The Fragrance and Flavor Industry

The Fragrance and Flavor Industry

WAYNE E. DORLAND

and

JAMES A. ROGERS, Jr.

Published by
WAYNE E. DORLAND COMPANY
Box 264, Mendham, New Jersey 07945
1977

Table of Contents

This book is respectfully dedicated to

Dr. Ernest Guenther

Steffen Arctander

and Paul Bedoukian

who have been so generous in sharing the
results of their technical studies with
others in the Fragrance and Flavor Industry

List of Illustrations

Publisher's Foreword

I should probably admit, for openers, that I have always had in mind that I would some day publish a book on the essential oil and aroma chemical business. Having been on the perimeter of this industry for almost fifty years, having known many of the leading figures in the business in their early days, and having always considered the field of fragrance and flavor raw materials a highly interesting one, I have been at least subconsciously storing ideas for such a book in my head for a long time. Thus, when I heard of the plan of the Essential Oil Association of U.S.A. to publish a 50th anniversary book about the industry, I was delighted to accept their invitation to become its publisher.

Subsequently the EOA abandoned plans for their publication, but by that time I was so firmly committed to so many friends who had gone to considerable trouble to help me dig out content for the book, that I determined to publish the volume independently. Too much work had already gone into the project to consider abandoning it. In its completion I have been fortunate in having the assistance of an exceptionally capable co-author, James A. Rogers, Jr., group vice-president and director of fragrance and flavor research for Fritzsche Dodge & Olcott, who brought to his role thirty years of experience with every aspect of the technical side of the flavor and fragrance industry.

From the start, my plan had been to include in "The Fragrance and Flavor Industry" a broad spectrum of the type information I thought would interest not only those working in the industry, but also those who use fragrance and flavor materials in the manufacture of consumer products. I believed that both groups of readers would have a lively interest in knowing more about the nature of essential oils and aroma chemicals, the categories into which fragrances and flavors fall, the type concerns which manufacture, import, process and sell these materials, the organizations to which they belong, and the reference sources from which those in the industry derive information about both raw materials and finished aroma products.

I felt too that, particularly those working in the fragrance and flavor industry, would profit from having a better historical background on their industry. Workers in any industry are, I am convinced, more productive members of that industry if they have a sense of pride in what they are doing; but it is obviously difficult to build up pride in your calling unless you first know more than a smattering about it. Thus we have included an ample historical chapter, tracing the background of many of the major firms which deal in fragrance and flavor materials, and have also singled out for mention some of the strong individuals who built the industry through their vision and ability.

In this effort it was recognized that we would inevitably face criticism, because space limitations made it impossible to deal in depth with every company and individual. The decision was made to confine our historical treatment principally to companies which were in business fifty years ago, and are still active today, plus a few others which were major factors in the 20's, but which have disappeared from the scene since then. If we have failed to include in this section a reference to some company or person whose history should logically have been reviewed, we regret the omission. Space limitations made it impossible to include histories of the firms that entered the industry more recently, many of which have made major contributions to the industry.

"The Fragrance and Flavor Industry" was designed to be a popular, non-technical study of the industry. We start by reviewing and emphasizing the highly important contribution that a superior fragrance or flavor formulation can make to the commercial success of a consumer product. Next we trace some of the steps the marketer must take, working closely with his perfumer or flavorist, in developing a superior formulation, pausing to remind the buyer of the fragrance or flavor compound that he must give the supply house his eager assistance, his complete confidence, and his full cooperation if he expects that supplier to be in position to do the best possible job.

There is a chapter on definitions, and others on raw materials, both natural and synthetic. There are chapters on fragrance and flavor families, the make-up of the industry, and its history. There are also several chapters for the more technically minded reader. One covers production processes, another the important contributions that science has made to the development of the F/F industry along the lines of instrumentation and testing.

Another chapter deals with product safety and government regulation. Still another tells what type people the industry employs, where they come from, what training and abilities they must have, and what their responsibilities are in the various jobs they fill.

So much for subject matter; now to the happy subject of credits. I leaned strongly for advice on individuals whom I have known for many years, and they cooperated generously. For the historical chapter, I got highly interesting details from such industry veterans as Mike Lemmermeyer, Sidney Friend, Col. Joseph Magnus, Grant Dorland, Dr. and Mrs. Felton, Ed Bush, Fred Theile, Bob Engel, Kenneth Voorhees, George Uhe, Charles Fischbeck, George Cadgene, Ed Strobl, Louis Allstadt, Howard Isermann, Herbert Halpern and many others.

The one individual who helped more than any other to carry the project on to completion was co-author Jim Rogers, who wrote the chapters on Definitions, Natural Raw Materials, Production and Quality Control, and also made major contributions to the chapter, Anatomy of the Industry, as well as to the one on Synthetic Raw Materials. He also offered many valuable suggestions for the correction and amplification of other major parts of the book. Without his assistance, it is questionable whether the book would ever have been completed.

George Cadgene, now retired, but who formerly worked for Givaudan, Trubek, and IFF, made major contributions to the chapters on synthetic raw materials, the anatomy of the industry and its history. Dr. Ernst Theimer, former vice-president in charge of research and development for IFF, wrote the chapter, The Scientific Side.

In the chapter on formula development we had generous assistance from the late Ernest Shiftan of IFF, and in the sections on flavors such industry experts as John Galuzzi of Norda, Inc., Charles Grimm and William Downey of Fritzsche D & O, and Mike Mason of IFF gave helpful advice.

Expert assistance with the chapter on industry personnel came from Thomas Lotito of Givaudan Corp., William A. Meyers of IFF and Bill Rosenberger of Fritzsche D & O. Also in this chapter I was delighted to have Henri Hoffman of Monsanto Flavor/Essence contribute an essay on "The Training of a Perfumer."

In writing the chapter on product safety and government regulation, I was assisted to a considerable extent by suggestions made by Robert Eiserle of Fritzsche D & O. Dr. Kurt Kulka of the same company made a number of helpful suggestions on the content of the synthetic raw material and other chapters. Steve Manheimer of J. Manheimer, Inc. helped with informed comments on the role of the essential oil dealer in the industry. J. F. Sansoldi Jr. of Charabot & Cie. gave me a tour of the Charabot fragrance manufacturing plant in Grasse, France, and offered many practical suggestions for material to be included in several chapters.

Delphi Marketing Services, New York, gave permission to use in the statistical chapter some of the data their organization had assembled for their book, "A Study of Flavors, Fragrances and Essential Oils." The same permission, much appreciated, was granted by C. H. Kline, Fairfield, N.J., to quote from their book, "The Kline Guide to the Chemical Industry."

I appreciate the permission granted by Ms. Jill Jessee to refer at length to her codifications of fragrance and flavor families. Naarden UOP Fragrances also kindly gave permission to condense and re-print an invaluable Perfume Check List originally published in the "Naarden Fragrance Magazine." Similarly, Haarmann & Reimer allowed us to condense and reprint their Chart of Fragrance Families from the H & R "Genealogy of Perfumes."

Illustrations were supplied by Ungerer & Co., Norda, Inc., Monsanto Flavor/Essence, V. Mane Fils, Inc., IFF, Fritzsche D & O, Danco Intl., Charabot & Cie. and Roure Bertrand Dupont.

When I started on this project I recognized that, in trying to cover a mass of detail in which I would be forced, particularly in the historical section, to rely on old and sometimes confused records, as well as fallible memories, including my own, a considerable number of errors must be anticipated. To minimize mistakes, I submitted portions of the manuscript to individuals I thought were in best position to recognize the errors and correct them, knowing that in spite of all such precautions numerous inaccuracies would inevitably find their way into the finished book. I regret such errors and apologize for them in advance.

I hope our book will have thousands of readers, that they will perhaps be indulgent in overlooking its shortcomings, and that they will find this account of a fascinating industry half as interesting as I have found that industry to be over the past half century.

Wayne E. Dorland

Mendham, N.J.
September 10, 1977

Chapter I

The Nature of Fragrance and Flavor

Fragrance and flavor are very closely related to one another. They are, we can safely say, more closely related than any other pair of the human senses. And, the fact that they are so closely related explains why such apparently divergent materials as flavors for foodstuffs and fragrances for perfumes, cosmetics and chemical specialties are sold by the same group of companies, the importers and processors of essential oils and the producers of aroma chemicals. It is the close relationship in humans between the senses of smell and taste that accounts for these two groups of raw material suppliers sharing a common industry.

Many psychologists have investigated and reported on how the senses of smell and taste function, how they interact, and how the mind and the body respond to taste and olfactory stimuli. However we doubt that even the experts have sorted out all the facts thoroughly enough so they can risk being didactic in this area. Certainly we cannot. The mechanisms of organoleptic detection and determination are complex, and we doubt they are thoroughly understood be even some leading psychologists. And, when experienced fragrance and flavor chemists attempt to correlate the properties or molecular structure of chemicals with the odor or flavor responses they elicit, and find how difficult it is, we begin to realize how much still remains to be learned in this area.

One psychological fact on which most experts agree is that human taste responds to only four basic types of stimuli: sweet, sour, salt and bitter. The experts also generally accept the fact that what we think of as flavor consists perhaps to the extent of 80% or more of aroma or smell. Incidentally, the aroma or fragrance connected with taste most often comes from the back of the nasal cavity, and through the mouth rather than through the nose. To check the validity of this observation it is suggested that the reader take a strong dash of hot mustard. The work of the psychologists on threshold values and electrical responses in nerves to discrete chemical stimuli has shown tremendous progress in developing correlation between stimuli and responses, but it would be presumptuous of the authors of this book to try to speak as experts in this area.

It is a fact, however, that a great deal of empirical information on the nature of taste and smell has been collected by flavor and perfume chemists and food technologists through their practical experience in dealing with fragrances and flavors on a day-to-day basis. Some of the experts have shared their experience with us, and we pass along selections from their views in this chapter.

With most foodstuffs, there may be — and often is — an overlapping effect to smell and taste. Frozen orange juice offers a good example of this overlapping effect of fragrance and flavor. It is probably not aroma alone that makes the consumer give such ready acceptance to this beverage, for its fragrance is by no means irresistible. But, added to the orange fragrance, is the sweet taste of the product, its tang, and the cooling and thirst-quenching effect. Flavor and fragrance combine to create a high appeal for orange juice reconstituted from frozen concentrate, or orange drink.

While most foods base their appeal on aroma, other food items such as soda crackers, pickles, sharp cheese, etc.

build acceptance on the basis of their taste rather than their aroma.

In still other cases, fragrance and flavor may act independently in a specific product. While one sense may give a favorable response to a particular material, the other sense may send a different message to the brain. Thus, garlic, onions or Limburger cheese are considered by many to have an appealing taste, but an odor which repels. Civet, on the other hand, has a smell which in dilution is sweet and quite acceptable, but its taste is unpalatable.

In spite of the overlapping effects in flavor and fragrance, there are certain exceptions — some basic differences which exist between the two senses. Almost without exception, every flavor is based on an item that is a natural food. An appealing fragrance, on the other hand, can be anything, either natural or synthetic, that is pleasing to the sense of smell. While originally fragrances were based on natural materials such as flowers, gums, herbs, bark and leaves, the capacity of the sense of smell is so great and so acute that any pleasant smelling material, natural or synthetic, can exhibit odor appeal, although originally aroma appeal was based exclusively on natural materials.

There is no direct parallel between the appeal of fragrance and flavor. There are many materials — such as neroli and lavender, for example — which have an appealing fragrance, but which offer no appeal whatever so far as taste is concerned.

One function of the senses of smell and taste is to protect animals, and today to a lesser extent man, against inimical materials which might otherwise be ingested.* The *odor* of

*A British researcher, Michel Treisman, an experimental psychologist at Oxford University in England, observed in a recent report ("New York Times", Aug. 13, 1977) that taste is a basic defense against intake of toxic substances. Animals reject anything bitter. If a poison is eaten, vomiting to eliminate the substance is often set off by a reaction on the stomach lining, or by reactions after the toxin has been absorbed into the body. Mr. Treisman said there is considerable evidence that sickness after ingestion of a poison can cause a conditioned aversion to the tastes and smells accompanying initial exposure to the substance.

carrion or other contaminated or germ-carrying materials is repugnant to humans, and tells us to avoid them. And, in cases where no repellent *odor* is present to warn us, our sense of *taste* protects us against such dangerous items as strychnine or moldy nuts by rejecting them. Unfortunately neither sense is infallible. Mushrooms, whether poisonous or not, have an interesting flavor, and even some spoiled canned products, such as fish or soup which may have developed botulism, are occasionally eaten without immediate rejection. However it is the sensations *pleasing* to the senses of taste and smell that are of primary interest to the flavor and fragrance industry, and thus the major thrust of this book.

Since there is so much subjective appeal to fragrance and flavor, in which all of us participate, it should not surprise anyone that fragrance and flavor play an extremely important part in affecting the decisions of millions of buyers on which of so many competing products to buy. This is the challenge of this most intriguing and complex industry, which represents a coalescence of trained experts, creativity, art, science and business. How these elements work together to help in the development and marketing of a host of successful consumer products will be the subject of succeeding chapters.

(Further observations on the nature of fragrance and flavor by Dr. Ernst Theimer may be found in Chapter XII — On The Scientific Side.)

Chapter II

Fragrance and Flavor Their Role in Product Success

Most marketers of consumer products are quick to recognize that the *fragrance* selected is the key factor in the success of any soap, cosmetic, or household product, and that *flavor* is the key to the market success of any food, beverage, or smoking product. The particular fragrance or flavor selected can and often does, spell success or failure at the point of sale. At the end of this chapter we will include a few of the many success stories, spotlighting instances where the fragrance or flavor selected by a marketer contributed in a major way to the success of his product.

A great deal of research has gone into cosmetic fragrancing in an attempt to determine just what types of fragrances have the greatest appeal for certain products. At one time, a rose fragrance, not too strong, nor too vague, was felt to be the most suitable for products like face powders or rouges, and fruit flavors were preferred for lipsticks. But, with the passing of time, many cosmetic lines have now taken their fragrance cues from successful, proven perfumes. Complete lines of cosmetics are now often fragrance-keyed to make a full product family.

At the moment there is a strong interest in fruity

fragrances, which are highly popular in shampoos and other hair preparations, and in various types of household products. They can range from the uncomplicated citrus or raspberry notes to very lush and sophisticated fruit fragrances in the peach or apricot family.

Usually, a perfume and cosmetic manufacturer introduces a perfume, followed shortly thereafter by a toilet water (by whatever name) and a cologne. If the fragrance is a success, then such items as bath powder, oil, soap, bubbles, etc. widen the family line. Thus, a woman who favors a certain perfume can support its fragrance with several other similarly scented products, rather than risk the possibility of a fragrance clash if fragrances of different character are used on other personal items. In another chapter (see pp 256-7) we will review the various steps in the marketing of a fragranced product from evaluation of the candidate formulas on through to introduction of the finished line to the market place.

To the cosmetic marketer, the problem of giving a full line of cosmetics the same fragrance may seem to be a very simple one. Actually, it can be very complex. You don't just add a dash or a dribble of the same fragrance oil used in the perfume to a soap or a powder, a foundation or night cream, and hope to have them all come out smelling the same. Each type product requires reformulation to suit the varied demands of the base materials.

We think of a certain popular bath line where the fragrance is greatly admired, but we don't think the manufacturer ever really succeeded in making the bath powder smell the same as the original cologne. It's sort of sweet, instead of being sharp, clear and clean. But it's probably the closest they could come. Sometimes, there just has to be a compromise. And, when it comes to soaps, there is another whole kettle of problems. We will cover the problems of soap perfuming in greater detail later (see pp. 46-49).

Why is one perfume and its fragrance companions a smash, while another may be a disaster? There is no single answer. If there were, we would have a succession of successes instead of an occasional success and a lot of "also rans." Fragrance is a "high risk" business. Time, often years, and considerable talent usually go into every fragrance introduction, but after they have been invested about all anybody can do is pray.

The perfumer assigned to the creation of the fragrance may be as talented as another. The ingredients in one fragrance may be as expensive as those in another. The package designers may be equally creative. The names of the two products may be equally appropriate and intriguing. The advertising budgets may be approximately the same. Distribution may be as carefully planned. Yet one product may sit neglected on the shelf, while shoppers jostle one another to buy the other.

The basic difference between the two products may be, and often is, simply a difference in fragrance appeal. And, while it is extremely difficult to understand just why one fragrance will be preferred so strongly over another, and there is no one, simple answer to the riddle, we are inclined, after years of studying fragrance successes, to put "innovativeness" at the head of the list of essential characteristics of a successful new cosmetic product. Perhaps it is the discovery of a new and exciting ingredient, or the combination of that ingredient with others in the composition, that makes the product a "smash" success. Perhaps it is a name and/or concept that is innovative. Perhaps the packaging represents a new departure. "Innovative" is an idea that combines many elements.

Equally important with innovation is timeliness. There seems to be a right time for a certain fragrance, and a wrong time. Only someone with a sixth sense, plus a sensitive finger on the social pulse, can say with relative certainty "This is the exact moment for this fragrance!" Sometimes a perfume introduced twenty years too soon can be re-introduced, promoted, and enjoy postponed popularity.

The prestige of the marketer also has a great deal to do with success at the counter. The marketer's name creates an advance ripple of acceptance. Yet, no rules were ever made that didn't have their exceptions. A curious anomaly is that quite often the very first perfume introduced by a cosmetic house may remain the perennial favorite, no matter how many new fragrances follow it. The maker might have had prestige to begin with, or might be an unknown name, yet an initial success could have provided the impetus leading to continued success.

We now have a trio of indicators of success: innovation, timeliness, and prestige. But there are still other factors. We would rank name of product next. Again, a name must be a timely one. An appropriate name of one year may be old-hat in another. But if one is innovative and venturesome, a name that goes counter to the social climate of the time may create a considerable stir. And, if a perfume has been a consistent success for many years, its name becomes so much a part of the fragrance culture that it is accepted without question.

We must certainly also mention advertising and packaging as important factors in product success, although it is a mistake to try to be didactic in these areas. One advertising campaign may send a new product into orbit, while another which might seem on its face to be equally well conceived is a failure. One perfume flacon with pure, simple, clean lines may be a success; another may be very elaborate and highly decorative, and also be a success. But both must be in good taste. A color scheme for a package may be feminine and romantic, or bold and daring, but both packaging and color scheme must be appropriate to the name of the product and the style of the fragrance. If all this is highly confusing, perhaps it is because the subject is confusing.

The importance of perfume as a major factor in a woman's decision on what toiletry to purchase was clearly demonstrated in a survey conducted in the latter part of 1976 by Louis Harris & Associates, New York, N.Y., for their client, PPL Fragrances, Maywood, N.J. The survey was based on 790 in-person interviews intended to measure consumer attitudes toward the perfuming of twelve product types, with particular focus on shampoos, toilet soaps, deodorants, anti-perspirants, hand creams, lotions and bath products.

The women queried indicated that in their minds fragrance is the second most important consideration in their decision on what toilet soap, bath oil or hand cream to purchase, apart from basic product performance. The only factor they rated more important than fragrance was feel or texture. Fragrance was rated even ahead of price in affecting the prospective purchaser's decision of what product to buy. Perfume was rated as the third most important factor affecting the purchaser's decision on shampoos or deodorants.

Most women said they favor delicate perfumes. The qualities in a perfume that influence them most are whether or not it is: clean, refreshing, pleasant, healthy, and delicate. Women's perfume preferences depend to a certain extent on personality types, said the researchers. "Sophisticated" women and teen-agers both like highly perfumed products, while "dull" women or the type that might be classified as activists for women's rights were reported likely to prefer no perfume at all.

Other conclusions drawn by the researchers were that:

1. Black women are least inhibited in their attitude toward perfume.
2. Lower-income women are receptive to perfumed products.
3. Younger women (18-34) are most open to perfume.
4. The younger woman may look upon perfume as a way of expressing herself, while the older woman tends to view perfume in the traditional sense as a way of enhancing her femininity.
5. Middle-aged women are not as much interested in perfume as younger women.

There is a large group of fragranced products such as shampoo, soap, deodorant or anti-perspirant that have a functional purpose, and with which fragrance, although important, is necessarily only a secondary consideration. For example, the marketer of a well-established shampoo based his entire advertising campaign for the product on the single fact that the fragrance was new and superior. But, keep in mind that the shampoo had already made its reputation as an effective product, gentle on the hair but strong against dandruff. This claim was retained and emphasized, and the appeal of the new fragrance simply added to the basic claim for functional excellence. The position of the product in the marketplace was just enhanced by the fragrance appeal.

By contrast, another new shampoo introduced a few years ago, which based its name and its entire promotion on its superior fragrance, had very rough going at the counter. A woman wants a shampoo first of all to get her hair clean. Then, she wants her hair left soft and shining, full of bounce and body. All other things being equal, the shampoo which has a delightful fragrance will have an edge on its competitors. But the primary function of the product must be fulfilled first.

Since shampoos are such an important part of the total hair preparations market, accounting for almost a third of overall volume, we will add a few specific thoughts here on the perfuming of shampoos. Not only have herbal scents recently become very popular for shampoos, but other innovations in perfuming are gaining favor, especially the use of fruity fragrances such as strawberry, green apple, honeydew, apricot, tangerine, etc. One shampoo perfume is an effective combination of cinnamon, sage, nutmeg and ginger. There seems to be no limit either to one-note fragrances for shampoos, or a combination of notes.

One producer of fragrance compounds has suggested that the single floral note is too often overlooked, although some one-note fragrances such as honeysuckle are now appearing on the market. This company feels that today's marketers are neglecting the rose as obvious (shades of the past), and counsels that we should not pass by such other single note fragrances as magnolia and violet. And it provocatively suggests that other fragrance notes in nature should be explored, such as green peas, bell peppers, coffee, peanuts, etc.

It then might be possible for an enterprising young lady seeking to make her hair fragrance a conversation piece to arrive at her office smelling like coffee, go out to lunch with a fragrance of peas and peppers emanating from her hair, then keep a cocktail date smelling of peanuts.

A fragrancing problem that often faces the perfume chemist is developing a fragrance that will successfully cover (mask) unpleasant odors innate in a product. There are many such problems, but a specific example from the hair preparations field can complete our immediate discussion of the perfuming of hair products.

A new gradual hair coloring entered the market recently in competition with a brand that had been No. 1. Apparently both products were about equal functionally, but the marketer of the new product featured a "smells better" test in its TV promotion. A woman was shown sniffing the hair of each of two look-alike male models. The head that "smelled better" was, of course, the one that had been treated with the advertiser's product. In hair colorings there is usually an initial and a residual odor following application. So, kudos to any product that can overcome such a negative characteristic.

Like the functional aspects of hair preparations, a deodorant must, first of all, deodorize; and an anti-perspirant must retard wetness. If one product is not as effective as its competitors, even the most marvelous fragrance in the world won't put it in a leading position. But, if it is functionally equal, then an appealing fragrance may be just the bonus that will give it a sales boost. There are some deodorants and anti-perspirants that are unscented, both because some people prefer such a product that way, and also because fragrance can be allergic to a very small percentage of people. But most people, consciously or not, are attracted by a pleasant scent in even the most functional of products.

A current campaign stresses the "pretty" fragrance of an anti-perspirant that has been improved both in its strength and scent. And, a deodorant in powder form, which utilizes baking soda for its deodorizing properties, comes both "regular" and in an herbal scent, giving the buyer a choice.

Deodorant soaps have their place in this discussion too. *Lifebuoy* was a pioneer, with its carbolic smell and its contribution of the phrase "B.O." to our language (See also p. 23). But today deodorant soaps don't have to come on that strong. Years ago the essential ingredient in a deodorant soap was phenol, which has a foul odor, difficult for the perfumer to cover. For a time hexachlorophene was widely used to provide the deodorant function, but problems with toxicity a few years ago led to its demise. Now there are numerous halogenated compounds which find use as bacteriostats. They are much less of a problem to mask. The deodorizing ingredients employed today don't telegraph their functional role so blatantly. Many of these modern day deodorizing soaps use fragrance as an added selling feature. Still, some people seem to retain the lingering impression that deodorant soaps must necessarily be deficient in scent appeal.

A current advertising campaign for a deodorant soap has one young person saying to a member of the opposite sex, when told the person uses a deodorant soap, "But you don't *smell* like a deodorant soap." Even if this copy line implies a negative view of deodorant soaps in general, and denigrates competitive products, it has no doubt contributed in a major way to the sales success of that particular brand.

Feminine hygiene deodorants emphasize fragrance as a selling feature. And a certain disposable, pre-measured douche is offered in a choice of two fragrances. All of these products are highly functional, but must, as with the hair coloring product mentioned before, be expertly formulated to retain full function while still being attractively scented.

But fragrance is far from limited just to fragrance items, cosmetics and related preparations. It plays a leading role, too, in a wide gamut of other products. Household products, for example, make extensive use of fragrance. Once, pine was the most popular scent in many household cleaners and disinfectants, but today lemon and herbal scents have largely replaced pine, although by no means completely. The clean, fresh, wholesome, outdoorsy fragrance of pine will undoubtedly always be with us.

From waxes and polishes to detergents and disinfectants, from whiteners to fabric softeners, a woman buys fragrance with function. She selects lemony scents in many of these products, including such products as *Lemon Pledge, Lemon-Freshened Borax in Fab, Lemon Fresh Joy, Old English* furniture polish with a lemony scent, and so many other products.

There is really nothing new in lemon-scented floor polishes. Our mothers and grandmothers used to make their homes smell fragrant and inviting by using lemon oil on their furniture. But lemon has been given a new twist today with its appearance in so many products. There has been a renewed emphasis on this old-time favorite.

We are all familiar with that "April fresh smell" of *Downy,* and with the new, softer fragrance of *Charmin* bathroom tissue. But fragrance goes still further in making product successes. Sometimes, fragrance isn't even mentioned, but it's still there playing a subliminal selling role.

Tests show that fragrance is often a sales factor even when the buyer fails to recognize its presence. It has been demonstrated that scented stationery, stockings and lingerie will out-sell unscented products, assuming that all other characteristics are similar. Textiles and leather, typewriter ribbons and ledgers, rubber bands and carbon paper, paint and plastics will often, unbeknownst to the prospective buyer, be scented to attract him.

Advertising brochures, circulars and blotters are sometimes scented to promote the sale of a product. Most of our readers must be familiar with the "scratch and sniff" technique used to acquaint people with a particular fragrance. Blotters are highly effective in selling a fragrance, and have the added advantage of serving as a sachet.

Back in the 30's, Givaudan Corp. used to market a specialty item called "Ink-O-Dors" for perfuming printing ink. The writer of these lines had the experience of printing several issues of a trade magazine, "The American Ink Maker" with these products. A raspberry fragrance was the most successful of those used. In addition to imparting a highly recognizable fragrance to the printed page, it also made the entire pressroom smell like a raspberry patch.

The development of microencapsulation in recent years has made more practical the incorporation of fragrance into a wide range of consumer articles. One we heard of just about a year ago is "Learning Shirts," introduced by Learning Apparel International, 350 Fifth Ave., New York, N.Y. T-shirts for children are printed with ink which contains a microencapsulated fragrance. A letter of the alphabet is printed on each shirt, together with a fruit or article of food beginning with that letter — thus A for apple, B for banana, C for coconut, etc. The ink carries the appropriate fragrance in microcapsules. Minnesota Mining & Manufacturing Co. developed the scented inks used with the process. Mr. Herbert Cohen of Learning Apparel International is reported to have a patent pending on employment of microencapsulated printing on apparel.

Fragrance has lots of other uses, but inasmuch as our present discussion is limited to products, we will only mention in passing that fragrance has environmental uses too. Fragrance can create an ambiance in movie theaters, or make workers happier and more efficient in offices and factories.

The ability of appealing fragrance to create sales is clearly apparent when a shopper goes into a bakery shop. With delicious aromas filling her nostrils, she may well be persuaded to buy more than just the items for which she was originally shopping. She may not even realize that her sales resistance has been broken down by the enticing scent of the bakery. We know of one lady who employed aroma successfully as a sales aid in moving real estate. Whenever she expected a prospective buyer to look at her house, she put a loaf of bread into the oven. The aroma of the baking bread, she insisted, predisposed the prospective buyer to be pleased with the house.

Another example of using aroma to help sell a product is the scenting of synthetic whipped cream to make it smell more like the real thing. Or, consider the cigar manufacturer who has the insides of its cigar boxes sprayed with an aromatic tobacco fragrance. When the box is opened, the heady aroma of good tobacco is so appealing the buyer can hardly wait to light up. Cigarettes have a very definite olfactory attraction, showing once more how closely smell and taste are related.

Space sprays are used to reodorize public buildings, even buses and subways. Scented sprays are often used in garbage cans and bath rooms, and a scent has been devised as a shark repellent. When synthetic rubber was introduced into general use back in the World War II days of the 40's, its objectionable odor had to be covered. One acute problem of reodorizing women's girdles made from synthetic rubber was solved by incorporating directly into the rubber during the molding process a fragrance which relied heavily on musk.

FLAVOR

Turning to flavor, we must keep in mind that, as pointed out in Chapter I, while fragrances can be based on anything pleasing to the sense of smell, flavors, almost without exception, are based on a natural (food) product, or combination of food products. The flavors used most widely are the fruit flavors. They find their major use in beverages, being used in heavy volume in soft drinks, cordials, etc. Among the flavor leaders in the fruit group are Apple, Banana and Cherry. After the ABC's come the citrus flavors, lime, lemon and orange. Next grape and the berry flavors, especially strawberry and raspberry. Then there are pineapple, peach, apricot and dozens of others.

In recent years the demand for all fruits, and particularly grapes, cherries and strawberries, has become so great that it is now impossible to grow enough of these crops to satisfy the demand. Thus it has been necessary to develop artificial flavors. The flavor industry has met this need by developing numerous flavor-identical compounds. The building blocks for these synthetic flavors are normally materials which match the exact chemical structure found in the natural products. For example, methyl anthranilate and ethyl acetate which are found in grape juice, are employed to make artificial grape flavors. Isoamyl acetate, the main component of banana essence, is the synthetic most widely used in making artificial banana flavor.

But, compounding a successful flavor compound is a much more difficult problem than simply mixing a few items together. The creation of a successful flavor is probably more difficult than compounding a fragrance. A flavor compound must not only smell the same, but also taste the same as the item it seeks to duplicate. Complicating the problem further is the fact that almost everyone has in his or her head a memory reference as to what a good cherry flavor, a good strawberry, or a good lemon should be. Flavor has a more universal set of standards, and any attempt to fool the public, or mother nature, in this area, is normally doomed to failure.

The components available for compounding a flavor are also more limited, particularly in the USA, because of Federal regulations. The area within which the flavor chemist must work is narrower than for the perfumer.

Finally, as with fragrance, there is a similar problem that must be solved of compatibility of the flavor compound with the base. A cherry flavor that is satisfactory for spray-dry will probably not serve for a hard candy, a soft drink, a pie filling, a cough drop, or a cherry scented chocolate. Cherry with pit would differ from cherry without pit, etc.

Whether the food product we are flavoring is a solid or a liquid, to be eaten hot or cold, as an appetizer or an after-dinner drink, a snack or a main course, flavor is the essence of its popularity, and a criterion often employed to judge its quality. Consumer approval will be accorded or withheld based on the flavor of such products as diet foods, natural foods, gourmet foods or trademarked foods, and the choice of flavor for a particular product is probably the most important consideration in the success of any food product.

Earlier in this chapter we said that innovativeness was the most important factor in determining the success of a *fragrance*. In the case of *flavors* the situation is different.

Certainly innovation may be important to the marketer, but the most important element in a flavor is that it be a true replica of what the average person expects that flavor to be. This is a somewhat more difficult standard to meet. Next, the flavor must have impact, and it must be reasonably consistent over an appreciable time period — it must have lasting power.

The development by the flavor industry of a wide selection of fruit flavors, both natural and artificial, has contributed in a major way to the tremendous expansion which has occurred in recent years in the sale of soft drinks, ice cream, canned juices, preserves, jams and jellies, candy, syrups and baked goods.

Of the nut flavors, perhaps coconut is the most popular, along with almond, both of which have wide use in the baking industry and have helped to expand the sale of baked goods. In addition, other nut flavors such as walnut, hazelnut, hickory, pistachio, etc. have helped to widen markets for the sale of both ice cream and baked goods. Vanilla is another top seller in the flavor list, for it is widely esteemed as a flavor in addition to its role as a fragrance.

An interesting group of flavors that have helped to build a large volume of sales for food stores are those which give a smoky flavor to processed meats and sausages. In addition, certain barbeque sauces, steak sauces and tomato ketchup now frequently have the smoke note built into them, widening still further their already very substantial market.

The flavoring of all forms of tobacco has done a great deal to expand the market for these products. Menthol is perhaps the ingredient used most widely in cigarettes, but many other flavors have been added to tobacco over the years to give an individual brand of cigarettes a distinct flavor and thus expand its sale. Menthol, of course, has a cooling effect, which carries through even as the cigarette burns; it is not degraded by burning as readily as other additives are, and many smokers find its cooling effect highly acceptable in cigarettes. Rum and maple were at one time extremely popular flavors added to pipe and chewing tobacco, and still sell in volume for this purpose. Vanillin is currently very popular, and is sometimes used in high concentration. The sweet, heavy odor of vanillin is very obvious, not only to the smoker, but also to all others in the immediate area.

Because of the recognition that smoking offers a medical hazard, it is only natural that any flavor additive to a tobacco product must pass close scrutiny before being considered for use. Some brands of cigarettes now produced, however, are still flavored with items which are unpleasant and perhaps undesirable in tobacco smoke. In recent years considerable emphasis has been placed on the development of filters which will permit selected flavor items to pass through the filter in sufficient quantity to still permit a good smoke flavor, while at the same time allowing the filter to remove tars and other materials which are undesirable. This approach has enjoyed wide acceptance, and may be an indication of the direction tobacco technology will take in the future, thus avoiding a future decline in the market for cigarettes.

The use of mint flavors continues to expand. The frequently used peppermint/spearmint combination is a highly popular flavor for tooth paste and breath fresheners, and has helped increase the sale of these products. Those who use mint flavored oral products avoid that horrid condition which advertising writers label "medicine breath." So the flavor of mint expands another market. In addition to having an agreeable taste, mint supplies another important ingredient for these products in that it is a mild "sense shocker." It provides "impact," making the user of the product aware immediately that it has gone into action. Even *Chap Stick* has now added a mint flavor to its two former standards, cherry and grape.

Mint of course also retains its role as an essential ingredient in those cooling after-dinner candies, either chocolate-covered or plain. It contributes something special to lamb in the form of mint sauce, is one of the essential ingredients in a mint julep, and continues to be the most accepted flavor for chewing gum.

PRODUCT SUCCESS STORIES

But, let's turn from the general to the specific with a series of "cameo success stories" on products that have achieved fame and glory in the market place, where the specific flavor or fragrance of the product has played an important role. First, fragrance. Then, flavor. We think you'll find these examples fascinating, not only for the intrinsic merits of the products discussed, but also for the human interest element in each story.

Coty's Rose Jacqueminot

Coty's *Rose Jacqueminot** is a success story if there ever was one. This is the perfume that launched young Francois Coty on his fragrance career back in 1900. Born a Corsican with the name of Spoturno, this young perfumer offered his first fragrance to an exclusive Parisian merchant, who said he wasn't interested. This so angered the developer of the product, who had renamed himself Coty, that he deliberately broke a small bottle of *Rose Jacqueminot* on the main staircase of the shop. The customers were enchanted with the fragrance, and besieged the shop owner with demands for it. He had to recall Coty and place a substantial order.

Rose Jacqueminot was excitingly different from any rose fragrance being sold at the time because of its warm ionone note. It was a fragrance triumph. M. Coty continued to introduce new and often daring fragrances. He also exhibited artistry in the flacons he selected.

The story of *Rose Jacqueminot* is one that belongs in the annals of perfume history. We repeat it here to insure its being kept alive. The current revival of interest in single floral perfumes should mean a rebirth of this beautiful fragrance. Perhaps, too, M. Coty taught his competitors the value of "demonstrations."

*A Chronicle of Perfumery, Firmenich S.A., 1973

Yardley's Lavender

Yardley of London's *Lavender** goes much further back in fragrance history. The Yardley firm was founded in London in 1770 by Thomas Yardley, who received a charter from the King (George III). Yardley introduced his *Lavender* in England about 1780, and the product was the first one to sell on the American market. This typically English fragrance continues to be popular today, especially as a fragrance for toilet soap. Fresh and clean smelling, it meets today's desire for the simple and natural, nostalgic and old fashioned. Although Yardley has introduced many newer fragrances in more recent times, many sophisticated and "modern," the Yardley name will always be associated automatically with *Lavender,* a family fragrance favorite.

Shulton's Old Spice

In the category of perfumes and colognes, we shouldn't pass by without mentioning *Old Spice.* Its fragrance was certainly a major factor in making it a best seller. The story of how this particular fragrance was developed is one of those bits of gossip around the cosmetic industry that makes a great human interest story.

William Lightfoot Schultz, who had sold his original company, the Lightfoot Schultz Co., a Hoboken, N.J. soap and shaving cream manufacturer, to the American Safety Razor Co. in 1932, started in business again under his own name in a company that was later to be renamed Shulton, Inc. He planned to introduce a ladies fragrance under the name *Early American Old Spice,* and wanted for this product an old fashioned perfume with a modern touch. He outlined his ideas to several fragrance houses, and asked them to make suggestions. Dodge & Olcott had developed a fragrance a year or two earlier which the perfumer, a Mr. Hauch, thought was exceptional, and which he believed was reasonably close to the type fragrance Mr. Schultz indicated he wanted. Though the fragrance had been

**Perfume Who's Who,* "Beauty Fashion," Sept., 1952

offered to several other potential buyers, it was still "on the shelf." D&O decided to let Mr. Schultz have a look at it. He liked what he saw and after he and the perfumer had made a few modifications the fragrance was accepted.

It was the story of the "Ugly Duckling" all over again, the winning lottery ticket that nobody wanted to buy, or the batch of chemicals mixed by mistake and left on the lab shelf overnight, which turns out to be the miracle product for which scientists have been seeking unsuccessfully for years. The fragrance was an instant success.

Following the introduction of *Early American Old Spice* for the women's market in 1937, Schultz brought out an after shave using the same fragrance about a year later. He shortened the name of this product to *Old Spice* and packaged it in a replica of an old fashioned rose petal jar. (The women's line had been packed in boxes which were covered with wood-veneer paper). The only change in the fragrance for the aftershave, which became the first item in Shulton's later complete men's line, was a little more citrus in the bouquet, which gave the product a more masculine tone. Otherwise the final fragrance was very close to the original compound that any one of several other companies might have had for the asking.

Schultz took the fragrance and made millions out of it. Perhaps the greatest factor in his success was that he had, or at least he developed, a concept to go with the fragrance. And he followed through with packaging and promotion geared to the theme he selected. They reinforced the effect of the fragrance, and helped to make *Old Spice* one of the great marketing successes in the cosmetic business.

Jean Nate's Fragrance Line

Jean Nate's *Friction pour le Bain* was her first bath preparation, and the start of probably the most popular of all lines of bath products. The concept, packaging and scent were established in the 1930's by Jean Nate (later Mrs. William Hunnefield), who was critical of a bath preparation she was using and suggested that the company make certain changes in it. Rebuffed, she decided to put out her own product. Later her husband joined her in the business.

In time, more bath products were added to the line. The company was eventually sold to Lanvin Perfumes, which later merged with Charles of the Ritz. Until the line was sold, no national advertising was done, yet the name, *Friction pour le Bain,* was familiar to countless men as well as women. Why? The secret was in the superb fragrance, a bright, tingling citrus blend. The formula has been carefully guarded all these years. There have been many imitators, but not one of them has succeeded in creating an exact copy.

There is another dramatic "demonstration" story involving Jean Nate's Bath Bubbles. A particular buyer showed reluctance to buy this item because there were many other similar bath products. Alexander Koswick, then vice-president and general manager of the Nate Division, filled up a sink with water and poured in some of the bath bubbles. The buyer was impressed and a sale followed. Later, it was jokingly suggested that a model should take a Jean Nate bubble bath in the store window. It happened. Shades of M. Coty.

The Jean Nate bath line was certainly not the first to use a citrus note in its fragrance, but it was the predecessor of innumerable citrus types to follow.

Colgate's Cashmere Bouquet

Turning to the soap category, we can think of a number of soaps in which fragrance was a major factor in building sales. For example, Colgate considered the perfume for one of its toilet soaps so important that they adopted as the name of the product, *Cashmere Bouquet,* the type of fragrance used in perfuming it.

Lever's Lifebuoy Soap

A completely different aroma helped make *Lifebuoy* immediately recognizable. Lever Brothers coupled a carbolic odor with an octagonal shape and a coral color to make their "health soap" a fantastic seller all over the world. So successful was Lever in getting the public to accept a carbolic odor for soap during the 30's, that a whole family of imitators of *Lifebuoy* was spawned. Dr. E. R. Thomssen, in an article on "Carbolic Soaps" in the December, 1935 issue of "Soap" magazine, remarked on the fact that the huge success of *Lifebuoy* had attracted dozens of imitators who copied the carbolic odor, and in some cases the color and shape of Lever's best seller. During this period Lever Brothers engaged in extensive litigation in an attempt to protect their exclusive right to market a soap of carbolic odor, pink color and octagonal shape. Their efforts enriched their patent attorneys, but did very little to stop the small soap makers who were determined to market a "me too" cake of "health soap."

Billy B. Van's Pine Tree Soap

Another toilet soap which built up from nothing to a very respectable sales figure in the space of a few short years back in the late 20's was Billy B. Van's *Pine Tree* soap.* Van was a baggy pants, slapstick refugee from the burlesque stage who wandered into the soap business quite by chance. The story he told was that he had been forced into retirement by ill health and selected a resort in New Hampshire in which to recuperate. There, according to Billy, he was restored to good health by the refreshing aroma of the pine woods.

Back on his feet again, he took it upon himself to come to the rescue of the great American soap buying public by "incorporating the pine fragrance of the New England woods into a cake of soap." The public bought the hokey story, and the soap. In fairly short order Van built up a sales volume which probably would not have impressed P&G, Lever or Colgate, but it kept "Sliding Billy" from having to go back to work. At the peak, he was selling his soap to 1,100 hospitals, 200 hotels and hotel chains, and eleven steamship lines, as well as millions of cakes to the public by direct mail. He gave all the credit for success of his soap to the piney fragrance.

*"Soap," November 1929, page 23.

S.C. Johnson's Lemon Pledge

The Johnson Wax* line of *Pledge* furniture care products has been the market leader virtually from its beginning. There are a variety of *Pledges,* the first introduced in 1958, *Lemon Pledge* in 1967, and later *Natural Wood Scent Pledge* and *Liquid Pledge.* The marketer, S.C. Johnson & Son, Inc., Racine, Wis., now in the hands of the fourth generation of the Johnson family, is one of the largest privately held corporations in the United States.

From the first Samuel C. Johnson, who began peddling parquet flooring in 1886, and then branched out into paste wax, the company has built its growth on a steady parade of new products. Some have done better than others, but across the board the Johnson people have had a good batting average. The company ranks first in four of its major product types (floor care, furniture care, insecticides and air fresheners), and second in its other two product lines (auto wax and laundry care).

An early believer in national advertising, Johnson lost no time in "being on the spot" in both radio and television. A favorite commercial tells the story of a housewife daydreaming herself out of her living room into a California lemon grove. The "lemon grove" commercial emphasizes how easy it is to use the product and how beautiful fine furniture looks when treated with *Lemon Pledge.* Lemon is a natural fragrance for a furniture care product. It is not only pleasant for the user, but also makes the home smell inviting and friendly for family and guests. Johnson's air freshener, *Glade,* is offered in a lime fragrance, in addition to lemon.

GREAT FLAVOR SUCCESSES

Just as there have been hosts of products that built their commercial success on an appealing fragrance, so there have been countless others that have continued to sell over the years because of unique and appealing flavor.

**Business Week,* July 6, 1968, "How a family keeps its verve."

Angostura Bitters

One of the most closely guarded flavor formulas is the one for *Angostura* Aromatic Bitters. This popular aromatic flavoring ingredient for both food and drinks was first compounded in 1824 by Dr. Johann Siegert, Surgeon-General in the army of Simon Bolivar, the liberator of South America. Dr. Siegert's headquarters were in the port of Angostura in Venezuela, a city now known as Ciudad Bolivar. Seeking to find a preparation which would make the food consumed by Bolivar's fighting men more appetizing, Dr. Siegert developed *Angostura* bitters. Sailors putting into port there discovered the product and took bottles home with them. Thus *Angostura* bitters became famous all over the world. Originally developed to make food more tasty, *Angostura* became widely used in drinks, because it improved the taste of many liquors and mixed drinks.

Beyond saying that the product contains tropical herbs and spices, and that it does *not* contain any bark of the Angostura tree, the formula for *Angostura* bitters has been very closely guarded. A company publication says that only three individuals know the formula. One copy of the list of ingredients is kept on the island of Trinidad, where *Angostura* is now manufactured. Another was deposited in a bank vault in Barclay's Bank in New York City. About twenty years ago, when the ink on the slip of paper carrying the New York copy of the formula started to fade, a second copy was made of the original formula kept in Trinidad. The new copy was cut into four segments, and the four parts sealed with sealing wax into four separate envelopes. These were then sent by registered mail — separately — to New York. Each section was mailed only after word had been received that the earlier segment had arrived. The four sealed envelopes were placed unopened in the vault. Thus does the manufacturer of this widely used flavoring ingredient guard the secret of *Angostura*'s formula.

Sole distributor of *Angostura* Aromatic Bitters in the United States is A-W Brands, Inc., 1200 Milik Street, Carteret, N.J.

Benedictine

Pre-eminent among the liqueurs is *Benedictine*, made originally by the Benedictine monks, but now manufactured commercially in France.* It has proved to be one of the flavor successes of all times. We don't know even the century in which it was first produced, but *Benedictine* goes back to the middle ages. It may seem strange that a religious order was engaged in such a side line, but there was justification for their interest.

Before modern medicine was developed, there was a constant search for curatives, almost always based on secret formulas. The Benedictine monks considered alcohol to have great powers and set about mastering the art of distillation. They developed *Benedictine*, and the Carthusian order created *Chartreuse*, two cordials which might never have been discovered if the monks had not been such great humanitarians.

These two liqueurs were made by macerating selected flavoring materials in strong spirits, then distilling the solution to obtain an alcoholate. The composition of *Benedictine* is still a closely held secret, but it is believed to contain lemon peel, angelica, and a number of spicy herbs. When made by some bottlers today, only a few flavors are used, with one note predominating, while in others the flavoring is more complex. Obviously the key ingredients are the flavors selected.

Worcestershire Sauce

The "original" *Worcestershire Sauce* was first made in Worcester, England, about 1834. It is still being made there. Lea & Perrins, Inc., Fair Lawn, N.J., were sole distributors for the United States in the late 1830's. They began manufacturing the product in the U.S. in 1876 or 1877, and now have full rights to the recipe.

The company offers a free cookbook which tells something about the ingredients in *Worcestershire Sauce*. They include: red onions, cloves, tamarinds, garlic, chili

*"The Givaudan Flavorist," No. 4, 1972

peppers, anchovies and eschalots. The dictionary calls this a "pungent" sauce. No wonder.

Lea & Perrins insist that the recipe for their product hasn't been changed in over 130 years, since it was first concocted by John W. Lea and William Perrins of Worcester, England, "from the recipe of a nobleman in the county." To this day, they say, it contains no "artificial" ingredients, only natural products of the choicest grades. And, they add, "it is aged in wooden casks, like fine wine." Sales of the product are in excess of $25 million annually to the retail trade alone. Here is another product where flavor of a unique sort created product success, and adherence to the same unchanging formulation has guaranteed a continued market for almost a hundred and fifty years.

Wrigley's Chewing Gums

The story of Wrigley's chewing gums is a fascinating one which illustrates American enterprise and a Horatio Alger-like success.

The Wrigley organization was founded in Chicago in 1892 by William Wrigley Jr., who had come to Chicago from Philadelphia at age 29, with only $32 in his pocket. He was ambitious to start his own business. Since his previous experience had been in selling soap, he decided to continue in this line.

Wrigley was a firm believer in the use of premiums. To help sell his soap he used a premium of a small can of baking powder. When this became more popular than the soap, he switched to the baking powder business. Again he relied on a premium, and started giving a couple packages of chewing gum with each can of baking powder. Before long he was in the chewing gum business.

Wrigley's first brands of gum were *Lotta Gum* and *Vassar*. He introduced *Juicy Fruit* in 1893 and *Spearmint* later the same year. There were many other brands at various times, including *Doublemint,* introduced in 1914, and *P.K.* in 1921. *Freedent* was introduced more recently to overcome the problem of chewing gum sticking to dentures. Wrigley's latest brand is cinnamon-flavored *Big Red,* introduced in March, 1976.

Wrigley was one of the pioneers in advertising. He used newspapers, magazines, car cards and posters as well as other media. Because *Spearmint* was a slow seller in its early years, it was advertised in 1906 in three eastern cities. Results were very encouraging. And, when his competitors decided to retrench during the general business slump of 1907, Wrigley seized the opportunity to make *Spearmint* a best seller by expanding his advertising program. It was America's favorite brand of gum by 1910, and *Juicy Fruit* and *Doublemint* were also on their way up.

Wrigley's had become an international business by 1910, and was exporting its products world-wide. Factories were established in Canada, Great Britain and Australia. The Wm. Wrigley Jr. Company has been a publicly owned company since 1919 and now has approximately 11,500 stockholders. The company has remained a "family business," however, and the third generation of the Wrigley family is still prominent in the management.

As their names imply, *Spearmint* and *Doublemint* gums are flavored with spearmint and peppermint. Because of the great success of the two products, mint has come to be accepted as almost a universal flavor for chewing gum. *Juicy Fruit* uses a different flavor, containing a number of natural fruit essences as well as some synthetic flavors. In all three of these best selling chewing gums, flavor was one of the primary ingredients in this extraordinary success story.

Life Savers

Another product success story built around the flavor of mint is this one. *Life Saver Mints* were first made in Cleveland in 1913 by Clarence Crane, maker of Mary Garden and Crane chocolates. Crane considered the mints a summertime specialty to supplement his existing line of candies. Known first as *Crane's Life Savers,* they have retained their distinctive shape and name since they were first introduced.

Unfortunately for Crane, he didn't share in the success of the product he started on its way. An energetic young salesman of car card space tried to convince Crane to advertise his product, but Crane was more interested in selling his company than investing more money in it. The advertising salesman, Edward J. Noble, together with a boyhood friend, bought the company from Crane for $2,900 and put some novel promotional ideas into operation.

Their first disappointment came when they realized that, while the candy Noble had tasted was fresh and appealing, there were hundreds of thousands of rolls on store shelves which were stale and flavorless because the cardboard package had absorbed the volatile peppermint flavor. Noble promptly devised a tinfoil wrapping to protect the flavor, but the storekeepers were sceptical and wouldn't buy the new package unless they were allowed to return their old stock for credit.

The new owners refused to acknowledge defeat and found new outlets. When making a sale, Noble would suggest to the store operator that he put the *Life Savers* carton near the cash register to cash in on impulse buying as the customer got his change. This and other ideas worked, for within a few years the Mint Products Company made a quarter of a million dollars.

The pioneering Ed Noble had many more marketing ideas. In their first year in business, he and his partner used to sit up nights putting samples into glassine bags. These were distributed on the streets and in public places by costumed girls. This idea of a "gift *without* purchase" wasn't the first time the idea had been used, but it proved effective in popularizing *Life Savers.* When the partners dressed up men in metal tubes painted like rolls of *Life Savers,* however, and sent them out on the streets of New York City one day, consternation was created when frolicsome students at Columbia University turned these tubular sandwichmen on their sides and rolled them down hill. There were other such humorous incidents which greeted the company's marketing efforts, but the public just laughed and went on buying *Life Savers.*

There are two kinds of roll candy made by Life Savers, one a pressed sugar mint, and the other a boiled drop, with a hole in the center. It was in 1929 that the company developed the technique for making the latter product, which had not been previously known. *Life Savers'* flavor story has been a kaleidoscopic one, flavors being added or dropped as consumer demand changed. In 1920 the flavors included peppermint, wintergreen, cinnamon, clove and licorice. Spearmint was added in 1936, by which time some of the other flavors had been discontinued.

The idea also evolved of offering multi-flavor as well as single flavor rolls. Lemon, Lime, Orange, Wild Cherry and Pineapple make up one of the most popular combinations. This first five-flavor roll was introduced in 1936. There are now three other assorted flavor rolls: Tropical Fruit, Fancy Fruit, and Assorted Sours. Single flavor rolls being offered today include Cryst-O-Mint, Stick-O-Pep, Butter Rum, Butterscotch, Wild Cherry and Tangerine. There are also *Life Saver* Lollipops in three flavors. Multiple sales are stimulated by offering a Christmas collection of 12 rolls of assorted flavors and a Halloween collection of 20 miniature rolls for "trick or treaters."

There are a million places in the U.S. today where you can buy *Life Savers*. Like everything else, inflation has moved prices up, and they no longer sell for the nickel a roll we remember as a child. But they are still a bargain in flavor, and another resounding tribute to the important role of flavor in product success.

Coca-Cola

Coca-Cola, the world's most popular soft drink, has a storybook history which began in 1886 when, according to legend, Dr. John S. Pemberton, a pharmacist in Atlanta, first produced the syrup in a three-legged pot in his back-yard. His partner, Frank M. Robinson, suggested the name, *Coca-Cola,* taking it from two of the constituents, flavoring extracts from the cocoa leaf and the kola nut. It was Robinson who wrote the name in the flowing Spencerian script of the day, which is preserved in the *Coca-Cola* trademark today.

Originally sold as a soda fountain drink for five cents a glass at Jacobs' Pharmacy in downtown Atlanta, *Coca-Cola* was first advertised on May 29, 1886, in the "Atlanta Journal" as "the new and popular soda fountain drink." The slogan, "Delicious and Refreshing" is just as meaningful today as it was then. In the first eight months, sales averaged 13 drinks a day to a handful of daring customers who must have liked what they tasted.

What they tasted was a compounding of many natural essential oils and flavors, with pure caramel used for color. The water was carefully treated to be neutral in taste and odor, and the sugar was highly refined. Of course, the exact formula has remained secret for all these many years.

But we left the struggling Dr. Pemberton behind with his new invention. Facing difficult times, he sold a two-thirds interest in his enterprise for $1200 to two friends, and later sold all remaining rights to Asa G. Candler, a native Georgian who had come to Atlanta 15 years before with assets of $1.75. Mr. Candler acquired complete control of the business by paying a total of approximately $2300. A born merchandiser, sales began to grow and distribution to broaden under his management of the company.

This astute businessman gave away thousands of complimentary tickets for a free glass of *Coca-Cola* as well as such souvenirs as fans, calendars, clocks, urns, and countless novelties. Advertising appeared on hundreds of different objects such as the Tiffany lamp at the turn of the century. Many of us can also remember the "pretty girl" change trays used in serving *Coca-Cola* in drug and confectionery stores.

In 1919, the Coca-Cola Company was sold for $25 million by the Candler interests to Ernest Woodruff, Atlanta banker, and a group he had organized. In spite of intense competition from other marketers of cola drinks as well as other types of soft drinks, sales of *Coca-Cola* have continued to grow. Consumption has now reached over 175 million drinks per day worldwide, marketed by over 1,500 bottlers in 135 countries. Not long ago, the 10th billion gallon of *Coca-Cola* was produced.

As flavor success stories go, this must be one of the most impressive on record, and *Coca-Cola* has become the favorite ambassador for the U.S. all over the world.

Chapter III

Developing That Magic Formula

In a number of the case histories we reviewed in the previous chapter, in which a superior fragrance or flavor played a major role in the marketing triumph of one product or another, the formula for that highly successful product was often the creation of the man or woman who started the company, the success of a single individual. But in recent years product successes are more liable to be successful team efforts. New fragrance or flavor introductions may now come from the marketing department more often than from the technical department or the perfume and flavor staff.

There is a growing dialogue between the perfumer, cosmetic chemist, flavorist and food technologist on the one hand, and the marketing expert on the other. This dialogue involves an exchange of ideas and planning. Those who formulate fragrances or flavors can't work in a vacuum, and those who know the market place need the stimulating contact with creative and technical people in the laboratory. In addition to those at the conference table, there are many others representing both sides working behind the scenes.

Norman Greif, Givaudan Corporation, in a talk on the "Application of Essential Oils and Compounds" delivered at the VIth International Congress of Essential Oils in San Francisco, September, 1974, put the marketing expert right in the driver's seat when he said "Finished goods manufacturers have generally in the past, and now almost exclusively, put the full responsibility for fragrance selection for their products in the hands of marketing people rather than technical people or fragrance compounders. This trend has made for more commercial and somehow more successful fragranced products having been launched in recent years."

Mr. Greif observed that in the past few years a large number of green floral perfumes, dozens of lemon-based products, and numerous highly scented toilet soap bars have been some of the results of marketing concepts, rather than concepts inspired by the perfume compounder. Granted that the marketing executive should have his finger on the pulse of trends — what people want and what they are likely to want in the near future — but he can contribute much more than this. He or she knows where the company's product should be positioned in the marketplace, what the competition is, the best time of year to introduce it, whether it should be test-marketed or released everywhere simultaneously, what promotion (including advertising) needs to be executed, etc. Very often, a name must be selected, so, creative members of the marketing or the advertising staff should also be present at meetings. After a name is chosen, a promotion and advertising program must be planned.

There is also the matter of cost, so obviously someone who knows what percentage of product cost should go into raw materials, is essential to the dialogue. Knowing the positioning of the product helps to determine what the retail price must be, and the group can work backwards or forwards in arriving at a breakdown. All other costs of the finished package must be estimated too, including labor and overhead, none of which concerns the fragrance compounder. But he needs to know what his share is, else he runs the risk of underestimating or overestimating his selection of ingredients.

Mr. Greif says it this way: "In selecting fine fragrances, the perfumer has the most knowledge as to what new compounds and materials are available, but I don't believe he is close enough to the marketplace or the consumer to apply this knowledge correctly. On the other hand, the marketing person doesn't know enough about new fragrance materials with which to develop new concepts. There must be more interchange of information between the compounder and the finished goods marketer if more fragranced products are to succeed."

Speaking for the representatives on the other side of the table, Mr. Greif puts the responsibility on the perfumer to find and develop new fragrance concepts, whether by using new chemicals more creatively, or by finding new ways of utilizing existing natural materials. The creative perfumer must come up with unique, special fragrances not only for an established type of product, but ones that will expand the use of fragrance into newer and more desirable consumer products.

Mr. Greif speaks of using naturals, creatively combined with new research chemicals, the former for fragrance beauty and the latter for lasting power and diffusion. These two must be correctly balanced for more uniformity of product and higher oil concentrations. Today, there is more emphasis than ever before on the lasting quality of a fragrance. This has resulted in higher concentrations being employed, especially in the United States and in South America.

Speaking further on the responsibility of the perfumer, Mr. Greif uses a phrase we think is the heart of the matter, which is that the perfumer is of the most help to the marketer in "selecting the style of fragrance." "Style?" Of course! A fragrance must be styled just as fashions are, or buildings, or table settings, or anything else. Present styles in fragrance, according to Mr. Greif, are sweeter, heavier, more feminine, sexier, all marked by nostalgia or a return to the past. Fragrance styles, like fashion styles, are cyclical, but new developments in fabrics or chemicals mean a return-with-a-difference.

In a speech delivered by the late Ernest Shiftan, International Fragrances and Flavors, Inc., at the 7th Annual Industry Conference of The California Cosmetic Association in Los Angeles, Calif., April 4, 1968, he too, spoke of trends and of the reciprocal relationship between the marketer and the perfumer: "I have often been asked and I have asked myself, if there is such a thing as a general trend in perfumery. We definitely know that trends exist. Whether we call them directions, developments, or evolutions, it all comes out the same. Nobody can deny that with passing times the conditions of life have improved, and even if certain people cherish the products of the past in other fields, I am sure that where perfumery is concerned they enjoy the improvements which have been made in the quality and the choice of fragrances. In the United States especially, the taste for good living has increased intensely in the last 10 years. On one side people have learned to be aware of better things and to appreciate finer perfumes as well as cosmetic products. On the other side the tremendous development in aroma chemical research has given us perfumers the possibility to improve the qualities of all perfumes, even inexpensive ones. The variety of fragrances which we can produce today may be 100 times as great as it was 50 years ago."

Although this talk was given almost ten years ago, what Mr. Shiftan said is still valid except that we might add "even more so." But to his conviction that the "quality of life" is responsible for new trends as well as greater discrimination in the choice of fragrances, we'd like to add some thoughts of our own. Supplementing the undoubted "quality of life" factor, there are also many social changes that make for changes in our life style. These also affect trends.

This is not the place for a long discussion on sociology, and we needn't detail these social changes. Briefly, however, there is much greater social freedom, more casualness, fewer prejudices, and restraints, a breaking down of class barriers, a recognition of minority rights, etc. There is a yearning, especially among the younger members of our society, for "things as they were," a craving for the simple and natural, the basic, the fundamental. We spoke of nostalgia before, but there are social reasons for this other than the better "quality of life."

Mr. Shiftan, like Mr. Greif, had further comments on today's raw materials situation. He said, "Trends are also dependent on the availability of raw materials. If some materials are less available, or get very expensive, they will be used less and less. Others which are available in larger quantities and, therefore, go down in price, will be used more. As a result, their character will be important at a specific time in the trends in perfumery. We are also looking forward to receiving even newer natural raw materials from different countries, in the not too distant future."

Then, on the subject of an effective dialogue between perfumer and marketer, Mr. Shiftan offers his thoughts: "Often the head of a fragrance or cosmetic house, or a director of an advertising agency, or a person involved in the promotion of a new product, would like to convey his own ideas to the perfumer. In order to do that he can either speak of natural odors with which he is familiar, as mentioned before, or he can speak of well-known perfumes on the market which he might like to use in a different way. By this I mean that someone might have the idea to give a lipstick the fragrance of a famous perfume or of a famous soap. In this case, the explanation is easy.

"On the other hand, if this person has an abstract idea, how can he convey that to a perfumer? He then has to have practical knowledge of an odor vocabulary. He should be familiar with most of the natural products, meaning the odors of all the flowers, the leaves, the barks, the roots, the gums, the grasses, the herbs, the fruits, the spices, the seeds, etc. Every person who is really aware of odors must also be aware of tastes and he will know many natural fragrances through the food he eats and the beverages he drinks. He should know and understand the meaning of 'dry' and 'sweet,' two expressions which are often misunderstood or, let us better say, understood in different ways by different people."

In addition to having a knowledge of and a vocabulary on the scents of natural fragrances, Mr. Shiftan rightfully expects the inquirer and seeker to use descriptive words, of which he gives a number of examples. We will use some of his and some of our own to illustrate: modern, full, gay, romantic, soft, intense, brilliant, sexy, sophisticated, brisk, fresh, complex, lilting, sprightly, magnificent, elegant. We could go on forever! Aside from the style of any particular fragrance, are there certain criteria by which it is possible to determine the superiority of a fragrance generally? These criteria, if so there be, might be expressed differently by a perfumer than by a manufacturer or by a customer, but in the final analysis they come to much the same thing. A perfumer might talk about the top note, body, and bottom note, or he might express the three phases in the development of a fragrance in other terms. Certainly, a perfume should be a thing of beauty, but beauty is largely subjective. What is beautiful to one person may leave another indifferent and unmoved. Beauty is an emotional thing, depending on nostalgia, association, aesthetic concepts, etc. It's personal.

When a finished goods house goes to a manufacturer of fragrance compounds seeking a fragrance, there is a great deal of information that must be supplied to the formulator of the fragrance before he can proceed about his task intelligently, and with any assurance that the fragrance, when ready for submission, will even approach what the prospective buyer has in mind. An excellent outline of the type information which should be supplied by the finished product marketer is contained in an article, "How Does One Describe A Fragrance?" in the 1/76 issue of "Fragrance Magazine Naarden" published by Naarden International, Naarden-Bussum, Holland. The procedure developed by Naarden's Fragrance Advisory Board is to provide a checklist for a perfume briefing in which the prospective customer is asked to answer thirty-three questions, divided into four principal groups related to: a Product Profile; a Market Profile; a Perfume Profile; and a Procedural Profile.

With Naarden's permission, which we gratefully acknowledge, we offer a condensed and reworded version of this checklist.

PERFUME CHECK LIST

1. For what product is the perfume to be created? (e.g. a shampoo, cream, toilet soap, scouring compound, etc.)
2. Is the product new, or is it an existing one being revamped?
3. Is it a single product, or does it form part of a line? If the latter, what are the other members of the line?
4. Are unperfumed samples of the product available?
5. If not, please give the physico-chemical properties of the product, e.g. — is it solid, semi-liquid, or liquid? If it is an emulsion, what type?
6. If the base contains essential ingredients which could influence the fragrance, please specify.

   In the case of lotions, is either ethanol or isopropanol employed? In what concentration?

   If an aerosol, what type propellant is used?

   Does the product contain chemically reactive ingredients? If so, what are they?
7. What is the pH of the product?
8. What is the color of the product? (As some perfume ingredients can cause discoloration, it is important for the perfumer to be aware of how his ingredient selection can be limited by the possibility of affecting the appearance of the product.)
9. What type packaging is employed?

   Aerosol or non-aerosol?
   Transparent or opaque?
10. Of what material is the package made?
11. Can samples of the packaging material be supplied?

Market Profile

12. In which area of the world will the product be marketed?
13. What is the attitude of consumers in this area toward this type of product? Is it regarded as a boon, or a necessary evil?
14. Has the general attitude toward the product type

changed in recent years? If so, in a positive or negative way?

15. Does the attitude in the market area differ greatly from that in the surrounding areas? If so, why?
16. Describe the attitude among consumers in the target area by (a) age, (b) sex, (c) social status, (d) type apparatus that must be present in a household in which the product is to be used.
17. Who are your principal competitors?
18. What are (and were) their shares of the market? Yours?
19. What are the demographic positions of their products, and yours, in terms of the four factors mentioned in question 16?
20. List the strongest and weakest points of your own product and those of your competitors' products.
21. What is your target group? Which consumer category is it your primary aim to reach?
22. How do you plan to position your product?

    In what senses must it distinguish itself from its competitors?

    If you are revamping an old product, in what senses must the new product distinguish itself from the existing one?

    Which properties will you call to the attention of potential buyers in your advertising?

Perfume Profile

23. Describe the type fragrance desired, as compared with products now on the market.
24. Define the desired fragrance more closely.

Fragrance Class	Examples
Fresh	Menthol, Petitgrain, Citronella, Eucalyptus
Green	New-mown grass, Liguster, Cucumber, Runner Bean
Citrus	Bergamot, Lemon, Orange, Lime, Tangerine
Floral	Lily-of-the-Valley, Lilac, Rose, Hyacinth, Jasmine, Carnation
Herbal	Camomile, Sage
Lavender-like	Lavender, Rosemary, Spike lavender

Fruity	Peach, Pineapple, Apple, Strawberry, Raspberry
Animal	Civet, Leather, Faeces, Amber, Musk
Spicy	Pepper, Mixed spices, Cinnamon, Clove, Nutmeg
Woody	Cedar, Sandalwood, Vetiver, Patchouli
Balsamic	Olibanum (incense), Opoponax, Labdanum
Coniferous	Pine, Juniper berry, Cypress
Mossy	Oak moss

25. Again relate the fragrance you desire to those of the principal competitive products.
26. Does the perfume play an essential role in your marketing mix?
27. What is the proposed concentration of perfume?
28. What is the target price for the perfume?

Procedural Profile

29. How many perfume suggestions do you want submitted?
30. What procedures do you employ to test?
    Consumer appreciation of the perfume?
    The technical properties of the perfume?
31. What criteria do you employ for selection, and what is the relative importance of each?
32. Will your test results be available to us for examination and comment?
33. What is your schedule?
    Date of briefing........
    Deadline for submission of suggestions........
    Deadline for screening test........
    Deadline for final choice........

Many marketers of finished products make no attempt to "tell" the customer that she should find their fragrances beautiful. They depend on the prestige of their name, perhaps adding a slogan to a picture of their flacon. Sometimes, a few phrases are added to give the customer an image. Happily, more and more manufacturers are describing their fragrance in terms of its ingredients.

We can remember when many manufacturers felt it

would be a mistake to do this, on the premise that if a potential buyer didn't happen to like lily of the valley or rose or carnation or whatever, she might not buy the fragrance. But, there are plus as well as minus factors in describing the ingredients of a fragrance. For one customer who might be lost, a dozen others could be gained. Perhaps the reluctance of some marketers to describe the nature and ingredients of the fragrances of their products needs to be re-examined.

Now, a still different approach has been taken by Houbigant in advertising their *Essence Rare*. It comes closer to a perfumer's description, but it's obviously a copywriter's language. The caption says, "We Searched Until We Found" followed by the name of the fragrance. Then this search culminated in finding a scent "that starts softly, develops beautifully, and never seems to end." Here we have the initial impression of a fragrance, its main theme, and its lasting quality. Just what a woman wants. And just what a perfumer promises. This seems to us to be a "real meeting of the minds," illustrating just what we were talking about earlier in this chapter.

Another approach that intrigues us is that of Lanvin in researching consumers' appreciation of fragrances as being in four different categories: Elegant, Romantic, Life Style and Sensual. Instead of thinking of the anatomy of a fragrance, they think of it in terms of "effect." That makes sense too.

Lanvin's research was undertaken in anticipating the advertising of *My Sin*. It was found that consumers regarded this fragrance as a "sensual fragrance with a sense of style, as an earthy scent with a certain elegance." In other words, *My Sin* fulfills three of the criteria set by the user herself. No perfume need be limited to fitting into only a single category, but can embrace two or more. Perhaps the results of this research will cause a change of direction in advertising thinking throughout the industry. We detect marketing behind this kind of research.

Going back to aesthetics for a moment, the response to a fragrance is certainly personal, as we said before, and there is no one standard of beauty. If there were, then we'd need only one perfect fragrance to please everybody. But a woman would be narrow, indeed, if she chose to wear only one fragrance constantly. A dozen different fragrances could be beautiful to her in different ways. Comparing the creation of a fragrance as an art form to those of other art forms, we excerpt again from Mr. Shiftan's talk as follows: "There are two basic methods for a perfumer to proceed in practicing his art. 'Inductively,' meaning to add ingredient by ingredient until a desired note is achieved, or 'deductively,' in making a compound of many products and then taking away, little by little, from this original mixture like a sculptor, until the intended effect comes out clear and clean.

"For instance, if it has been decided to create a perfume which is fresh in the beginning, not very heavy, and comparatively related to the active way of life, like sporty, or more adapted to a collective life, meaning that the perfume should not shock a neighbor, one which is not too typical nor too violent, then the perfumer has to think of creations which will produce that effect. His work consists of making several combinations until he finds one which is pleasant, comparatively new and which has a character of its own. He might take one of his formulas and then work, as mentioned before, like a sculptor, taking out certain ingredients to come to a new combination which has the desired characteristics for a special product."

A perfumer is a highly disciplined person who must exercise self-criticism and trained judgment, but Mr. Shiftan accepts and applauds the fact that he also needs critics. What artist doesn't! Just as critics in all fields of art, fragrance critics need to possess creative imaginations even if they, themselves, do not create. Mr. Shiftan proposes that fragrance critics should consist of both women and men of different ages and of different experience, but that all should have a cultured background and an acquaintance with fragrances over a long period of time. They should be able to recognize patterns in fragrance preferences so as to anticipate what might be the "coming thing."

Norman Greif sets these criteria for a good perfume: (1) most desirable out of the bottle; (2) it has to stay true from top note to dry down; (3) it should diffuse around the body and elicit compliments; (4) it should last long and remain recognizable to the end, and (5) it should have a style of its own, whether floral, sexy, green, sporty or just clean and fresh.

Whether we use the suppliers' words or the manufacturers' words or the consumers' words, we need the point of view of all three in achieving a superior fragrance formulation.

If not exactly spelled out, it is obvious that so far we have been limiting ourselves for the most part to such forms of fragrance as perfume, toilet water and cologne. But there are a number of other vast areas of perfumed products, including bath preparations, hair preparations, treatment items and make-up, household and industrial products. The perfuming of soaps alone presents complex and unique problems, so we will treat this subject separately at the end of this fragrance discussion.

Certain generalizations can be made that apply equally to all of the above categories of perfumed products. Manufacturers must, if they are to produce superior formulations, supply the perfumer with his base materials whether they be powders, oils, liquids, foams, creams, soaps, whatever. If the perfumer is simply translating an accepted perfume for use in other products, or if he is developing an entirely new fragrance, he must make many different tests for such things as stability, discoloration, irritation, etc. Even after several weeks of testing, he may fall short of perfection, and then he must begin all over again. A fragrance that is beautiful in a cologne, for instance, might not come out true in an entirely different base. It might discolor the finished product. Or it might be found, after testing, that it has a short shelf life. Corrections must be made, fragrances may need to be reformulated, ingredients substituted, etc. This may well be a long and tedious process and everyone concerned must be patient.

In a talk given by J. Scott Birnie of Lever Brothers Co. at

the International Congress of Essential Oils in San Francisco in September, 1974 on "The Role of Perfuming Materials in Finished Goods," Mr. Birnie makes many observations that apply not only to soaps and detergents, but to other kinds of products as well. We are excerpting some of these statements, and later will quote him in a more specific context.

Mr. Birnie reminds us that the consumer is the final reference on the acceptability of any product and that the perfume ingredient must reflect the consumer's preference. He says, "Unless marketers are familiar with current attitudes and preferences of the customer, we may fail to provide her with what she needs and wants. Twenty years ago, in England, a strong pine note was the almost universal perfume for household cleaners. Today cleansers are sold with lemon fragrances and other citrus notes, because women prefer them. Life styles are changing, and suppliers must reflect these changes."

Mr. Birnie spoke of the recent trend in the United States towards the naturals, not as opposed to synthetics, but to the type of herbal, outdoors, fresh fragrances that have been developed so successfully in the hair care, shampoo and skin care areas. Thus, the perfumer has been challenged to develop new types of perfumes to meet these consumer trends.

For a successful collaboration between perfumer and manufacturer, Mr. Birnie outlines three necessary elements. First, the manufacturer must define precisely what his product is supposed to do, he must define the target consumer group, and specify the function of the perfume. Birnie suggests a detailed briefing to the perfume supplier. Secondly, there should be a clear understanding of the nature of the testing to be used to determine the final product. And the third element lies in the service that the perfume house can provide. Even after the perfume has been selected, the supplier should be able to advise the customer on trends in other fields which could well apply to that customer's product. Just as the marketer can guide the perfumer on trends, the reverse is also true.

We have quoted from Mr. Shiftan earlier and will add some additional comments in this connection. He points out that the fragrance of functional products such as soaps, detergents, creams or tissue paper is usually smelled by the individual user. This is in contrast to colognes and perfumes which are shared with others. The perfumer performs accordingly. Not only must he think of stability, discoloration and irritation, but sometimes of solubility of the ingredients in the formulation, resistance to extreme temperatures, reactions with alkaline or acid ingredients present in the product, compatibility in aerosol formulations, etc.

Mr. Shiftan says, "Fortunately, the number of aroma chemicals and also of special rectified natural materials is so plentiful today that an experienced perfumer who has a competent research department behind him has no difficulty whatsoever in adapting his formula to the desired product. The most important thing is, naturally, that the perfumer create a pleasant perfume which helps to sell the final product."

In Mr. Birnie's talk, referred to earlier, he gives special attention to the perfuming of soaps, citing the recent trend for incorporating a higher percentage of perfume into soaps. Where 1% was once standard, now the usual soap bar contains 1.5% or even 2% of perfume. And to quote at greater length: "Nor is there any doubt that the future will see a greater involvement of perfuming materials suppliers in the methodology of the soap and detergent industry, — first technically. Suppliers must be familiar with the formulas of our products, their processing, packaging, and estimated shelf-life. Raw material shortages have produced notable changes in the formulas of many products which affect the performance of the perfume. Changes in packaging materials can directly affect the impact of fragrances. Legislation, like the ban on phosphates, can lead to the need for radical reformulation, requiring new consideration of perfume type and concentration."

A tremendous amount of scientific data on soap perfuming has emerged over the years which is of constant concern to perfumers and perfume chemists as well as to soap manufacturers. Of course, the customer is concerned only with the final result. Is the fragrance of the soap delightful, leaving her skin pleasantly scented, without interfering with the effect of her perfume and related toiletries? We are speaking, in this context, of fragrance only, without reference to the other qualities of a superior soap — lathering power, ability to cleanse, lasting characteristics, etc. Naturally, a successful soap must score well in all these respects, but the fragrance can well be an important factor affecting the customer's decision of what brand to buy.

One of the best studies of soap perfuming ever written was an article by the late Philip Chaleyer, which appeared in the 1934 *Soap Blue Book.* Although published over 40 years ago, much of what Mr. Chaleyer said then is still valid today. We base the following discussion of soap perfuming on Mr. Chaleyer's article, with a few revisions by modern experts in the field to bring the author's comments into line with current practice in the art of soap perfuming.

Perfuming a white bar of soap places more restrictions on the perfuming materials used than the perfuming of colored soap, especially as regards discoloration. In the case of white soaps, the candidate fragrance must be tested in the sunlight, to see if discoloration takes place on exposure. The cakes under test should be observed for at least two months. It can happen that if two perfuming materials are tested separately, neither will cause discoloration, yet when the two are mixed together discoloration may occur.

Discoloration tends to be caused by certain groups of chemicals, but may also result from impurities in any single ingredient, so sources of supply must be carefully checked. While some aromatic chemicals or essential oils are excellent in soaps, others must be rejected, or their use limited to certain types of soap.

The cost of the perfume ingredient is obviously an important factor in perfuming soaps, especially the mass market kind, so the soap manufacturer is always strongly concerned with the price factor. Too great a concern over "price-per-pound" however can be a mistake, because the more expensive of two perfuming ingredients may have not only a finer fragrance, but may also deliver twice the perfuming value. "Cheap" can sometimes be expensive.

Commercial grades of essential oils can sometimes be irregular in quality, so the goal must be to procure those of highest purity. Quality also varies with climatic conditions, country of origin, and methods and processes used in distillation. Impurities may be picked up during storage which can cause discoloration. For example, oil of lavender can develop acidity during ageing, causing it to attack drums or cans in which it is stored. Traces of metal from the containers can cause discoloration. Or, if iron kettles or milling machines are not kept in prime condition, again there can be traces of metal in the oil which will cause discoloration.

Soluble resins and balsams in large proportions are recommended for giving a good odor background. Styrax, balsam peru and balsam tolu form a good background for all types of soap fragrances, but are particularly good fixatives for lily of the valley, lilac, chypre, rose and in general, all floral bouquets. Benzoin is very good for rose and oriental bouquets; orris for violet; castoreum for oriental bouquets and, in general, whenever an animal fixative is required. With the exception of castoreum, soluble resins can be used in large amounts, constituting up to 20 to 30% of the perfume in the formula. Some soluble resins discolor in white soap and must be limited to colored soaps. The colorless essences do not cause discoloration and can be substituted in the case of white soap. They are excellent fixatives, especially for floral bouquets. Distilled essences are many times stronger than soluble resins, and although more expensive they represent excellent value in terms of fragrance delivered.

Discoloration, with certain materials, may be so infinitesimal as to be unimportant, as in the case of castoreum liquid absolute which is usually used in minute quantities. Oak moss, already fairly dark in color, sometimes contains large amounts of chlorophyll which can turn this perfuming ingredient even darker because of the combined action of light and alkalis. If olibanum is pure, properly cleaned and purified, it will not discolor. Orris not only does not discolor, but is very stable. All the styrax products are also stable and can't be faulted for discoloration.

Results obtained with aroma chemicals are generally good, because they are well standardized and of uniform quality, although, of course, there are exceptions. For instance, the chlorine in benzyl acetate will form yellow spots in white soaps. Esters and phenols stored in drums or tins will frequently pick up traces of iron, causing spots. Certain artificial musks may cause trouble, but can be tested for purity.

An analysis of the determining factors in making a selection of these materials made by Mr. Chaleyer takes the following into account: (1) Discoloration; (2) Odor Character Change; (3 and 4) Lasting Index for Wrapped and Unwrapped Soaps; (5) Good as Anti-Oxidant; (6) Intensity of Odor; (7) Good for all Soaps; (8) Good for Colored Soaps Only.

The meaning of these headings is mostly self-evident, but we will add a few words of explanation which may be helpful. On odor change, certain products may change their character through the combined action of alkali and humidity. For instance, terpenes take on a turpentine-like odor, clove becomes sweet and loses its spicy character, etc. Materials that act as anti-oxidants help to prevent rancidity; others may have the reverse effect. Stability is the key in factors (7) and (8) and that refers both to discoloration and character of odor.

Even this brief resume of soap perfuming problems is indicative of what we said earlier about this phase of perfuming being highly complex.

Developing a Flavor Formula

From this glimpse of some of the problems the perfumer must solve in creating a successful fragrance for any one of a long list of consumer products,.we'll take an even more rapid glance at some of the problems the flavorist faces in developing a successful flavor formula. Both the perfumer and the flavorist are highly creative, but their targets are normally different. Where the perfumer doesn't just repeat or try to imitate Nature, but often works with abstract concepts in mind, the flavorist directs his efforts almost entirely to reproducing natural flavors.

As suggested in the previous chapter, creating a successful flavor compound is often a much more difficult problem than compounding a fragrance. A flavor compound must not only *smell* the same as the item it is attempting to duplicate, but must also *taste* the same. A further complication is the fact that, because it must be extended and diluted for use in the finished food product, the flavor compound must initially be more intense than the flavor as it exists in nature.

Thus, the natural character of an artificial cherry flavor is enhanced by adding benzaldehyde, a strong aldehyde found naturally in cherry juice. Flavor chemists also often add homologs of benzaldehyde, such as tolyl aldehyde, to produce a more natural flavor. Similarly, decyl aldehyde and ethyl butyrate, natural components of orange essence and orange juice, are the enhancers used to compound a more intense orange flavor when required.

It is difficult to produce sufficient strawberry concentrate in a natural strawberry flavor to build up the sugar content high enough to make an acceptable finished product. Thus another natural food flavor, such as apple concentrate, is sometimes used to enhance the sweetness of the strawberry. Apple is high in natural fruit sugar and acids, but has a relatively weak flavor of its own. It can be incorporated in a strawberry flavor to make it sweeter, without materially affecting the flavor of the strawberry. However, when so used, the resulting flavor must be labeled "artificial strawberry."

Consumer activists have tried to make food buyers suspicious of "artificial" ingredients in foods, including flavors. In doing so, they have probably done them a disservice. The general public should realize that, without the availability of synthetic flavors, which help the flavor industry to extend the limited stocks of natural flavoring materials, many products would have to go to market unflavored, and the prices of flavors would go to the astronomical levels at which herbs and spices used to sell in Europe in the Middle Ages, before opening of the sea routes to the East Indies in the 15th century made these materials more freely available in the European market.

Those unacquainted with the problems of the flavor compounder often gain little but misinformation from reading labels. It would be to the advantage of all concerned — to the general public, to flavor buyers and food marketers, as well as to flavor compounders, if more people were educated to appreciate the common uses of such terms as "artificial."

Flavorists and food technologists have outdone themselves over recent years in developing new taste and flavor sensations, — for products old and new. And, with every successful new formula they develop, they continue steadily to expand the market for flavoring materials, just as marketers of perfumes and cosmetics, soaps, and household products are steadily expanding the market for fragrances.

In a recent interview with William J. Downey, vice president and chief flavor chemist of Fritszche Dodge & Olcott Inc., he told us about an imminent meeting with a customer at which both technical and marketing people would be present. His company had created a new flavor which they felt would not only be custom-made to suit the product, but Mr. Downey explained, sometimes an especially designed flavor can serve as a springboard for use in an entirely different product. This is where the representatives for the manufacturer can prove very valuable.

This was confirmed by Michael Mason, vice president and general manager of the Flavor Division of International Flavors and Fragrances, Inc., a company that maintains its own marketing department. Dr. Mason feels that much of the success of a flavor depends upon the competence and experience of the flavorist, along with the availability of the raw materials with which to work. However a flavorant is not a "flavor" until it performs in the end food product in a manner consistent with the expectations of the customer. Thus the food technologist also plays an important role in the development of a successful flavor.

We will not explore the multitude of flavors that have emerged from the laboratory, but will talk instead of marketing trends and new products coincident with changing life styles in our eating habits.

Quoting briefly from a paper by Mr. Downey, "Few areas have expanded in so many directions as the food industry. A more and more diverse array of new products continues to pour forth, all of which require flavors and many flavor specialties with unique properties and characteristics. To satisfy this need, the flavor chemist has been called upon to balance his artistry with the employment of highly sophisticated instruments."

Think of the overwhelming growth of snack foods; the modern need for convenience foods of all kinds, including frozen dinners; the multitude of beverages, especially the powders to be mixed with water; the different ethnic mixes of vegetables to be combined with meats; the instant soups; the gravy powders; and on and on. We can be sure that there will be many more to come!

We cite some of the recent innovations in the food industry as examples. These few by no means encompass the enormous expansion that has occurred in the food industry over recent years, but perhaps will serve to make our point. We are not going to try to track down who invented what, nor assign the shares played by the flavorist, the food technologist, and the marketing people, by the supplier and the manufacturer. But certainly these recent food innovations seem designed to satisfy consumer wants, to give the consumer what she wants — often before she may know she wants it.

For example, we are intrigued by all the new taste

accents given to coffee. A hostess no longer asks simply whether or not a guest takes cream or sugar. She may have a whole coffee menu to offer, whether the several *International Coffees* of General Foods, or coffee with a mint or an orange flavor, or whatever. We have long been accustomed to mint or lemon with tea, but coffee is getting into the act too.

If you're a tea fancier, Nestea does it up both brown and fancy in a delicious blend of *Nestea* with tangy fruit juices and bubbling ginger ale, to which you can add a fresh fruit garnish if you like. Nestea offers a booklet of several dozen recipes called "Nestea Recipes For All Seasons." Hot or cold, tea is a gourmet thing these days. If you abhor monotony in your beverage, here's an adventure for you. There's literally no end to variety in cup or glass, whether coffee, tea, or any number of other beverages.

A number of lipsticks have been given flavors in the past. But it's something new when Bonne Bell flavors their lip gloss with *7 Up,* introducing the *Uncola Lipsmacker* as a winning taste treat. We anticipate more tasty lips to come.

Croutons are welcome additions to soups and/or salads. A newly introduced product in this category is *Bac-Os,* which are crunchy and have a bacon-like flavor. In the salad field is *Buttermilk Salad Dressing,* marketed by Hidden Valley Ranch, a really delicious and different dressing. It is offered in five flavors, and comes in a mix form in an envelope, convenient to store.

Another taste treat, recently introduced, is *Eggo Brand* Strawberry Waffles. These waffles are frozen, real time-savers for a rush breakfast, fast and easy to fix, and at the same time delicious to eat. Still another novel taste sensation is McCormick/Schilling's *Sloppy Joe*'s seasoning mix to add flavor to hamburgers or ground meat.

Another innovation, which is really not an innovation at all in a manner of speaking, is a line of strong drinks marketed under the name, *Hereford's Cow.* There is a *Strawberry Cow,* a *Mocha Cow,* a *Banana Cow,* and a *Chocolate Mint Cow,* each a blend of natural flavor with grain neutral spirits. The advertising employs the punch line, "A Cow-on-the-Rocks is not a bum steer." We have been told that this recent introduction is a smash success, far outselling all the milk punches and egg nogs of the past. Or, you might prefer *The Tomcat,* a peach sour flavored mix offered by Early Times to add to bourbon whiskey. There are a number of other mixes, one of the best sellers around the New Jersey area being *Tree-Ripe* lemon juice mix, marketed by Tree-Ripe Products, Inc., Fairfield, N.J.

When we come to ice cream, there are a multitude of fresh, new and delicious flavors being offered today. No longer are we limited to vanilla, chocolate or strawberry, as in our youth. Some marketers now offer over a hundred flavors including such rarities as Pumpkin Pie, Cheese Cake, Bubble Gum, Licorice, Coconut Crunch and Peanut Brittle.

Every flavor compounding problem that the flavorists and food technologists solve carries with it the potential for a further expansion in this country's exploding market for foods of all kinds.

Chapter IV

Definitions

Every trade has its own vocabulary — the language which those "in the industry" speak to one another. It normally consists of trade jargon, colloquialisms, idioms, abbreviations, extensions. contractions and combinations of conventional words which take on a special meaning for those who understand this "in" language. Every specialty industry has its own specialists, who need their own special terms, which they readily invent, adapt or adopt to suit their special needs.

The fragrance and flavor industry is no exception; and since it encompasses several highly specialized fields, which have many unique characteristics, it has developed a highly colorful and quite descriptive language. That language has been shaped by numerous influences, including the fact that the industry is not only ancient but also very modern, it is both an art and a science, it is multi-lingual, both subjective and objective, and since it is world-wide in scope it necessarily has roots in both the Occidental and the Oriental spheres.

We find old words used, along with the most modern; English, French, German and Italian words; highly technical terms and far-out, arty phrases. In addition to a language for fragrance, there is another for taste and flavor. One result of the confused nature and origin of the industry's language is that even members of the industry often disagree as to what a specific term means, and there are inevitable errors and contradictions. We will try to clear up the picture a bit with the following discussion of a group of the most important basic terms used in the F/F industry. We hope that, as a result, readers of this book will be in better position to understand the content to follow, and will also be better equipped for the future to understand the language which industry members use in talking to one another.

Our vocabulary will describe the raw materials with which our industry works, the processes it uses, chemical names, trade names and classifications of flavors and fragrances, the finished products in which flavors and fragrances are such important ingredients, and finally the effects on subjective human senses exercised by fragrances and flavors. Because of a common derivation, it will be found that many of the definitions used in the F/F industry are similar to those used to describe products of the drug and pharmaceutical industry.

Actually, in many cases, we will not attempt to *define* some of our terms precisely, but will rather *discuss* them. Nomenclature is always a thorny subject, inviting controversy, and in cases where we are aware there is considerable disagreement as to exactly how certain words should be defined, we will not push our own opinions aggressively. We will also reject the usual course of taking up terms in strict alphabetical order, and start right out with a discussion of "essential oils," the basic products with which the fragrance and flavor industry started. Several years ago when the American Society of Perfumers tried to agree upon definitions of fragrance terms, they suggested the following for **Essential oils:**

> Volatile oils obtained by various processes from leaves, petals, twigs, roots, bark, seeds, woods, fruit peels, etc.

We are not satisfied with this definition. A better one we believe might be:

> Aromatic oils derived from raw materials of botanical origin by such processes as distillation, expression, extraction, maceration, etc. which isolate and concentrate the basic aroma principle of the raw material.

But this definition doesn't completely satisfy us either, so we will compromise by just discussing what "essential oils" are rather than attempting to *define* them. The word, "essential," does not translate as "most necessary," although this phrase certainly describes the value of essential oils. Rather the term "essential" is derived from "quintessence," the vital or significant portion of a material, its descriptive odor or flavor component, the concentrated characteristic of a natural flavor or fragrance raw material.

The raw materials from which essential oils are produced include leaves (peppermint), fruit (lemon), bark (cinnamon), roots (vetiver), grasses (lemongrass), wood (sandalwood), heart wood (cedar), gums (olibanum), balsams (Peru), berries (juniper), seeds (celery), flowers (rose), twigs (clove stems), and buds (clove). These are processed to yield their quintessences, which are mainly water-insoluble oils composed of complex mixtures of organic compounds of a variety of chemical structures. The oils are mostly devoid of cellulose, glycerides, starches, sugars, tannins, salts, etc. which occur in the natural product.

The method most commonly used for separating the essential oil from the natural raw material is steam distillation, but some oils are dry distilled (no water) under vacuum (labdanum), expressed (bergamot), or extracted (jasmin). Most expressed oils, particularly the citrus types, were formerly produced by hand pressing, sponge pressing, or by ecuelling (another manual method used on limes); but the expressed oils are now produced on modern, high-speed equipment. Many of the flower oils are extracted, using a purified petroleum ether. Enfleurage, the old process of extracting the floral aromatic principle on a neutral fat, is no longer a viable production method.

Yields of essential oils from the raw materials vary widely. Clove buds, for example, yield 15-18% by weight of essential oil, while garlic bulbs yield as low as 0.1% of oil.

Most essential oils are liquid at room temperature; however some are semi-solid *(Mentha arvensis,* also called cornmint or Brazilian mint); while still others are solids (oil of guaiacwood).

Proceeding with other industry products, which we will take up alphabetically, we will continue the procedure of discussing the various product types in detail rather than attempting to offer rigid definitions. Our first product type is the Absolutes.

Absolutes — are processed products that have been at least doubly extracted, first using non-polar solvents such as petroleum ether, followed by a polar solvent which is usually alcohol. Absolutes are thus equivalent to a resolubilization of a concrete, which further concentrates the vital components of the concrete, removing the waxes, glycerides, and di- and tri-terpenes. This produces a clear, highly concentrated material of improved solubility, strength and color. In some cases a neutral solvent such as benzyl benzoate may be used to improve fluidity.

Aroma Chemicals — are any natural isolates or synthetics which have an aroma. The natural isolates are removed mechanically (by distillation, etc.), or chemically (by hydrolysis, salt formation etc.) from a natural essential oil or product. The isolates are further modified. For example, bois de rose may be distilled to produce linalool, which is then acetylated to make linalyl acetate; cedrene and thujopsense are fractionated out of oil cedarwood Texas, then acetylated to produce acetyl cedrene; citronellal is fractionated from citronella oil, then reacted to produce hydroxycitronellal.

Synthetic aroma chemicals duplicate chemical structures found in nature, such as ethyl butyrate from butyric acid, and ethyl alcohol or citral, either from pinene (a sulfite turpentine source), or from isoprene (a petroleum source). A synthetic may have an aroma reminiscent of a natural product, although being entirely different in chemical structure. Typical examples would be hydroxycitronellal which smells like lily-of-the-valley, or the nitro-musks, which on dilution smell to some extent like natural musk.

Balsams — are the resinous exudates of trees and bushes, similar to gum resins except that balsams contain cinnamic and/or benzoic acids or their ester derivatives such as benzyl benzoate, cinnamyl cinnamate, etc. Typical balsams are balsam tolu, copaiba balsam, or benzoin Siam. Balsam fir Canada is a misnomer, as this material is a gum rather than a true balsam. Because of their composition, balsams have a warm, spicy note, described obviously as "balsamic."

Concentrated Oils — are processed terpeneless or sesquiterpeneless oils or other oils such as folded citrus oils in which the innocuous and unstable components of the oil such as terpenes and sesquiterpenes (which are flavorless hydrocarbons) have been removed in whole or in part by fractional distillation, solvent extraction, counter-current dual solvent extraction, thin film evaporation, molecular distillation, or recently by preparative gas chromatography. In some cases both fractional distillation and solvent extraction are used. Industry-wide specifications do not apply to concentrated oils, as they are specialty products of individual manufacturers which have been processed to improve stability, solubility and/or flavor concentration. Such products as oil bergamot terpeneless is approximately a 2.5 fold concentration; oil grapefruit terpeneless is approximately a 30-fold concentration; and oil orange sesquiterpeneless is about a 40-fold concentrate.

Concretes — are processed products resulting from extracting the natural raw material (jasmin *flowers,* orris *root,* labdanum *leaves,* etc.) with purified petroleum ether or other non-polar solvents. The solvents are then removed by vacuum distillation. The remaining materials are mostly solids or semi-solids at room temperature, as they contain in addition to the essential oil also di- and tri-terpenes, waxes, glycerides, and other similar type materials. This process is applied primarily to flower oils which cannot be steam distilled and also to other natural raw materials of botanical origin. Concretes have limited solubility in finished products because of their components, but may be further concentrated to absolutes. Extraction with petroleum ether separates out the cellulose, starches, sugars, tannins and other water-soluble components.

Essences — are products which endeavor to capture the highly volatile top notes of natural products, or which attempt to emphasize the top notes by a physical process such as developing a product with high evaporation rate. An example of the latter type essence is the simple solution of an essential oil in dilute alcohol.

When citrus fruits such as orange, grapefruit and lemon are processed to make frozen concentrate, essence oils of these juices are collected in the essence recovery units as the juice is being concentrated for solids content. The essences are separated from the water distillate of the juice portion after the expressed oil and the juice are separated by centrifuging (which takes place earlier in the process).

Other essences are made by water distillation and alcohol addition, or dilute alcoholic distillation, of such items as coffee or cocoa. In many cases the word "essence" is used to describe products which have a strong topnote, regardless of the method of production.

Extracts — are products in which the essential components of a natural raw material are taken up by a solvent such as alcohol, water, propylene glycol, or combinations of solvents. Hot or cold percolation or steeping are the methods used. After filtration, the finished extract may be used as is, or diluted or concentrated according to the requirements of the finished product. The most widely used extract is vanilla extract.

Extraits — are the alcoholic washes of enfleuraged fat, rarely encountered today.

Fixatives — are items of low volatility and preferably exotic odor (mainly the animal products, ambergris, castoreum, and the odoriferous gums such as galbanum, the woody oils such as sandalwood, synthetic musks, and in some cases the odorless benzyl benzoate, etc.) which have the ability by chemical or physical absorption to reduce the volatility of the other ingredients in a compound so that they will be released consistently over a longer period of time.

Flower waters — are aqueous distillates which remain after flower oils have been separated off, and no apparent essential oil results from further distillation of the flower charge. Flower waters are solutions in water of the more hydrophylic components of an oil. They are produced when there is still usable fragrance in the still charge, yet no more essential oil can be removed economically from the distillate.

Folded Oils — are essential oils which have been processed to concentrate the flavor components. For example, citrus oils contain high concentrations of terpenes, mainly limonene, which in addition to oxidizing readily, makes no addition to flavor but rather dilutes the flavor components. These oils are concentrated, or "folded," in an effort to reduce the terpene content. A 2x (two-fold) oil is one in which 100 lbs. of the original oil has been reduced to 50 lbs. by precise vacuum fractionation. A 5x (five-fold) oil is one in which 100 lbs. has been reduced to 20 lbs. Each type of oil (grapefruit, orange, etc.) must be treated differently in concentrating, since they have different flavor components and component analyses. Folding concentrates the significant components almost directly proportional to fold strength, however there is inevitably some loss of vital items even with the most efficient operation. Oils up to 25-fold have been manufactured, but a law of diminishing returns (yield vs. flavor concentration) normally limits folding to 7 to 10 times. The folded oils are more stable than single fold oils.

Gums — are the resinous exudates of the bark, twigs or leaves of trees or shrubs. These exudations may be forced by injuring the tree by incisions (as with olibanum, opoponax, myrrh, etc.) and allowing the sun to dry the exudate into tears or hardened droppings. In the case of labdanum, the leaves and the twigs surrounding the leaves exude the resin. When leaves and twigs are put into boiling water, they release the crude resin which is then skimmed off the top of the liquid in the vessel. The crude resin may subsequently be refined to reduce color.

The gums described above are water-insoluble, and are gum resins. There are other water-soluble gums such as gum arabic, tragacanth, agar, etc. which are rarely used in fragrance applications as is, but do find employment as thickeners and spray dry carriers in the manufacture of water-soluble fragrance and flavor compounds.

Oleoresins — are extractions, usually of natural food or flavoring raw materials, using a selected solvent to remove the vital components. The oleoresins contain the essential oils, plus other vital non-volatile components which enhance the flavor of the essential oil or act as fixatives. Thus, red pepper or capsicum may be extracted to concentrate the heat principle, capsaicin, and its derivatives. Black pepper is extracted to produce piperine and its homologs which contribute to the characteristic bite of that spice. Paprika and turmeric are extracted to separate the color components in high concentration. Other oleoresins such as sage, mace, tarragon, etc. give truer representations of the whole spice itself, and exhibit improved heat stability in the finished products in which they are used.

Pomades — are combinations of purified fat and flower oils produced by the enfleurage process. These are no longer articles of commerce.

Rectified Oils — are essential oils which have been processed by redistillation to improve a particular property or characteristic by removal of objectionable notes or concentration of preferred components. This is usually accomplished by vacuum distillation. Oil bergamot II rectified is bergamot oil that has been distilled to remove color and phototoxic coumarins and psoralens. Peppermint II rectified has been treated to remove dimethyl sulfide, an objectionable and fairly volatile compound. Oil eucalyptus II rectified has been treated to upgrade the concentration of the preferred ingredient, cineol, from 85 to 95%, at the same time removing harsh notes. Oil cassia II rectified has been steam distilled a second time to remove color and high boiling residual components.

Resinoids — are extracts of gums, balsams or resins and certain other products such as orris root, which consist in whole or in part of resinous material. They are used as fixatives. Resinoids may also be employed to make the natural product more usable by fluidizing it with a neutral solvent such as benzyl benzoate, diethyl phthalate, or dipropylene glycol. This also often improves solubility. Where color is a problem, as with labdanum and oakmoss, the resinoids may be decolorized.

Resins — are water-insoluble, solid or semi-solid exudates of trees and shrubs. As a class they include both gums and balsams. Also included are products such as amber and copalin (resins of fossilized vegetable origin). Not all resins contain essential oils. The lichens (oakmoss and pinemoss) are extracted to produce resins. However these extracted resins are not true exudates.

Specialties — are natural oils, natural isolates or synthetics, either alone or in combination, which are used as building blocks for flavor and/or fragrance compounds. They are less complex than a finished F/F compound, and may be the product of special processing treatments or unique raw materials. They are usually supplied by a single company under a trade name (such as *Wardia, Sandela, Lyral,* etc.).

Tinctures and Infusions — are natural materials which have been processed by extended solvent treatment. If the solvent is maintained at room temperature or warm, the resulting extract is called a tincture; if hot, > 60°C. or boiling, it is called an infusion. The solvent used is most often alcohol or water-diluted alcohol. The resulting product is filtered either before being put into storage, or before use. Usually the significant flavor/fragrance components of these natural products are non-volatile, or poorly volatile. They include such constituents as animal fixatives, certain gums and balsams. Some absolutes such as ambergris, castoreum, labdanum, opoponax and orris root are made into tinctures. Few if any products are labeled as "infusions," but some infusions may be called "tinctures."

When products are made into tinctures, storage qualities are improved, and finished products may readily be made from the tinctures merely by dilution. Also, as the tinctures age, there is a synergistic effect which produces a fixed alcoholic solution. This results in a superior finished product.

PRODUCTION PROCESSES

We move now from the types of *products* used and manufactured by the F/F industry to a discussion of the *processes* employed to produce these items:

Distillation — is a physical process which achieves separation of the components of a mixture of two or more liquids based on the difference in their vapor pressures. As the vapor pressure of a substance is exceeded, it will vaporize out of a mixture. The vapor that passes off at a selected temperature can then be condensed back to a liquid by cooling below the vapor pressure of the material and collected as a purified liquid. There are two general types of distillation used in the F/F industry: (1) distillation of non-miscible mixtures (water and oil) or hydrodistillation, and (2) distillation of miscible mixtures, which includes rectification and fractionation, usually by vacuum distillation.

Vacuum distillation is accomplished by reducing the pressure within the distillation apparatus pot, column or condenser and receiver. When a material is subjected to vacuum distillation, its vapor pressure is increased. Thus it will vaporize at a lower temperature, consequently less heat is required to distill the charge.

With steam distillation, not only does the steam rupture the oil cells and release the volatile oil, but the steam also exerts a partial vapor pressure on volatile organic liquids to permit the liquids to exceed the vapor pressure required for conversion to vapors at a lower temperature. Thus a mixture of water and oil is condensed, liquefied and separated.

Hydrodistillation or steam distillation involves several modifications. Steam from an external source can be blown through a charge of herbs, twigs, roots, etc., and the steam and oil vaporize out and are condensed.

In water distillation, the herbs are immersed in water and the charge is heated to the vapor pressure required. Steam and oil are vaporized, and condensed to water and oil.

Another steam distillation method employs dry steam, with a heated jacket keeping the contents of the still above the vapor pressure of the combination of oil and water. No water is present, just vapor (steam), and the combination of oil vapor and steam is condensed and separated.

Expression — is the removal of essential oil or other liquid content from a substance by exerting pressure, then collecting the expressed oil. Many different types of pressing equipment are used to remove essential oils from natural products, notably the citrus fruits. They range from the original primitive screw-press and elementary hand presses on down to modern, high-speed equipment such as the FMC and Brown presses.

Extraction — involves release of the oil content of a substance by direct contact with a solvent selected to remove the desired components from the material being treated. Many different types of equipment may be used, a number of different processes, and there are numerous physical modifications of the method employed, i.e. hot or cold solvent, percolation or mixing, countercurrent dual solvent, or solid and solvent extraction, etc. The purpose of all these modifications is to get the desired components of a substance separated from the unwanted components.

FINISHED PRODUCTS

Our next group of definitions are those that apply to finished products. About ten years ago the Nomenclature Committee of the American Society of Perfumers tried to agree upon a set of definitions of terms commonly used in the fragrance industry. A group of these definitions was published in the January, 1968, issue of "Soap, Perfumery and Cosmetics," a British trade publication. We are not too highly impressed by the adequacy of some of the definitions they proposed — particularly for such words as "perfume," "cologne," or "toilet water." Thus we offer our own definitions for these three important product categories:

Perfume oil is a mixture of fragrance raw materials chosen and blended by a skillful artist to present an alluring, consistent rendition of a natural fragrance item (for instance a flower such as rose) or a combination of natural fragrance items (cinnamon, sandalwood, etc.) or an entirely unique aroma (green, aldehydic, etc.) which appeals to the sense of smell.

Perfume (Extract) is a solution of a perfume oil in cologne spirits, a purified ethyl alcohol 95% by volume, sometimes treated for fixation. The rate of use for a standard perfume is normally 22-32 ozs. of perfume oil per gallon (128 ozs.) with alcohol. A perfume is many times the strength of the natural or selected item, since concentrates are used. Super perfumes or skin perfumes may contain from 35-70 ozs. of perfume oil per gallon cologne spirits.

Toilet Water is a solution of a perfume oil more dilute than a perfume. Normally 4-8 ozs. of perfume oil per gallon with diluted (circa 79%) aqueous cologne spirits or denatured alcohol is the recommended rate of dilution.

Cologne is a solution of perfume oil in diluted alcohol, normally used at a rate of 2-5 ozs. of oil per gallon with diluted alcohol (30-70% aqueous). Sometimes a trace of citrus (bergamot, lemon) may be added to give the cologne a fresh citrus character.

Fruit Concentrates — are processed fruit juices in which the water has been removed to concentrate the flavor essence and the solids (mainly sugars and acids). Special equipment is used to recover the essence and to remove only water, with the minimum amount of heat being applied. The concentrate is usually a 5 or 6-fold.

Fruit Essences — are the concentrated aroma components of fruit juices which are present in very low concentration (i.e. less than 0.1%) in the juices themselves. Although similar in chemical structure to the components as they occur in essential oils, these fruit essence homologs are more water-soluble. They are collected with difficulty as water solutions which contain small amounts of ethyl alcohol, naturally fermented in the juice.

DESCRIPTIVE TERMS

The definitions proposed by the Nomenclature Committee of the American Society of Perfumers are currently being revised, but there is no indication when this work will be completed and the new definitions ready for release. Pending such announcement, we reproduce below a number of the Committee's original definitions of Descriptive Terms used in the fragrance industry, which still seem quite acceptable:

Animal — Odours or notes with a sensuous character.

Blend — Any harmonious mixture of fragrance materials.

Body — Main fragrance theme.

Citrus — Odours reminiscent of fruits, such as orange, lemon, bergamot, grapefruit, etc., which impart a freshness to a fragrance.

Cloying — An odour that satiates the senses beyond a natural desire.

Diffusion — The ability of a fragrance to radiate from the wearer and permeate the environment.

Dry Out — Final phase of the main fragrance.

Evaporation — The process of changing from a liquid to a vapour.

Fixation — The property of a fragrance that prolongs the odour life and produces a continuity of odour.

Fruity — The impression of fruit odours within the fragrance theme.

Full Bodied — A well-rounded-out fragrance that possesses depth.

Green — Notes that recall fresh-cut grass, leaves and vines.

Harmonious — Order, accord and unity in a fragrance.

Heavy — A forceful and intense effect.

Lasting Qualities — The ability of a fragrance to retain its character over a given period of time.

Mossy — Typical note reminiscent of forest depths.

Sophisticated — An abstract creative interpretation.

Strength — The relative intensity of a fragrance impression.

Thin — The lack of body and richness.

Top Note — Initial fragrance impression.

Undertones — The subtle characteristics of the fragrance background.

Volatile — The property of being freely diffused in the atmosphere.

Woody — The impression of wood odours within the fragrance theme.

FLAVOR TERMS

An excellent set of definitions of terms associated with flavors has been assembled by the Committee on Sensory Evaluation of the Institute of Food Technologists. With their permission, which we acknowledge with appreciation, we reprint their suggested definitions of 86 frequently used terms:

Absolute Judgment — A psychophysical method in which no standard stimulus is used and which requires an estimate of the absolute intensity of a stimulus or a categorical judgment based on an observer's experience.

Absolute Threshold — (See *threshold*).

Acceptance — 1) An experience, or feature of experience characterized by a positive (approaching a pleasant) attitude. 2) Actual utilization (purchase, eating). May be measured by preference or liking for specific food item. The two are often highly correlated, but they are not the same.

Acrid — Sharp and harsh or bitterly pungent.

Acuity — Ability to discern or perceive stimuli; sharpness or acuteness.

Adaptation — Loss of sensitivity to a given stimulus as a result of continuous exposure to that stimulus or a similar one.

After-Taste — The experience which, under certain conditions, follows the removal of a taste stimulus; it may be continuous with the primary experience or may follow as a different quality after a period, during which swallowing, saliva, dilution and other influences may have affected the stimulus substance.

Ageusia — Lack or impairment of sensitivity to taste stimuli.

Anosmia — Lack or impairment of sensitivity to odor stimuli.

Appetite — Lust or desire for anything, but more especially for food. (If hunger be regarded as the specific sensory experience localized in the region of the stomach and presumably aroused by slow rhythmical contractions of the stomach walls, then appetite for food can exist without hunger, and in civilized man ordinarily does occur, especially as habituated behavior dependent upon past experience with food.)

Aroma — A distinctive characteristic suggestive of fragrance or odor.

Astringent — Quality perceived due to the complex of sensations caused by shrinking, drawing, or puckering of the skin surfaces of the mouth, accompanied by a driving of the blood from the tissues. Dry feeling in the mouth.

Basic Tastes — (See *taste*).

Bitter — A quality of taste sensation, the taste of quinine sulfate being a typical example. Perceived by the circumvallate papillae at the back of the tongue.

Body — The quality of a food or beverage, relating variously to its consistency, compactness of texture, fullness, or richness.

Burning — As related to taste is used most often to describe the sensation of heat caused by the application of inadequate stimulus (q.v.), to the sense system, e.g., that of pepper, mustard, or other strong spices. It arises from the skin senses, including that of pain as well as those of temperature. When the adequate stimulus (hot material) is applied, the term *hot* is more likely to be used.

Chewy — Tending to remain in the mouth without readily breaking up or dissolving. Requiring mastication.

Compensation — The result of interaction of the components in a mixture of stimuli in which each component is perceived as less intense than each would be alone.

Confidence Interval (Statistical) — A range of values which has a given probability (usually .95 or .99) of including the true value (parameter) of the quality attribute being rated.

Contrast — Juxtaposition of two different sensations which results in intensifying or emphasizing their contrary characteristics. It may be of two types: (a) simultaneous, or (b) successive.

Contrast Effect — A judgmental phenomenon which appears in evaluating food samples of different preference (or quality) levels where the presentation of one sample tends to make a following sample of the opposite quality rate either higher or lower than they would if they had been rated independently.

Cooling — A physical sensation in the mouth resulting from the presence of a cold liquid or solid. Also result of chemical action (menthol) sensed by the skin.

Difference Test — A comparison or test of quality variations without indication of preference.

Discrimination — 1) Perception or difference between two or more objects in respect to certain characteristics; 2) a differential response to two stimuli which differ quantitatively or qualitatively.

Dual Standard Method of Difference Testing — Two samples are identified and presented first as knowns, then they are given again as unknowns for the observer to identify.

Duo-Trio Method of Difference Testing. — One of a pair of samples is identified and presented first. Then the observer receives two more samples as unknowns in random order. The time interval can be varied as desired. The observer's task is to pick the *different* sample.

Enhancement — (See *contrast*).

Expert — Generally, an individual acknowledged to be experienced and skillful in a special practice; in the food and beverage field, a specialist who usually confines his diagnostic judgments to a specific product under specific conditions, even though others attribute to him special powers of discrimination, sensitivity and perspicacity applying to other products and conditions.

Fatigue — Condition of organs or organisms which have undergone excessive activity with resulting loss of power or capacity to respond to stimulation. (See *adaptation*).

Flavor — 1) A mingled but unitary experience which

includes sensations of taste, smell and pressure, and often other cutaneous sensations such as warmth, cold, or mild pain. 2) An attribute of foods, beverages and seasonings resulting from the stimulation of those senses which are grouped together at the entrance to the alimentary and respiratory tracts — especially odor and taste.

Flavoring — Any substance, such as an essence or extract, employed to give a particular flavor.

The Flavor Profile Technique — A method of qualitative description analysis of aroma and flavor. The method makes it possible to indicate degrees of difference between two samples on the basis of individual character notes, and the degree of blending, and the over-all impression of the product.

Forced Judgment — A reaction required by an experiment in which "don't know," or other indeterminate answers are not permitted.

Fragrant — A pleasing olfactory quality; odors which are distinctly pleasant smelling.

Gritty — A hard, stone-like sensation, usually caused by the presence of sand particles or stone cells. Not a sensation, but a judgment based on complex perception.

Gust — A unit of gustatory intensity relating to the threshold of a given substance.

Gustation — A sense (taste) whose receptors lie in the mucous membrane covering the tongue and whose stimuli consist of certain soluble chemicals; e.g., salts, acids, sugars, etc.

Haptic — Pertaining to the skin or to the sense of touch in its broadest sense.

Hedonic — Pertaining to feeling or feeling tone; hedonic tone is the pleasurable or unpleasurable accompaniment of characteristics of conscious experiences.

Inadequate Stimulus — When a reaction is obtained by the application of energy which is not "normal" to the sense system involved (such as an electrical current applied to the tongue producing a taste), the nonnormal stimulus is termed *inadequate*.

Judge (See *subject, observer*). — One who participates in a test by providing data which generally are verbal reports of his experience. The term connotes that the person has some special qualification; also, it connotes a situation where primary attention is directed toward the samples.

Limen — (See *threshold*).

Masking — In taste, odor, or flavor applications, it is a component quality within a mixture which dominates or overrides another quality or other qualities present, thus changing the quality of the perceived resultant without benefit of chemical interaction of the components themselves.

Matching — The process of equating or relating, pair by pair, for experimental purposes, usually to determine the degree of similarity between a standard and an unknown, or two unknowns.

Mealy — A quality of mouthfeel denoting a starch-like sensation. Friable.

Medicinal — An olfactory sensation denoting a usually unpleasant medicine-like taste or smell.

Multiple Comparison — An unlimited number (usually more than three) of samples is presented to the observer simultaneously in random arrangement or in accordance with a predetermined statistical design. Significance of results is usually calculated by the variance method, or a rapid approximation thereof.

Mouthfeel — The mingled experience deriving from the sensations of the skin in the mouth during and/or after ingestion of a food or beverage. It relates to density, viscosity, surface tension and other phsyical properties of the material being sampled.

Objective — 1) Capable of being recorded by physical instruments or as a consequence of a repeatable operation. 2) Not dependent upon the observations and reports of an individual, and thus verifiable by others.

Observer (see *subject, judge*) — One who participates in a test by providing data which generally are verbal reports of his own experience. This term, in contrast to *judge,* connotes a situation where primary attention is directed toward the person's response.

Odor — Sensation due to stimulation of the olfactory receptors in the nasal cavity by gaseous material.

Odorant — A chemical substance which stimulates the olfactory receptors.

Olfactometer — An instrument for controlled presentation of odor stimuli, used for measuring thresholds and other quantitative values.

Organoleptic — 1) Affecting or making an impression upon an organ or the whole organism; 2) capable of receiving an impression; 3) sometimes used as a synonym for *sensory* when referring to examination by taste and smell (obsolescent).

Paired Comparison, Method Of — A psychometric or psychophysical method in which stimuli (samples) are presented in pairs for comparison on the basis of some defined criterion, such as preference, intensity, degree of a defined quality, etc. The traditional form requires that each sample in a series be compared with every other member in the series.

Paired Preference — Paired comparison method using the preference criterion.

Palatable — Agreeable to the "taste," savory; hence acceptable or pleasing.

Panel — A group of people (observers, subjects, judges) comprising a test population which has been specially selected or designated in some manner, e.g., they may be trained, or have special knowledge or skills, or may merely be available and predesignated.

Preference — 1) Expression of higher degree of liking. 2) Choice of one object relative to others. 3) Psychological continuum of affectivity (pleasantness — unpleasantness) on which such choices are based. This continuum is also referred to as that of degree of liking or disliking.

Primary Qualities — Within a specific sense, those qualities which are considered basic and from which all other qualities can be compounded. There is general belief that salt, sweet, bitter and sour are the four primary taste qualities.

Psychometrics — The quantitative or mathematical aspects of psychological procedures. It includes determination of the mathematical relation between quantitatively varied stimuli and the frequencies of a given judgment upon them, usually by one of the several psychophysical methods.

Psychophysical Methods — A group of specific procedures used in psychophysical investigation. Fechner adopted three: the methods of limits, average error and constant stimuli. Guilford lists others and a number of variations of each procedure.

Psychophysics — The study of the quantitative relations between a stimulus and the ensuing sensation or other experiences.

Pure Odors or True Odors — Olfactory experience unaffected by the action of other senses of the skin and mouth, especially the trigeminal and taste receptors.

Quality (1) An aspect, attribute, characteristic, or fundamental dimension of experience, which involves variation in kind rather than in degree.

2) The composite of those characteristics that differentiate among individual units of a product and have significance in determining the degree of acceptability of that unit by the user.

Rancid — Having a rank odor or taste, as that of old oil.

Rank Order (Ranking) — A psychometric method that may be used in multiple comparisons where the subject considers all of the samples in a series at the same time and is required to rank them in order on some designated dimension, such as preference, intensity, quality, etc.

Rating Scale (see *absolute judgment*). A method for securing and recording a judgment concerning the degree to which a stimulus material possesses a specified attribute, usually by placing a mark at an appropriate position between the two extremes of a line that represents the possible range of degrees of that attribute.

Reference Sample (see *standard*) — In a multisample test, the sample designated as the one to which all others are to be compared.

Salty (Saline) — A quality of taste sensation of which the taste of sodium chloride is the typical example.

Score — 1) (*noun*) A value assigned to a specific response made to a test item; 2) (*verb*) To score a food is to rate its properties on a scale or according to some numerically defined set of criteria.

Sensory — Pertaining to the action of the sense organs.

Sharp — Characterizing an intense or painful, well localized reaction to a substance being eaten or smelled, e.g., various acids and alcohols.

Significance (Statistical) — Referring to the probability that differences (among samples or treatments) are real, and not due to chance variation. A difference which has a .05 probability of occurring by chance alone is usually considered a significant difference, and a highly significant difference is one which could occur by chance with .01 probability.

Single Stimuli, Method Of — Any psychophysical or psychometric method in which a judgment follows the presentation of one stimulus only.

Sour — A quality of taste sensation of which the taste of acid is the typical example.

Standard — A sample presented as a model or example. The standard sample conforms to a specified level or degree of quality.

Stimulus — A change of energy in the environment which affects a sense organ.

Subject (see *judge, observer*) — One who participates in a test by providing data. This is the generic term which is appropriate wherever any of the related terms may be used. Thus, it includes not only those who provide subjective or verbal data but also those who respond in any other way. Subject, Judge, and Observer are terms whose usage is supported by custom and, as such, are distinguished from such casual words as "tester," "taster," "panelist," "sniffer," etc.

Subjective — Pertaining to individual experience which can be observed and reported on only by the person involved, i.e., mental phenomena that are usually reported verbally.

Subliminal — Below the threshold (applied to stimuli which are not sufficiently intense to arouse definite sensations, but which, nevertheless, have some effect upon the responses of the individual).

Sweet — A quality of taste sensation of which the taste of sucrose is the typical example.

Tactual — Pertaining to the sense of touch.

Taste (see *gustation*) — One of the senses, the receptors for which are located in the mouth and are activated by a large variety of different compounds in solution. Most investigators usually limit gustatory qualities to four: saline, sweet, sour, bitter. Distinguished from *flavor,* the experience to which taste contributes.

Texture (see *mouthfeel*) — Those properties of a foodstuff, apprehended by the eyes and by the skin and muscle senses in the mouth, including roughness, smoothness, graininess, etc.

Threshold or Limen — A statistically determined point on the stimulus scale at which occurs a transition in a series of sensations or judgments. Thresholds are of three kinds: 1) The threshold of sensation, *stimulus threshold,* or *absolute threshold,* often designated as RL, is that magnitude of stimulus at which a transition occurs from no sensation to sensation. 2) The *difference threshold* is the least amount of change of a given stimulus necessary to produce a noticeable change in sensation. It is often designated as the DL, and the interval or unit as the j.n.d. (just noticeable difference). 3) The *terminal threshold* is that magnitude of a stimulus above which there is no increase in the perceived intensity of the appropriate quality for the stimulus (above this point, pain often occurs).

Triangle Method of Difference Testing — The test unit consists of three samples. All three are coded so that they are unknown, but the observer is told that two are alike and one is different. It is his task to pick the *different* sample.

Chapter V

Fragrance Families

No matter how hard you try to pin a label on them, fragrances resist being confined within rigid categories. They are willful. With few exceptions, they escape precise classification, and wander off into fellow classifications. The obvious reason is because in so many cases fragrances partake of the characteristics of more than one group. Fragrances are free!

And yet the struggle goes on to attempt to organize hundreds of fragrances. We don't know how many classification systems have been set up over the years, but we have a number of authoritative ones at hand. Trying to reconcile them with one another is as confusing as starting the project in the first place. It makes one wonder "Why all the bother?" However, there is just something in human nature that abhors chaos and craves order, so perhaps we must have classifications. At least the systems can serve as guidelines, if not as absolutes.

Fragrances may have started out years ago being relatively simple, but as time passed and science progressed, they became more complex. We doubt, however, that born classifiers will become discouraged. They will keep trying.

There is something tantalizing and challenging in the whole classification process that makes us refuse to rest even if we never reach our goal.

Recently Haarmann & Reimer Corp. offered a logical and carefully thought out division of fine fragrances into nine families which represents a sort of family tree of fine perfumes ("The Genealogy of Perfumes" prepared by Mr. Peter Worner). The families are:

green notes
flowery notes
aldehyde notes
Chypre notes
oriental notes
tobacco notes
leather notes
Fougere notes and
citrus notes

Mr. Worner subdivides some of his categories, and also suggests examples of perfumes which are typical of these classes as follows:

Green - the classic perfume in this category, he says, is

Vent Vert (Balmain, 1945)

other typical examples are

Grafitti (Capucci, 1963)
Mademoiselle Ricci (Nina Ricci, 1967)
Weil de Weil (Weil, 1971)
Espiegle (Atkinson, 1963)

Floral

Floral-Fresh

Diorissimo (Dior, 1956)

Floral-dry-herb

Carnation-powdery

L'Origan (Coty, 1905)
L'Heure Bleue (Guerlain, 1912)
Maderas de Oriente (Myrurgia, 1918)

Linden-like

Soir de Paris (Bourjois, 1929)

Herbal-spicy

Moment Supreme (Patou, 1933)
Blue Grass (Arden, 1933)
Aquamarine (Revlon, 1952)

Pure Floral

Quelques Fleurs (Houbigant, 1912)

L'Air Du Temps (Ricci, 1948)
Joy (Patou, 1935)

Aldehyde

Aldehydic-Floral
Chanel No. 5 (Chanel, 1921)
Arpege (Lanvin, 1927)
Mme. Rochas (Rochas, 1960)
Climat (Lancome, 1968)

Aldehydic-Floral-Woody-Powdery
Je Reviens (Worth, 1932)
Bois des Iles (Chanel, 1926)
Caleche (Hermes, 1961)
Calandre (Paco Rabanne, 1968)
Chamade (Guerlain, 1970)

Chypre

Aldehydic-Fresh-Mossy
Chypre (Coty, 1917)
Crepe de Chine (Millot, 1925)
Ma Griffe (Carven, 1944)

Aldehydic-Mossy-Animalic
Bandit (Piguet, 1944)
Miss Dior (Dior, 1947)
Intimate (Revlon, 1955)
Cabochard (Gres, 1958

Mossy-Fruity
Femme (Rochas, 1942)
Fruit Vert (Florel, 1930)
Rumeur (Lanvin, 1934)

Oriental
Shalimar (Guerlain, 1925)
Shocking (Schiaparelli, 1935)
Tabu (Dana, 1931)
Youth Dew (Lauder, 1952)

Leather and Tobacco
Cuir de Russie (Chanel, 1924)
Scandal (Lanvin, 1931)
Tabak Blond (Caron, 1919)

Fougere (Fresh-Aromatic-Spicy-Mossy)
Fougere Royal (Houbigant, 1882)
Canoe (Dana, 1935)
Moustache (Rochas, 1947)
Brut (Faberge, 1964)
Jade East (Swank, 1964)

Citrus — the classical Eau de Colognes are the most typical representatives of this class

Another student of perfumes, Ms. Jill Jessee, author of *Perfume Album,* has also tried her hand at classifying fragrances. Her system, she acknowledges, is a composite of several classifications that have been proposed by others, with some of her own ideas added. We like it well enough to reproduce it here as at least a convenient frame in which to picture the different fragrance categories.

I - *Single Florals*

There is one class or family of fragrances upon which everybody seems to agree — the *Single Florals.* By that, of course, is meant those fragrances that achieve the effect of a single kind of flower, no matter how obtained. We assure you that such an effect is seldom easily obtained, but that is a matter for another chapter.

Examples would be florals like Jasmine, Lilac, Rose, Sweet Pea, Gardenia, Lily of the Valley, Violet, Narcissus, Carnation, Heliotrope, and many others. Their popularity waxes and wanes, but we seem to be enjoying a revival of appreciation of these Johnny-One-Note fragrances just now. That may decline in due course, but if so, we venture to prophesy that they will return one day as fresh and lovely as ever.

II - *Floral Bouquets*

That sounds simple too. You just imagine a bouquet of a number of different flowers, harmoniously combined. Sometimes such a combination is called *Multi-floral.* Take your choice. They come to the same thing.

The "French Committee on Perfume" recently issued their classifications in which they make only one *Floral* group, which includes the *Single Note Floral, Rose and Jasmine,* and *Multi-Floral.* We certainly admit the supreme importance of Rose and Jasmine in all perfumery, but we still prefer to make the *Single Florals* a separate group from the *Floral Bouquets* or *Multi-Florals,* and to make *Rose* and *Jasmine* as a combination, one or the other. Our reasoning is that if these two flowers are of equal importance in a fragrance, then it is a *Floral Bouquet,* but if one is definitely dominant, then we'd call it a *Single Floral.*

But we go off into all kinds of more important tangents in this group. Some *Floral Bouquets* are aldehydic, and more and more are becoming so. Some may be Spicy, some Green, some Woody, some Fruity, etc. Then, although there is a melody of flower notes, we may choose to classify them under another heading altogether. Carnation, for instance, or a bouquet in which Carnation is the featured scent, is certainly Spicy, but our choice would be to leave it in a floral group. Often it's more explanatory to use several descriptive words and break out of boundaries altogether. For instance, Floral, Aldehydic, Woody.

Or, you could do as one classifier does and separate the Floral group into sub-groups such as *Floral Aldehyde, Floral Aldehyde Woody, Floral Aldehyde Mossy.* One company divides the *Multi-Florals* into *Floral Bouquets* and *Aldehydic Florals* even though there may well be aldehydes in a minor degree in some of the *Floral Bouquets.* But this company feels that if the aldehydes are emphatic enough, they should be thus separated.

III - *Modern Blends*

Our way of designating these strongly aldehydic fragrances which might be floral, woody, mossy, or whatever, is to call them *Modern Blends,* a term created some years ago for "McCall's" magazine which issued a "Perfume Guide" for their readers. It was felt that the word, aldehyde, would not be properly understood by the consumer. You can translate it in your mind to aldehyde blends if you wish, but as this is a twentieth century discovery, we think such blends might still be called "modern" for the sake of the novice and/or consumer.

IV - *Chypre Bouquets*

More problems! The "French Committee on Perfume" makes the *Chypres* a separate category, with four sub-divisions: (a) *Chypre;* (b) *Chypre fruited;* (c) *Chypre Green;*

and (d) *Chypre Animal.* But, first, we'd better know what Chypre means, a word that confuses even some people in industries related to our own.

Chypre is the French word for the island of Cyprus in the Mediterranean. This name was chosen by Francois Coty for one of his early masterpieces introduced in 1917 and the name became a generic term in our fragrance vocabulary. Coty's *Chypre* is highly complex and has been analyzed as follows: "mainly bergamot-oakmoss-amber 'accord' with a floral ensemble led by jasmine and including rose, ylang ylang, orange flower on an opoponax type of background with balsams, resins, labdanum, coumarin and the wood trio, patchouli-sandalwood-vetiver." Perhaps the dominant ingredient in this fragrance is oakmoss. But although all *Chypre* fragrances include a mossy note, not all mossy notes are necessarily chypre.

Now, whereas the "French Committee on Perfume" makes the *Chypres* with its subdivisions one whole family, the company referred to above makes *Chypre Bouquets* a part of their designated *Forest Blends* called "woody-mossy-leafy." This designation will be our next one, but we also discuss it here as this firm associates it with the *Chypres.*

We quote from this company's manual "On Fragrance" as follows: "The aroma chemical industry has given the perfumers a variety of woody notes which are characteristic of modern perfumery, often part of or associated with a chypre 'accord' lending itself to numerous variations. For some, green notes (different from that of oakmoss), which may be found also in aldehydic florals, have constituted a modern trend."

This company includes "fruity" perfumes in this class, because most of them are chypres with a distinct fruity note. And you will remember that the "French Committee on Perfume" also includes fruity fragrances in the *Chypre* category. But, for what we think is a clearer understanding, we are going to treat fruity fragrances separately and will also make a *Woody-Mossy-Leafy* category, (to follow) which we relate to the above-mentioned *Forest Blends.*

V - *Woody-Mossy-Leafy*

We relate this category to two of those of the "French Committee on Perfume," their *Woodsy* and *Green, Nature* (expressed in French of course). In their *Woodsy* group they include the scents of bark, precious woods, and spices. The principal components include cedar, patchouli, santal, and vetiver. We do not include spices in this category, as we have a separate classification for them, but we do include such warm aromatic woods as sandalwood, cedarwood, rosewood and such "woody" smells as patchouli and vetiver. We also include earthy notes and "green" notes such as oakmoss, fern, herbs, leaves, geranium, lavender, etc.

"Green" is big just now. This type note has long been in perfumery use, especially in men's fragrances, but as perfumes are currently becoming more and more unisex, women, too, are greatly enjoying the "green" fragrances. As women have become more liberated and more active in sports than ever before, it follows that this type fragrance is reflective of their life style.

As our fragrance vocabulary is inadequate, we must borrow extensively from other vocabularies. Green is a color. But it is an easy mental transition to think of it as a scent by associating the color with the smell of green, growing things. "Green" means something a bit different to each individual — perhaps, the early morning smell of dewy grass, of a lawn just mowed, of trees and bushes right after a soft rain, of a leafy lane of trees. It makes us think of a fresh spring day and, by implication, of youth, the outdoors, sports.

Green notes in a fragrance can vary in their quantity and in the variety of other notes with which they are combined, and the final fragrance should be described accordingly. Again we quote from "On Fragrance": ". . . we could say that the fragrance of a bouquet of flowers with only a few leaves is 'floral,' the fragrance of leaves and twigs without flowers is 'green,' the fragrance of a bouquet of flowers with leaves is 'floral green,' the fragrance of a bouquet with a few flowers among leaves, twigs and branches is 'green-floral.'

The essence of galbanum is sometimes used with "green" notes as in Balmain's *Vent Vert* which has been called a precursor of the green perfumes. Introduced in 1947, it is obvious the feeling for "green" is not newborn.

VI - *Spicy Bouquets*

Spices give both flavor and fragrance but, now, we are thinking of fragrance. Some flowers such as Carnation mentioned above are spicy. Carnation is clove-spicy, but your nose should be able to recall all the variations of spice on a kitchen spice shelf. The perfumer has borrowed cinnamon, vanilla, ginger, cardamom, cassia, caraway, pimento, pepper, celery, basil, tarragon, nutmeg in addition to clove.

So, a spicy bouquet featuring one or a combination of these stimulating smells serves up a very tempting fragrance. We think spicy bouquets deserve to be a group in themselves. It seems redundant to say that a spicy note may be only one of a number of other notes in a fragrance which might, then, be classified by an expert in some other group.

We go along with the "French Committee" on the next classification although it does not represent a large number of fragrances. But it does seem distinct enough in character to deserve a separate grouping.

VII - *Leather*

We needn't tell you that the French word is "cuir," familiar in fragrance naming. In the old days, high quality leathers were tanned in Russia with birch tar instead of oak used for ordinary leather. Birch tar is widely used in perfumery to give that so-called Russian leather note, this note sometimes being combined with floral and animal notes, as well as with chypre.

One of our reference charts has a *Floral Leathery* category and a *Tobacco Leathery* category. We have already mentioned floral plus leather, but tobacco plus leather is doubly appealing.

VIII - *Fruity Blends* excluding Citrus

True, fruity notes often belong to the *Chypre* family as previously mentioned, but, again, we would like to separate them for easier understanding. And for still easier understanding, we will put the citrus scents in another discussion. Think, rather, of fruits such as peach and apricot, pear and pineapple, cherry and grape, strawberry and raspberry, apple and avocado. Think of their fragrance rather than their taste, or perhaps the two senses are linked as they are with spices. Either way, you think "fruity."

It may be a smooth and mellow "fruity," or it may be a sweet and sunny "fruity." You may think in colors such as gold or red. You may think of an orchard or a plantation or a garden. You may have memories of your childhood, or of a recent trip to some fruit-growing spot. But whatever your association with "fruity," it has to be a pleasant one, one that you'd like to have perpetuated in a bottled fragrance.

Any fragrance that we'd put in this category would have to be obviously "fruity." Just a hint of fruitiness wouldn't be enough to satisfy us. Of course, we acknowledge other notes, but give them a low profile.

IX - *Citrus Blends*

The citrus fruits are so different in character and have grown to such tremendous proportion in the marketplace, that they deserve their own category and discussion. We see Lemon and Lime especially on dozens of product labels today (as a scent of course), but the other members of the citrus family such as orange and mandarin, neroli and bergamot are very active too.

The citrus scents may well be an expression of our present-day return to Nature, but we are returning to a long-ago day if we start at the beginning of their commercial use. We go back to about two hundred years ago when they were important ingredients in the first colognes, the so-called "classic" colognes.

The names Jean Marie Farina (Roger and Gallet) and The House of 4711 are old, familiar ones. The European

Eaux de Cologne (Koln is the German city where they were first made) were based on such citrus ingredients as bergamot, orange, lemon, neroli and lavender. They were admired for their superb refreshant qualities and were enjoyed by all members of the family, male and female, young and old. It isn't just today that the same fragrance appeals to both sexes. The story of Eau de Cologne is a long and fascinating one for which we haven't the space, but that these "originals" are still in demand is a tribute to their excellence.

There have been various departures from the Colognes mentioned such as Lavender Water, Hungary Water, Florida Water, etc. There have been such other additives besides the citrus and lavender ones as amber, florals, Russian Leather, etc., but the citrus character was always preserved.

Quite different are the popular type colognes developed in the United States, which derive their fragrance character from their parent perfume and are not what we are talking about here. Such colognes may well include citrus notes if the perfumes from which they are descended contained them, but the European colognes are real, citrus-based fragrances.

As we said earlier, citrus scents proliferate all around us in other forms from soaps and hair preparations to many types of household products, and the end isn't in sight. Nor must we forget that citrus scents have long been popular in men's fragrances, along with lavender, fougere, tobacco, herbs, woods, and others. In fact, citrus notes appear in any listing of men's fragrances in a majority of the names, even when other notes are included.

X - *Semi-Oriental and Oriental*

This is such a definite and accepted category that there can be little argument about it, as little perhaps as was true of the *Single Florals* with which we opened this chapter. We can't say as much for the categories in between. We doubt that any classifier would hesitate to call a fragrance a *semi-Oriental* or an *Oriental.*

These are the sophisticates of fragrance, as the *Single Florals* are the ingenues. Their inspiration is the Far East, also the source of many of the ingredients. They are rich, warm, heady, intense. They may have floral notes, spicy and woody notes, balsamic and resinous notes, and animal notes. Insistent notes such as musk, amber, civet, patchouli, vetiver, opoponax, etc. are characteristic of these blends.

As is self-evident, those fragrances that are more subdued but are still based on the Oriental motif have become known as *Semi-Orientals*. They were developed because Western women often found the typical *Oriental* fragrances too strong for their taste.

There has been a recent resurgence of interest in the *Orientals*. The Musk craze was only one highly visible example, but what was originally thought by many to be merely a fad grew into a prevailing trend. There is still speculation as to the reasons. We think that the pull of the Eastern mystic religions, which appealed particularly to the young, was a force. And, undoubtedly, the Sex Revolution played a part, with the Musk fragrances being advertised blatantly as frankly sensual.

Recently some of the earlier Oriental fragrances have been revived, using either a nostalgia theme or a Far Eastern theme. Certain Oriental fragrances never lost their popularity but, for a time were "swept under the rug". Now they have made a comeback.

As the reader has perhaps gathered, most fragrances are a combination of two or more families, and even the experts can disagree as to the category to which each should be assigned. It's all highly controversial. Numerous classifications have been made in the past by fragrance experts, and no doubt more will be made in the future.

****

Principal References:

The "French Committee on Perfumes" Classification

Givaudan Corporation's "On Fragrance".

McCall's "Perfume Guide".

Grayson Associates "A Genealogy of Fragrance".

TABLE 1

FLAVOUR CLASSES	REPRESENTATIVE EXAMPLE
Fruit Flavours	
Citrus-type flavours (terpenyl)	Grapefruit, orange
Berry-type flavours (non-terpenyl)	Apple, rasberry, banana
Vegetable Flavours	Lettuce, celery
Spice flavours	
Aromatic	Cinnamon, peppermint
Lachrymogenic	Onion, garlic
Hot	Pepper, ginger
Beverage flavours	
Unfermented	Juices, milk
Fermented	Wine, beer, tea
Compounded	Soft drinks, cordials
Meat flavours	
Mammal	Lean beef
Sea food	Fish, clams
Fat flavours	Olive oil, coconut fat pork fat, butter fat
Cooked flavours	
Broth	Beef bouillon
Vegetable	Legume, potatoes
Fruit	Marmalade
Empyreumatic flavours	
Smoky	Ham, kippers
Broiled, fried flavours	Processed meat products
Roasted, toasted, baked	Coffee, snack foods, processed cereals
Stench flavours	Cheese

Chapter VI

Flavor Families

We could eat and drink for hours on end without exhausting all the flavors there are in the world. From morning cereal to midnight snack, from breakfast cocoa to dinner demitasse, from cocktail to cordial, with all meals and every drink, we enjoy a constant parade of flavors. Nor are flavors limited just to edibles and potables. There are also flavors in many drugs we have to take, in chewing gums and breath fresheners, in tooth pastes and mouth washes, every product that can be made more pleasant by having flavor added to it.

In this chapter we will try to organize flavors into families, just as we did with fragrances. Unlike fragrances, however, where a number of other students of the subject have attempted to codify fragrances into logical systems, few such attempts have been made with flavors. Many manufacturers of flavor compounds classify products they sell by end use; thus they will offer soft drink flavors, ice cream flavors, candy flavors, chewing gum flavors, etc., but the only attempt we found in the literature to make a general classification of all flavors was in an article by Dr. G. Ohloff of Firmenich & Cie., Geneva, Switzerland,

"Classification and Genesis of Food Flavours," published in the October, 1972, issue of "The Flavour Industry."

Dr. Ohloff established nine general classes of food flavors, which he then further subdivided, as shown in Table 1. In describing some of these flavor classes he commented that "meat and fat flavours are produced by enzymatic activities, partially in the presence of oxygen. The action of heat yields cooked and empyreumatic flavours. In the first case this process provides mainly hydrolysis products, whereas the second case involves a sequence of drastic molecular modifications of part of the original food. Cheese flavour, the typical representative of stench flavour, is produced by a strong enzymatic activity on protein substrates."

Ohloff recognized that the classification he suggested is incomplete, saying "It is certainly possible to add some further flavour classes. In any event a finer subdivision will be necessary." For example, he suggested that the spice classification be further subdivided according to odor and distinguishable taste qualities. The subdivision, based on sensory impressions, would include: floral, phenolic, almond, herbaceous, caraway-like, minty, balsamic, and camphoraceous.

Another classification of flavors which seems to have considerable merit is one made by Ms. Jill Jessee, writer on flavor and fragrance. She had the advice and assistance of a number of leading U.S. flavorists in her pioneering effort to divide flavors into logical categories.

But before we review her seven general categories of flavors in detail, we ask our readers for a moment just to *think flavors.* Thinking flavors is almost like *tasting* flavors. Most of us have noticed that as we are about to eat a slice of cucumber, or slide a bite of lemon pie into our mouth, the taste buds are already tingling in anticipation. But, if you taste something that you mistook for something else, your taste buds are startled and it takes a second or two for them to recover. When you expect one flavor, but suddenly encounter another, a state of semi-shock can result. In the discussion to follow, however, when we talk about a certain kind of flavor, the reader should know from experience just what it tastes like. Ms. Jessee suggests seven major flavor classifications, as follows, based generally on the origin of the flavoring material:

1. Fruits - citrus and berries
2. Herbs and spices
3. Nuts, beans and seeds
4. Leaves, barks, bulbs and roots
5. Dairy-type flavors
6. Meat, poultry, and seafood
7. A "catch-all" category in which she puts all the flavors which don't fit into the other categories.

This is not necessarily the best of all classifications of flavors that could be established, or the most valid. Some experts to whom we showed the list suggested that the categories should be further broken down into natural and artificial. However, since in this stage of our discussion we are concerned only with the *flavor* of the various materials, and not their *derivation,* we have stayed with the original plan which has the merit of limiting the number of categories as closely as possible. We believe Ms. Jessee's categories are acceptable, and certainly serve as a convenient framework for the discussion of the individual flavors. So, we take up her first category.

1. *Fruit Flavors*

Examples: Lemon, Lime, Orange, Grapefruit; Strawberry, Blueberry, Cranberry; Cherry, Peach, Apple, Pineapple, Banana, Grape

Used in: Desserts, Gelatines, Confectionery, Baked Goods; Beverages (from Juices and Soft Drinks to Alcoholic Wines and Cordials); Tobaccos; Chewing Gums; Instant Breakfast Foods, Cereals; Marmalades and Jams; Pharmaceuticals

Citrus

Citrus fruits originated in the Orient about 5,000 years ago, then their culture moved through Persia to the

Mediterranean countries. Columbus is believed to have brought the first citrus seeds to the western hemisphere, and Ponce de Leon reportedly introduced citrus cultivation to Florida. Probably more oranges are consumed than any other fruit, while lemon is a highly popular flavor for many uses, vying with vanilla for top preference. The United States now produces some 40% of the world's lemons, mostly in California. Next comes Italy, which is also a center for bergamot production.

Berries

The most popular berries are the strawberry, raspberry and blueberry. Strawberries are a comparatively recent introduction, their cultivation having been started in Europe in the 15th century. In addition to a delectable flavor, the strawberry also has a beautiful shape and an exciting color, as does the raspberry. As color is believed to influence taste, the brilliant red of these two berries may well help explain why they are such popular flavor favorites. Blueberries are the second largest berry crop in North America, second only to strawberries.

Banana

Another fruit that stands high in the popularity ratings is the banana. It, we are told, is technically a berry, although seldom recognized as such by any but botanists. Bananas are said to have originated in southern Asia and were cultivated in India for centuries before Christ. They spread to Africa, then to southern Europe, and to North America about 1690. Today, almost 80% of the world's exports of bananas come from Latin America, with the biggest producing countries being Costa Rica, Panama, Honduras and Ecuador.

Grape

Another very popular fruit, and one of the most widely used fruit flavors, is grape. Cultivation of grapes and the production of wine from them dates far back into the origin of man's history. Grapevines may have originated in Persia, and certainly the grape was widely cultivated throughout the Near East. Grapes did not have to be introduced into the New World, for they were already growing here when the first explorers arrived. However American stock was considered inferior, so in time vines were imported from France and the native grapes improved in flavor by cross-hybridization. Grapes now grow in almost every country in the temperate zone, although there is a wide range in taste in the varieties produced in different areas, and thus a wide difference in the esteem in which they are held.

Our next flavor category is herbs and spices.

2. *Herbs and Spices*

Examples: Mint; Sage; Marjoram; Cinnamon; Cloves; Oregano; Thyme; Tarragon; Ginger; and so many others

Used in: Chewing Gums and other Oral Products; Salad Dressings, Sauces, Savories; Breading, Croutons; Snacks, Cereals; Beverages; Pharmaceuticals

Mint

In our discussion of herbs and spices we will start with one of the most common and widely used herbs, mint. The Random House Dictionary defines mint as "any aromatic, labiate herb of the genus *Mentha,* having opposite leaves and small, verticillate flowers." The two most common members of the family are spearmint and peppermint, but altogether there are at least 25 species of this perennial which grow throughout the temperate and sub-tropical countries.

The Egyptians were acquainted with mint long before the birth of Christ. Mint was also known very early in China, whence it traveled to Japan. There were, even then, two different types of mint. The Chinese mint contains an unusually high proportion of menthol, most of which is removed before the oil goes to market. The Japanese type is considered superior both in flavor and fragrance.

There are specimens of mint in the British Museum

from Hertfordshire collected in 1696 by one John Ray. The English variety of peppermint oils long had the finest reputation, but production is negligible compared with the U.S., China, Japan and Brazil. American mint is similar to the original Mitcham variety, but various types are now grown.

From garden mint, the first favorite of herb gardeners for kitchen use, to the ubiquitous peppermint and spearmint, mint flavor is high in popularity, especially in the United States. Chewing gum, mouth washes, breath fresheners and tooth pastes, all use the flavor of mint.

Clove

The word "clove," comes from the French "clou" which means nail. The clove resembles a nail in its shape. The clove is the unopened bud of a tropical evergreen tree. Cloves are so small in size that it takes 4,000 to 7,000 dried buds to make a pound of clove spice. Nevertheless the clove trade is still big business. The principal exporting countries are Madagascar, Zanzibar, Tanzania and the Malagasy Republic.

Cinnamon

Another widely used spice is cinnamon, one of the first spices enjoyed by man. Egyptians used to import cinnamon as far back as 2,000 B.C. Pungently sweet in aroma, cinnamon comes from the dried bark of certain evergreen trees. In Sri Lanka it is rolled into long, slender "quills." America imports most of its cinnamon from Indonesia and South Vietnam, then re-exports a fair amount to Mexico, where this spice is highly esteemed.

Our next spice family includes a number of highly popular nut, bean and seed flavors.

3. *Nuts, Beans, Seeds*

Examples: Vanilla; Cocoa, Coffee; Cola; Almond, Pecan, Peanut

Used in: Flavoring extracts; Desserts and Baked Goods; Confectionery; Tobaccos; Cereals; Snacks; Beverages

Vanilla

Vanilla is the most popular of all flavors. Natural vanilla has a very delicate flavor which shows a tendency to evaporate when exposed to the heat of baking. Vanilla extract needs ageing to develop its full flavor. It smells "creamy" and subtly sweet, and adds just that "lift" which keeps any food from being "flat" and lacking in flavor.

The vanilla bean, which is similar in appearance to a string bean, grows on a vine belonging to the orchid family. 60% of the total world supply of natural vanilla comes from Madagascar. The natives cultivate by far the larger part of the crop in garden plots and in the bush country; the balance is grown by Europeans on plantations. Pollination is a delicate process which must be performed by hand. Vanilla culture is thus expensive. With a limited supply, natural vanilla is a high-cost flavor, which explains why so much of the vanilla now used is "artificial," made from lignin; a natural by-product of the wood pulp industry.

Chocolate

Another favorite flavor is chocolate, which comes from the cocoa bean. Both vanilla and chocolate were first used by the Aztec Indians of Mexico. Chocolate was so prized that for a time it was used as a medium of exchange. Vanilla and spices were added to chocolate to make the favorite drink of the Aztecs.

The Spaniards liked the drink and introduced it to Spain about 1519, whence its use spread to other parts of Europe. The English began to drink cocoa about the middle of the 17th century, and it became so popular there that "cocoa houses" were established in which intellectual and society leaders discussed ideas over steaming cups of cocoa. The "cocoa houses" were the forerunners of the famous English Clubs.

It is difficult to describe the flavor of chocolate, perhaps because it is so unique. It tastes rich and velvety, smooth and mellow, soothing and satisfying. Chocolate is a superior dessert flavor in cakes, frostings, puddings, ice cream, candies, right down the line.

Coffee

The so-called coffee bean is actually not a bean at all, but rather a seed. It combines an appealing aroma with an irresistible taste, and although brewed by many different methods all over the world, and varying in color from a light brown to almost an outright black, this popular beverage seems to exert an almost universal appeal. There are more blends of coffee then one can readily keep track of, and in the United States dozens of competing brands, which differ widely in flavor.

As with cocoa, the adoption of coffee as a beverage by the British in the mid-17th century led to the appearance of English coffee houses, which became a center for gossip, discussion and occasionally intrigue. The thousands of American "coffee shops" today are nothing like their British counterparts, but still provide evidence of the continued high popularity of this exhilarating drink. Many coffee lovers also enjoy the flavor of coffee in such liqueurs as *Kahlua.* The chief exporting countries are Brazil, Colombia and Venezuela, some of the Central American republics, and the East Indies. Coffee lends itself to many adulterants, chicory being the most common.

Cola

The cola nut, sometimes called the guru nut, is not really a nut at all, but rather a seed kernel from which the rough seed coat has been stripped away. The nuts are about the size of horse chestnuts. The United States imports most of its requirements from the West Indies and Latin America.

The bitterness of the cola flavor results from the caffeine content of the nuts. Caffeine provides a stimulant effect, which no doubt is one reason for the exceptional popularity of cola drinks. It stimulates, as coffee does, and also is an unsurpassed thirst quencher.

Cola drinks are of course the most popular carbonated beverages. They have been known for many years, long before *Coca-Cola* was invented. A British patent was taken out for a carbonated beverage containing fruit juices and cola constituents back in 1866, and as early as 1700 there are references in the literature to the popularity of the bitter cola fruit among the blacks in Guinea.

Peanut

Certainly the peanut is the most popular of all nuts in the United States. But, again, it isn't technically a nut but a legume, a vegetable related to the pea and the bean. It grows, not on a tree, but underground, on the roots of a bushy, annual plant, about a foot and a half tall. But, we will treat the peanut here as a nut, for it has the character and appearance of a nut, and, most important, a "nutty" flavor.

When one handful of peanuts leads to another, peanuts have an obvious appeal as a universal flavor treat. What would patrons at a bar do without bowls of peanuts to keep them occupied while waiting for their next drink? And how would children ever reach maturity without countless school lunches of peanut butter sandwiches? Peanuts, or "ground nuts" as the British call them, are certainly nutritious, but the important thing about them with which we are concerned in this chapter is that they have a delicious flavor.

The peanut is widely cultivated all over the world in tropical and subtropical climates, and is grown in the sandy soils of many of the United States.

Coconut

Again, the word 'nut' is something of a misnomer. What we call a coconut is really the fruit of the cocoa palm. Coconut has become one of our most popular flavors comparatively recently. It is consumed either ripe or unripe, raw or cooked, and when shredded is widely employed to make cakes, pies, cookies, confectionery and puddings more attractive as well as more flavorful.

Another product of the coconut palm which has indirect flavor applications is coconut oil. When saponified and fractionated, purified coconut oil fatty acids are obtained which are basic raw materials for flavor esters.

Other fatty acid fractions are reacted with glycerol to produce standardized vegetable oils which serve as extenders for products such as turmeric or paprika.

The United States gets most of its supply of coconuts from the Philippines. Ceylon, British Malaya and the Dutch East Indies are also major producers.

Leaves, barks, bulbs and roots constitute a different kind of flavor family:

4. *Leaves, Barks, Bulbs, Roots*

Examples:	Tea; Licorice; Quinine; Onion, Garlic; Horse Radish
Used in:	Beverages; Candies; Tobaccos; Dips; Spreads; Sauces; Oral Products

Tea

Next to coffee, tea is probably the most widely used beverage, and in many parts of the world it is even more popular than coffee. The English of course take tea very seriously, and no visit to a Chinese restaurant would be complete without cup after cup of delicately flavored Chinese tea. The Japanese are great tea drinkers too, and make an art form as well as a delightful ceremony out of serving and drinking tea.

Although there are many fancy teas, including the herbal variety and even jasmine and rose hips, basically there are only two kinds of tea, green and black. That is not because they come from different plants but in the way the leaves are rolled and cured.

Chinese tradition says that tea was discovered as early as 27 B.C. In England, tea drinking started in the 17th century. Samuel Pepys mentions having his first cup of tea in 1666. In the early 19th century, the English introduced the growing of tea to India and Ceylon, and the Dutch into Java. The Asiatic countries are the largest producers of tea, although there is limited production in Russia as well.

Hot or iced, plain or fancy, with or without lemon or milk, with or without sugar, tea with its delicate aromatic taste is a favorite beverage all over the world.

Licorice

Licorice is a favorite flavor of our childhood, but actually goes back much further. It is one of the oldest and most popular of candy flavors. In Greek the word for licorice means "sweet root." Licorice contains a crystalline acid (glycyrrhizic) much sweeter than sucrose. The ability of licorice to quench thirst was recognized at an early date. Troops in the Greek and Roman armies were often supplied with licorice on their long marches. In Egypt licorice water was a favorite drink, which is still esteemed today. Licorice has long been used to combat hoarseness and sore throat, which accounts for the popularity of licorice-flavored cough drops.

Licorice is used extensively in tobacco products, about 40 million pounds a year. It is used as a flavoring and sweetening agent; as a blending ingredient to make tobaccos milder; and as a moisture-retaining agent. Licorice extract is also used to flavor chewing tobacco.

Quinine

Another odd flavor is that of quinine, which comes from cinchona bark. It is used medicinally for the treatment of malaria and other ills. We are most familiar with the bitter but refreshing taste of quinine in tonic waters, — as the perfect complement to gin and vodka. Quinine tonic is also a great thirst quencher straight.

Onion

We might have included onion and garlic in our category of spices, as they are found on the condiment shelf in powder, flake or salt versions. However we feel they really belong more accurately in the Leaf, Bark, Bulb and Root category as examples of bulbs.

Onions seem to grow in almost any climate, and with greatly increased demand they are now cultivated on a large scale much more scientifically than in earlier years. At one time onions were despised, the Latins considering onion eaters synonymous with barbarians. There are three general varieties of onions: the red ones are extra strong; the yellow, intermediate; while the white variety are the mildest. Spanish and Bermuda onions are delicately flavored.

Garlic

Garlic, of course, is a close relative of the onion. Like onion, it is an ancient seasoning, eaten by Egyptian workmen five to six thousand years ago.

These two strong and pungent flavors, onion and garlic, have no substitutes, although the cook can exercise an "either-or" decision when preparing certain dishes. Next, we move on to

5. *Dairy-type Flavors*

Examples:	Milk, Cream, Ice Cream; Butter and Margarine; Cheese; Yogurt
Used in:	Almost everything, both "as is" in some cases, as well as in cooking; Spreads and Dips; Salad Dressings; Desserts

Milk

We all depend on milk every day, whether for drinking or in other foods, so there is no need to discuss the flavor of milk. When we think of milk, we tend to think automatically of cows' milk, our accustomed kind, forgetting how many other animals give their milk to mankind. From asses to zebras, and including mares, goats, sheep, camel, reindeer and the llama, milk has provided nourishment to many peoples for many years.

If it weren't for the flavorist, we might well experience a shortage of milk in days to come as world population continues to grow. But we have a feeling that our industry's talented flavorists will be able to solve this problem and make it unnecessary to contemplate any tremendous expansion in the herds of milk-yielding animals.

Ice Cream

As for ice cream, that would take pages of flavor talk. Where we grew up with vanilla, chocolate and strawberry, today flavors have multiplied by the dozen. Now you can enjoy a different flavor every day and not repeat yourself for months.

Yogurt

Yogurt, too, has gotten into the flavor act. Once a joke, this sour milk product is now one of the most popular foods on the market, especially in urban areas. Literally millions of people now eat yogurt. True, yogurt is something of an acquired taste, but once acquired, one can readily become an addict. Many celebrities confess, or even boast, of their liking for yogurt; dieters applaud it; busy people find it quick and satisfying. Yogurt marketers hail their product as the new "fountain of youth," and credit yogurt with encouraging longevity. But, who can believe advertising?

With all those fruits added, yogurt has even become a dessert. It is a popular snack, as well as a whole diet meal. As a substitute for sour cream, yogurt not only has a similar taste but lots fewer calories. It is easily digestible, and is frequently recommended by doctors for its medicinal value.

Yogurt supposedly originated in the Middle East, then migrated to Europe, finding a welcome particularly in the Balkan countries. Like many foods, it probably was discovered by accident when bacteria somehow came into contact with jugs of souring milk. Those who tried it liked the taste, and the product didn't spoil, the way ordinary milk does when it sours. Initially, yogurt was probably made from sheep or goats' milk. Mares' milk is used in some countries. In the United States, yogurt is made from cows' milk from which most or all of the fat has been removed.

Butterscotch

Butterscotch and caramel are two perennially popular flavors which fall into the Dairy Product category. Both are what are referred to as "reaction type" flavors. In other words, they result from reacting together — normally we mean cooking — various natural ingredients which produce a flavor blend different from the flavor of any one of the ingredients alone, and more pronounced than the flavor of a mixture of the unreacted ingredients. To illustrate, if we simply mixed butter, vinegar and sugar, we would taste each ingredient separately. When we react (cook) the three materials together, we get a different and more pronounced flavor, a blend of the flavors of the three ingredients — butterscotch.

Cheese

This subject takes courage, for it is long and complicated. We won't try to compete with the cheese gourmets, but just stick to generalities. Cheeses are produced under more than 2,000 different names in all parts of the world. Best known are the Greek, Italian, Swiss, French, Scandinavian, English and American varieties, but every country takes pride in its native cheeses.

Like yogurt, cheese also probably happened by accident. One story is that it was discovered several thousand years before Christ by an Arabian traveler who put some milk into a pouch made of a sheep's stomach for his travels. During his journey the heat of the sun and the enzymes in the lining of the stomach changed the milk into a snowy white curd. The traveler tasted the resulting product and liked it, and the world has been eating cheese ever since.

Until the middle of the last century, cheesemaking remained a home or farm operation, but techniques started to vary in different geographic areas so that distinct flavors began to emerge. The basic principle, of course, is the same: "the coagulation of milk to produce curd by the separate or combined action of the enzyme rennin or lactic acid."

Whether you prefer your cheese soft or hard, mild or sharp, in any degree between, or all of them at different times, there are plenty of cheeses to please every palate. Cheese is a taste treat in sandwiches, with wine, *au gratin*, in salad dressings, omelets, fondues, rarebits, in *Quiche Lorraine*, or straight as a simple dessert. But there are dozens of other cheese uses: in soups; as an ingredient in bread, potato chips, pretzels and corn chips; as a flavor note in tomato sauces and instant gravies. Whole stores are devoted to cheeses, a paradise for cheese-lovers.

6. *Meat, Poultry, Fish and Seafood Flavors, and Flavorings*

Examples: Beef, Pork, Lamb, Chicken, Duck, Turkey, Tuna, Shrimp, Clams.
Used in: Entrees and Main Dishes; Salads; Soups; Appetizers; Dips; Snacks; Sauces; Cereals; Pet Foods

This classification certainly occupies an important place in the consumer concept of flavor. As most of us eat meat or fish or poultry almost every day, there is little to say to add to our personal flavor experience. But we suggest that texture as well as smell have much to do with our enjoyment of this class of flavors. The smell of sizzling bacon in the morning arouses the taste buds much as does the smell of coffee brewing. The smell of a turkey roasting or a steak on the barbecue grill, stimulates the appetite. Then there are the characteristic textures of these foods, — from the lighter textures such as chicken and fish to heavier and chewier such as beef and pork, often described as "something you can get your teeth into."

We will also talk here of flavorings added to meat, fish, poultry, et al., to enhance their natural flavor, contributing smoky or spicy accents. "Smoky" is a taste favorite, whether we are talking about smoked ham, bacon, dried beef, turkey, salmon, cod, kippers, oysters, pheasant or sausages. A smoky accent is a very special taste, subtle and delicious, which seems to soften the basic taste of the food, yet brings out its fullest flavor.

Then, there are the spicy accents such as those added to many of the sausage group of meats, — from breakfast links on to pepperoni and salami. Of course many spices are added when meats are cooked, but we are talking here specifically about the spices that are incorporated into meats *before* we buy them. There are also spicy sauces of all kinds, such as spaghetti sauces, often with meat included. Spices were probably used originally as a preservative for meats, or to disguise the flavor of meat that was a bit past its prime. Now they are an integral part of the meat flavor we are accustomed to expect.

Another spice often added to meat is curry, which we didn't mention in our direct coverage of spices. This is a condiment that makes an entirely different dish out of lamb or chicken, especially.

7. *Other Selected Flavors*

Examples: Potato, Tomato, Mushroom; Honey, Maple
Used in: Main Dishes; Soups; Salads; Omelets; Cheese and Egg Dishes; Snacks; Appetizers; Juices, Beverages; Condiments; Garnishes; Candies; Syrups; Pharmaceuticals

Potato

First, consider the potato, with its mealy, earthy, satisfying flavor, which can be prepared in so many versatile ways that it seems like a lot of different vegetables. Mashed or hashed; fried, scalloped, baked; and chipped, potatoes are not only demanded by the "meat and potatoes" man but by everyone. Unlike many other vegetables, we find no relics of potato culture left by primitive peoples.

The potato probably originated in South America in the Andes. The first written description of potato was in 1537 when a member of a Spanish expedition to this area noted: "Floury roots of good flavour, a gift very acceptable to Indians and a dainty dish even for Spaniards." Apparently, the potato then went to Spain around 1570. Ireland received the potato directly from the New World, and it soon became the staple food crop for Irish peasants.

Other countries at first found the potato objectionable, accusing it of being poisonous, capable of shortening the lives of those incautious enough to eat it. Not so in Prussia, however, where Frederick William commanded that potatoes be planted in the Lustgarten in Berlin in 1651. Seed potatoes were distributed free, and a royal circular was issued giving advice on plowing, fertilizing and planting. Even when potatoes were accepted as proper fare, however, eating them was accompanied by a social stigma, as the rich thought potatoes suitable only for the meaner sort of persons. In recent years, however, the lowly potato has come a long way!

Tomato

The tomato, which is vegetable, fruit and berry all at once, suffered an even worse reputation than the potato at one time. Now, however, its inimitable flavor has given it an improved image and an important place in our diet. Tomatoes range from the size of cherries to the huge beefsteak size, from yellow in color (once called "golden apples") to the brightest, most luscious red. The name, "love apple," by which the tomato was often known, may have come from its shady reputation as an aphrodisiac.

Most people today find the tomato a tart and saucy, juicy and zesty, versatile and indispensable food. Whether you drink its juice straight, or in your Bloody Mary, whether you serve tomatoes in soup or salad or on pizza, whether you like the tomato flavor in catsup or in any number of other sparkling sauces, whether you eat them raw or fried, or stuffed and baked, whether you like them pickled and in relishes, or in dozens of other ways, the flavor of tomatoes is delicious. We would all miss this unique taste if there was ever a tomato shortage.

Mushroom

Unlike the potato and the tomato, mushrooms have been venerated for centuries in spite of some of the strange powers with which they're credited. A tribute was paid to this tasty fungus by the Roman writer, Martial, who expressed his opinion about 2,000 years ago when he said, "Gold and silver and dresses may be trusted to a messenger, but not a mushroom, because he will eat it on the way." Not only did the Romans find the mushroom palatable, but the excellence of French cuisine owed much to this small but choice edible from the 17th century on. It is said that Marie de Medici's chefs brought mushrooms to France from her native Italy.

The mushroom comes in infinite varieties, some of which are poisonous. Pennsylvania, alone, has 1,500 kinds of both edible and inedible mushrooms that have been identified. There are mushrooms that cause extraordinary hallucinations.

Mushrooms make elegant appetizers, and are perfect in soups and sauces, in cheese and egg dishes, on meat and seafood and fowl, mixed with other vegetables, or just sauteed. They have an intriguing earthy flavor and a meaty taste that is an ideal complement to other foods or a proud dish all on its own.

Honey

From the sublime to the sweet sublime, honey and maple. Honey comes in a myriad of flavors, from alfalfa to persimmon, from thyme to sweet clover, from orange blossom to sage, and many, many others, all depending on the type of blossoms attractive to the bees. Honey has been loved for centuries in all countries and by all ages. Aristotle called honey the "dew distilled from the stars and the rainbow." A pound of honey requires the nectar essence of two million blossoms. That should justify calling honey liquid gold.

Honey is of course widely used in the kitchen, but it has also won an honored place as an ingredient in pharmaceuticals. Honey is a useful ingredient in tonics, cough medicines, and certain vitamins. It has been used for many purposes throughout history, from conferring immortality and bestowing prophetic powers, to its use in Egypt as an ingredient in embalming fluids.

Maple

Another flavor familiar to all Americans is maple. It adds a delightful taste pleasure to waffles and griddle cakes, and the delicious maple sugar candies so popular in New England. Maple is also a much-liked flavor in some alcoholic beverages, and in ice cream and cakes. It's an old-fashioned kind of flavor with nostalgic overtones, one that can never go out of fashion.

References:

A Glossary of Spices, American Spice Trade Association, copyright 1966.

P.&E.O.R., June 1955

The Givaudan Flavorist, Number 4 - 1968; Number 1, 2, 3, 4 - 1969; Number 2, 3, 4 - 1970; Number 2, 3 - 1971; Number 2, 3 - 1972; Number 2, 3 - 1973; Number 2 - 1974; Number 1, 2, 4 - 1975; Number 3 - 1976.

Chapter VII

Natural Raw Materials

The fragrance and flavor industry originally depended almost entirely on natural products for its raw materials. These items were the first to appeal to the human senses, having social as well as religious uses extending far back into ancient times. The vagaries of supply, and the steadily increasing costs of the natural products have in recent years reduced their use proportionately, as compared with synthetics, but naturals still tend to be the raw materials of choice for fragrance and flavor products.

Many natural raw materials are used in *both* flavors and fragrances: for example, oil clove bud, oil cinnamon, jasmine absolute, even rose oil and absolute civet. A general classification of the natural products, breaking down the many different products employed into categories, should help us to approach the subject of natural raw materials in orderly fashion. We can separate them into the following groups:

essential oils
concentrated and rectified oils
resinoids
gums, balsams and resins

- terpeneless and sesquiterpeneless oils
- natural flower oils
- flower waters
- concretes and absolutes
- tinctures and extracts
- animal fixatives
- oleoresins
- fruit juices and concentrates

Essential oils, our first category of raw materials, originate in all parts of the world. People and companies involved in importing and supplying them must become experts in world geography. Many essential oils are produced in the United States. The U.S. is the world's largest supplier of peppermint and spearmint oils. More than 100,000 acres are devoted to mint production in the U.S.

Mint Oils

Exports of peppermint oil from the U.S. in 1976 amounted to $31.3 million, to which can be added $12.9 million of spearmint oil. Total U.S. production of peppermint oil in 1976 was 3.7 million lbs., amounting to $55.3 million, and of spearmint, 1.68 million lbs., valued at $21.7 million. U.S. production of all essential oils in 1976 totaled $76.8 million.

The delightful aroma and the mildly stinging taste of the mint oils was appreciated far back in ancient history. The Israelites used to spread mint leaves on the hard dirt floors of their homes, so their cooling fragrance would be released when walked upon. Mints are mentioned in the Bible, a 10th century Japanese medical book describes the use of mint oil in an eye wash, and in the Middle Ages mint was used in medicines.

The peppermint oil we know today was first mentioned in 1696 by an Englishman named John Ray, and it is believed that commercial distillation of mint oil began in England shortly after that date. By 1797 there were 100 acres of peppermint planted at Mitcham, England, and production was estimated at between 2,000 and 3,000 lbs. annually. This is no longer a producing area.

The United States imported peppermint planting stock from England about 1812, the first plantings being made in Massachusetts. Mint culture then spread to New York state and Ohio, and in about 1835 to Michigan. Then, about 1919, peppermint growing was introduced to the states of Washington and Oregon. It expanded rapidly in this area, due to high sunlight intensity, favorable soil conditions and adequate water.

The history of spearmint is less well known, but we are told that it first appeared in European gardens in the 9th century. The two types now grown in the United States are called "common spearmint" *(Mentha spicata),* and "Scotch spearmint" *(Mentha cardiaca).* Scotch spearmint is more pubescent than either peppermint or common spearmint. It is not as frost-resistant as common spearmint, and is grown almost exclusively in Michigan and Indiana, because of its greater productivity and vigor, although some Scotch spearmint is also grown in the Far Western states.

The mint oils occur in minute glands, mainly on the undersides of the leaves. The oil globules are held in place by thin membranes which resemble pliable cellophane. When the membranes are ruptured by pressure and heat, as in steam distillation and comminuting, the oil is released and can be collected.

The price of mint oil has been forced upward in recent years by the competition of other crops for the use of suitable land, which has tended to reduce mint acreage. Nevertheless, new growing techniques, including specifically the introduction of selective herbicides such as the uracil compounds, has been responsible for significant yield increases. New strains of peppermint and spearmint have also made their contribution to raising yield levels.

The aroma of mint continues to enjoy high popularity. A recent study in Britain indicated that only lemon and cinnamon are more popular aromas than peppermint. The study showed that men prefer peppermint, while women choose spearmint. Introverts tend to prefer peppermint over spearmint, with the reverse preference for extroverts. Mint oils are not greatly in demand in fragrances, although peppermint is a good aroma for soaps, room fresheners and the like.

Cedarwood Oil

The United States is the main source of cedarwood oils, of which there are two principal varieties, *Virginiana* and *Mexicana* (or Texas). The *Virginiana* has a softer, more oily type note, one we associate with the red heartwood of the cedar tree. Highly aromatic and persistent, it is used mainly in perfumery, little or none in flavors. Oil cedarwood Texas has high impact, is camphoraceous and persistent. It is frequently used in room fresheners. This oil is fairly inexpensive. It serves as the raw material for a variety of cedar fragrance materials, including cedrol, cedryl acetate, cedrenol, cedrenone and acetyl cedrene.

Citrus Oils

The other domestic essential oils of most consequence are the citrus oils. As described elsewhere (see pp 279-81) these oils are expressed or "cold pressed," being produced as by-products of fresh fruit which is the primary monetary product of the grower. Orange, lemon, grapefruit and tangerine oils are the principal domestic citrus types, with small quantities of lime and tangela oils available. When juice concentrates or frozen juices are processed, the essence oils of these items may be sold rather than being used as additives in the concentrates.

About 10 million lbs. of orange oil are produced annually in the U.S., excluding distilled oil or limonene byproduct. Production of lemon oil, mainly from the California - Arizona area, amounted to about 1.5 million lbs. in 1976. The amount of lemon oil produced depends directly upon the amount of fresh fruit available in excess of that required for the fresh fruit market.

Citrus oils are used in flavors for soft drinks, in dry drink mixes, cake mixes, candies and adjunct or auxiliary flavors (such as iced tea with lemon). They are also used in fragrances for soaps, in eau de colognes to provide fresh notes, and in the formulation of other light, refreshing perfumes. Oils of orange, lemon and grapefruit are used in the same manner. Oil of tangerine, produced in much smaller quantity, has more specific uses, since its taste and aroma are somewhat different from those of the other citrus oils. The heavy body and unique topnote of tangerine can serve to enhance the character of fragrances and to give special notes to candies, drinks and such items.

Domestic orange oil is in adequate supply, although approximately 750,000 lbs. were imported in 1976, for price or variety reasons, but other citrus oils must be imported. Lemon oil also comes from Italy, Greece, Israel and Spain, as well as in increasing amounts from Argentina and Brazil. Imports of lemon oil were 635,000 lbs. in 1976. The imported oils are used in the same manner as domestic oil, that is mostly for soft drinks.

The dollar value of imports of distilled and expressed lime oil is quite substantial ($6.5 million). Most of it comes from Mexico, with smaller quantities from Haiti, the Dominican Republic, and now South and Central America. Distilled lime oil is used mainly for soft drinks and candies, and for after-shaves and light colognes in the fragrance area.

Expressed lime oil is a highly fragrant oil, heavier and more citrusy than lemon. It has limited flavor uses in soft drinks as an enhancer. Its use in fragrances for after-shaves, deodorants and like products emphasizes its distinctive odor.

Oil of bergamot is produced in southern Italy where it is expressed from a small citrus fruit which resembles an unripe lemon. U.S. imports in 1976 were 157,000 lbs. It has a powerful, flowery, citrus aroma, and is highly valued by the perfumer for use in fine fragrances, particularly colognes. Recently one of the minor non-volatile constituents of the expressed oil (bergamottin) has been cited as a phototoxic chemical. This same type chemical has been found in limes and lemon expressed oils. No action has been taken to restrict use of bergamot oil, but more oil bergamot, rectified, is now being used. The rectified oil unfortunately does not have the full body of the natural expressed oil.

We complete our discussion of citrus oils with oil of petitgrain, which is distilled from the leaves and twigs of the bitter orange tree which produces the fruit for oil bitter orange and the blossoms for oil of neroli. (U.S. imports in 1976 were 1,834 lbs. valued at $230,000.) The usual source of oil of neroli is France. Paraguay is the main source for oil bitter orange and oil petitgrain.

Spanish Jesuit missionaries first brought the orange tree to South America. The trees have continued to grow in a semi-wild state, spreading across the plains and through the jungle. The bitter orange tree has thus provided Paraguay with a flourishing essential oil industry. Started near Villa Rica in 1880 by a French botanist, about 440, 000 lbs. of oil are now exported annually. Most of the commercial oil of petitgrain which enters world commerce comes from Paraguay. It is used in soaps, colognes and toilet waters to impart a mellow note in conjunction with lemon, bergamot, orange, etc.

Citronella Oil

Two of the most important oils derived from grasses are citronella and lemongrass. While long an important article in the essential oil trade, production of oil citronella Ceylon estate has been declining over recent years. It has seldom been processed for derivatives, only occasionally being acetylated. Rather, it is used directly in fragrances. Oil citronella Java, the variety grown in Taiwan, Honduras and other Central American and West Indian islands, as well as in Argentina and Brazil, is processed to yield its content of geraniol, citronellal and citronellol. The chemical derivatives available from these isolates are discussed elsewhere (see pp 156-7). Citronella Java has been the most useful raw material in the perfumer's arsenal for many years. Despite the increasing production of synthetic aroma chemicals, imports of citronella Java are still impressive.

Oil of Lemongrass

Oil of lemongrass is frequently cultivated in the same areas which produce the other grasses, namely Guatemala, Indonesia, India, etc. The U.S. imported 244,000 lbs. of oil of lemongrass in 1976. At one time lemongrass was the only source of citral, ionones and vitamin A. Now synthetic citral has cut deeply into the market for the natural product. However, lemongrass provides a citral which can be labeled "natural," and therefore still enjoys wide use in artificial lemon products with natural flavors. The top notes of lemongrass, myrcene and methyl heptenone, still make this oil a useful product for deodorants and industrial fragrances.

Oil of Palmarosa

Oil of palmarosa was at one time an important oil, when pure geraniol containing no citronellol was required. It is still a useful product in soap perfumery, but limited production (21,000 lbs.) has made it a more expensive source of its derivatives which can be made from other less expensive oils.

Oil of Clove

The several varieties of clove oils are valuable natural F/F raw materials. The clove industry originated in the Molucca Islands, and this spice was one of the objects of the voyages of Columbus and the spice traders of the 15th century. Plants were taken clandestinely out of the Moluccas in the mid-18th century, and cultivation begun in Reunion and Mauritius (formerly Ile de France). In 1818 the Arabs introduced cloves to Penang and Zanzibar, and for many years Zanzibar was the only major source of clove products. Recently Indonesia has again become involved, and the volume of oil shipped from there, particularly clove leaf oil, is increasing. Zanzibar, however is still the major world supplier, and demand for cloves is high enough to insure the economic soundness of the island.

There are three types of oils derived from the clove tree. The finest, and the only oil used almost exclusively *per se* in perfumes and flavors, is oil of clove bud, which is distilled from the dried, unopened buds. This full, spicy, persistent odor and flavor is used for all spicy notes, carnation, and many oriental fragrances. It is also used in toilet waters, especially in conjunction with bay oil. Oil clove bud has many flavor uses, frequently being employed in dentifrices, mouthwashes and the like as an enhancer. Because of its relatively high cost, it is often replaced to some extent in these products by eugenol from clove leaf.

Oil clove stem is available in only limited volume, as it can be produced only when twigs and stems are available for processing. When available, it is of intermediate fragrance quality, but possesses a good spicy note. It is rigidly controlled by the Malagasy government and now most production is probably blended into clove bud oil. The general improvement in the quality of clove leaf oil limits the demand for stem oil.

Oil clove leaf is rarely used as such in formulations. It is normally rectified to improve the quality, and is used mainly as a source of eugenol.

Oil of Cinnamon

Several types of oil of cinnamon are available. Oil of cinnamon Ceylon (Sri Lanka) is made from the bark of the shoots which grow from the stumps of cinnamon trees. The natives strip the bark off the shoots in long quills, dry it, and roll one strip inside the other. The old fashioned cinnamon stick served with hot buttered rum is such a quill. These quills, perhaps three to four feet long, are wrapped in bundles for export. Most oil cinnamon Ceylon is distilled in the United States. It has a beautiful, warm, spicy odor, and is a favorite for cola drinks or candies. It also has many applications in fragrances, particularly in orientals.

Another type of cinnamon oil (called cinnamon Saigon) comes from Vietnam and the surrounding area, where it grows wild. To produce this oil, bark from the branches, the branches themselves, and bark from the tree trunks is used. This is the main source of cinnamon spice as we know it, but only a limited amount is distilled for oil for use in flavors.

Oil of cinnamon from the Seychelles Islands is not quite as fine an oil as is the Ceylon variety, but is less expensive.

The various cinnamon leaf oils are used as sources of eugenol, as they contain high amounts of this item. They are warm, spicy oils, and have uses similar to clove oils.

Oil of Cassia

Another species of cinnamon grows just over the mountains from Vietnam in southern China, in Kwangsi Province. Oil of cassia is distilled in this locale. With a very spicy, full bodied cinnamon fragrance, it has established a unique place for itself in the flavor list. It has been used for many years to flavor soft drinks, particularly colas. Oil cassia once had wide use in soaps, but its high price now precludes this application.

Cassia was unavailable for a long period, when our trade with Red China was interrupted. Only since trade has resumed has cassia oil again become available, but with rising demand the price has risen sharply in a most uncommunistic-like fashion.

Oil of Ginger

Another spice oil, oil of ginger, is distilled from the rhizomes of the ginger plant. Jamaica ginger has the finest reputation, but ginger is also grown in Africa, India, China and Japan. Fields of ginger resemble potatoes in the manner in which they are planted. After the stems of the plants wither, the rhizomes are dug and prepared for distillation. The distillate is pale yellow; it has the characteristic smell of ginger, but is not as pungent as the finished spice. Much of the oil is distilled in the U.S. The perfumer uses ginger to give an interesting flair to his oriental blends. Oil ginger has limited use in flavors.

Oil of Nutmeg

This is a very valuable flavor and fragrance material. There are two types of nutmeg oil, both from the same tree species, but different harvesting methods and different climatic conditions yield two different oils. Oil of nutmeg East Indian has a heavy, woody-phenolic dryout and is more spicy. Oil of nutmeg West Indian, mainly from the island of Grenada, is thinner, more terpeney, and finds greater use as a ground spice than as an oil. Nutmeg oil is often used in after shaves and men's fragrances. Its principal flavor uses are in condiments, cola type beverages, baked goods, pickles, and in sausage and other prepared meat flavors.

Other Spice Oils

There are many other spice oils which are excellent raw materials for the flavor and fragrance industry, but space limitations make it possible to do no more than mention a few of them by name. They include: mace, coriander, thyme, origanum, tarragon, cumin, marjoram, pennyroyal, black pepper, pimenta, bay, rosemary, sage, etc. These oils have their own individual attributes and are useful tools to both the flavorist and the perfumer. Those seeking a more detailed knowledge are referred to the numerous books on the subject, some of which are listed in Chapter XVI (see pp. 388-391).

Sandalwood Oil

We mentioned cedarwood oil earlier. There are several other valuable woody oils which are among the mainstays of fragrance fixation. These include oils of sandalwood, amyris and rosewood. Sandalwood is a tough, compact, close-grained wood which gives off a delightful fragrance long prized by the perfumer for its warm, woody fixation.

Most sandalwood oil comes from Mysore in the southern part of India. In earlier years the trees had been cut so freely that sandalwood had become an endangered species. The remaining sandalwood forests are now under government control. Trees are not harvested until they reach full maturity, and government officials mark the trees to be cut. Sandalwood is an exceptionally hard wood, so hard that it must first be chipped by a machine rasping process before the chips are charged into stills. In recovering the oil, steam is blown through the still charge for about 36 hours, since the wood contains mostly sesquiterpenes and sesquiterpene alcohols which are comparatively non-volatile.

Sandalwood oil is a frequently used ingredient in perfumes, particularly in orientals, wherever the price can tolerate it. Nearly all orientals contain some sandalwood, and many of them a great deal. Sandalwood also is often combined with violet, cassie, reseda and rose. It blends well with heliotropine, coumarin and vetiver. It is very popular in soaps. In addition to its own unique fragrance, it also adds fixative values to a compound in which it is employed.

Oil of Amyris

Oil of amyris, or West Indian sandalwood, comes from Central and South America and the West Indies. The tree from which it comes is actually unrelated to the Indian sandalwood tree except for the fact that it is a heavy, oil-laden wood. It is used for making furniture, and has a soft, balsamic character. It is called candle wood by the natives, because it is so full of oil that it lights up like a torch. The oil is reasonable in cost and has been used for many years to extend East Indian sandalwood or vetiver oils. It is an excellent fixative in applications where its overpowering fatty-woody character does not preclude its use. It exhibits weaker intensity than East Indian sandalwood or vetiver.

Rosewood Oil

Here is another oil whose history dates back to Biblical times. It is highly prized by the cabinet maker and builder for its beauty and durability. From it linalool may be recovered. The complete oil is frequently used in rose compounds. The oil comes from Brazil, Peru, Honduras and French Guiana, where the rosewood tree grows.

Other oils which have woody notes as part of their character, although they are not derived from wood, include oils of vetiver root and patchouli leaves.

Oil of Vetiver

Vetiver grass growing in sandy soil develops a heavy bundle of fibrous roots. If allowed to remain in the ground for at least a year and then slowly steamed-distilled, a rich, woody-earthy character develops in the oil. Vetiver oil is used extensively in fragrances, not only as a fixative, but also as a major contributor to chypre and fougere bases. It blends well with almost any fragrance material. Vetiver is also used as a source of other valuable fragrance materials such as vetiverol and vetiveryl acetate. Vetiver is grown in the island of Reunion, in Haiti and India. The Reunion type oil is preferred, but is the most expensive.

Oil of Patchouli

There has been a tremendous revival of interest in patchouli in recent years. This powerful fragrance oil, distilled from the dried leaves of the patchouli plant, an East Indian mint, resembles somewhat our garden variety of sage, except that the leaves are less fleshy. The current popularity of patchouli is perhaps related to the current vogue for "woody" odors.

Patchouli has a strange sort of fragrance, persistent and haunting. It is one of the most intense in the entire plant kingdom. Besides its woody character, patchouli might also be described as "dry and powdery" or "earthy." Along with sandalwood, it is used in many heavy oriental bouquets.

Oil of Geranium

The oil distilled from the leaves and stems of the geranium plant is one of the most important essential oils used in perfumery. It has a powerful, yet soft and lasting rosy odor. It blends well with rose bouquets and orientals, is used in fine perfumes of all types, and is indispensable for powder fragrances or perfuming fine soaps.

In harvesting, the plants are gathered shortly before the flowers open, and allowed to stand for about a day before being distilled. Sometimes rose petals are added to the still charge to intensify the rose note, but the geranium oil itself has a rose fragrance.

Geranium oil may be saponified and fractionated to make rhodinol, an aroma chemical which is even more fresh and flowery than geranium oil itself. Rhodinol is in effect the purified natural alcohols of the geranium plant, including geraniol, citronellol and linalool in perfect combination. Rhodinol is used in all rose perfumes because of its highly refined character and because it imparts considerable depth and heaviness to the fragrance.

Geraniums are grown for conversion into essential oil in many different countries, including Reunion, Algeria, Spain, Russia and France. Worldwide, many thousands of acres are devoted to geranium culture. Reunion alone produces over two hundred thousand pounds of geranium oil a year. The United States imported 230,000 lbs of geranium oil in 1976.

Oil of Celery Seed

The layman would not consider oil of celery seed a fragrance material, and only a few might realize it is used in flavors. The oil has both applications. Its odor is warm and spicy, sweet and rich, and very diffusive. Oil of celery seed has an extremely low threshold value, and therefore must be carefully applied. The best grade seed comes from France. It is exported and distilled in many countries, including the U.S. This oil is used primarily in condiments, and in trace amounts in fine fragrances.

There are several other seed oils which are very useful. Ambrette, carrot seed and parsley seed oils impart green, vegetal notes to compositions in which they are used.

Oil of Orris Root

Another powerful F/F raw material which requires careful use is oil of orris (iris) root. The finest orris root is found near Florence, Italy, where the coat of arms of the city is a white iris on a red shield, assumed to be in honor of the iris flower. The orris roots are peeled by hand, cleansed, then dried outdoors on straw mats, or indoors in warm, well-ventilated chambers. Some of the roots are pulverized, with the fine powder used in sachets, dusting powders, or incorporated into soap. Most of the crop, however, is distilled to yield oil of orris, the production center being in Grasse. In addition to the distilled essential oil, an oleo resin is also produced by extraction. Orris is an excellent fixative for lavender water, violet perfumes and fine soaps.

There is a French orris root which produces an oil with a more delicate aroma, more intense, with even more of a violet character. As might be anticipated, it is also more expensive. Other root oils which were important at one time, but are now dwindling in supply, are oils of lovage root and angelica root.

Specially Processed Oils

In the chapter on Definitions (pp. 59, 63) the purpose in making concentrated oils, rectified oils, terpeneless or sesquiterpeneless oils was discussed. Many essential oils can be processed to increase odor or flavor intensity, remove terpenes, stabilize the oil, improve solubility, remove waxes, fixed oils or objectionable top-notes, or to improve color. The processed oils are used for generally the same purposes as the oils from which they are made, taking advantage of the selected physical or chemical improvements in character.

Flower Waters

The natural flower oils are extremely important in the creation of fine fragrances. Depending on the heat sensitivity or the oil content of the individual flower, these products are hydrodistilled or extracted, and are therefore available in a variety of forms, as volatile oils, absolutes, concretes or flower waters.

Orange flower water and rose water have been produced for generations in the Grasse area of southern France, and are generally produced wherever roses or neroli are steam distilled. These products are much used in all the Mediterranean countries — France, Spain and Italy — for baked goods, liqueurs and fondants. They are regularly called for in many old recipes. Flower waters have also been used in the past for medicinal purposes, and in the fragrance area are popular products for incorporating in toilet waters, colognes and lotions where the true flower note is desired. Flower waters are, of course, the distillation waters after the volatile oils have been separated off.

Concretes and Absolutes

Of all the flowers processed, about the only ones treated by distillation are rose, orange blossoms, lavender, lavandin and ylang ylang. The others are extracted with solvents. The combination of heat and moisture inevitably produces some chemical and physical reactions, and as a result otto of rose and oil of neroli (orange blossoms) do not fully represent the true fragrance of the living flower. More true to life are the absolutes and concretes of cassie, champaca, carnation, gardenia, genet, jasmine, geranium, jonquille, lavender, lily, mimosa, narcissus, orange flower, rose, sage clary, tuberose, violet flower, violet leaf and ylang ylang.

Rose Oil

Bulgaria is the center of rose culture. During the rose harvest, in May and early June, the peasant women pick the flowers during the early morning hours while they are still fresh with dew. As it takes about 4,000 lbs. of roses to make a pound of rose oil, an idea can be obtained of the profusion of the rose harvest. The rose grown in the "Valley of Roses" in the Balkans is called *Rosa damascena.* It has a sweet, full, honey-like bouquet. Turkey is also now a major supplier of a variety of *Damascena* rose oil.

In Grasse, in southern France, the *Rose de Mai* grows *(Rosa centifolia).* Egypt is also now producing a large volume of otto of rose, used as is or blended with other rose oils. There is no one characteristic rose fragrance. Some roses smell fruity, others have a violet-tinged fragrance, while still others have a hint of musk or myrrh. The perfumer has scores of roses fragrances from which to make his selection, and in almost every fine perfume there is sure to be a scent of roses. The rose has clearly earned its title, "The Queen of Flowers."

Jasmine

An oil even more precious than rose, and even more essential in making a fine fragrance is jasmine. Actually, these two flower oils, rose and jasmine, are absolutely indispensable to the perfumer. The jasmine has been known to man perhaps longer than any other flower. There are many jasmine-growing areas around the world — Egypt, and other parts of Africa, Madeira, the Italian Riviera, Arabia, India, Taiwan — and, of course, Grasse. Jasmine imparts smoothness and gives "life" to the perfumers' creations. It is vital to the beauty of a perfume, and is considered an essential ingredient in a fine fragrance even if present in only a minute quantity.

Orange Blossom (Neroli)

The rose, jasmine and orange flower are sometimes referred to as the "three graces" of perfumery. Of the three, the orange blossom is the most fragile and dainty. The orange flower got its name, neroli, back in 1680, because it was used to perfume the gloves of the wife of Flavio Orsini, prince of Neroli. Oil of neroli is obtained from the freshly picked flowers of the bitter orange and sweet orange trees. The former yields, by far, the finer and more volatile oil. The oil from the blossoms of the bitter orange tree is called *neroli bigarade,* that from the sweet orange tree, *neroli Portugal.* The perfumer uses the term, neroli, to describe only the oil obtained by distillation. The extracted product is called orange flower oil. In France, almost all oil is extracted, but in many other areas, including the Italian Riviera, it is distilled.

Neroli gives a full, flowery note to perfumes. It blends well with floral bouquets of the heavier type, and almost all fine perfumes require some percentage of this essence of orange blossoms. The fragrance is light, evanescent and ambrosial.

Oil Ylang Ylang

The most intense flower oil has perhaps the strangest name, — ylang ylang. This means "flower of flowers" and also suggests that it is a drooping flower. The heavy hanging blossoms, greenish-yellow with a red center, have an intense, beautiful fragrance.

The finest ylang ylang oil is said to come from the Philippines, but it is also produced in substantial volume on Madagascar, Nossi Be and Reunion. Ylang ylang gives a lift to oriental perfumes, and is also frequently blended with jasmine, lily and violet. It is used not only for its own delightful fragrance but also as a modifier of other floral notes. Ylang ylang is almost indispensable in building a lilac fragrance. It softens and rounds out any harsh notes in a composition.

Lavender

Lavender is a low-growing evergreen plant belonging to the mint family which grows wild like heather. Originally lavender grew principally in the high altitudes of Spain and northern Africa, but is now native to England, Norway, France and the United States. We think of the word, lavender, as almost synonymous with "clean" and "fresh," and indeed the botanical word for the flower, *lavendula,* is probably derived from the Latin, *lavare,* meaning "to wash."

Most of the commercial lavender, now a hybrid and grown on plantations rather than found wild, is produced in southern France. Lavender grown in England, but not a commercial product, is said to have a finer fragrance than French lavender, induced by the moderate, moist climate. A true lavender, similar in composition to the original French variety, has been grown in Tasmania.

Oil of lavandin, a hybrid developed by crossing true lavender *(Lavandula officinalis)* with spike or aspic lavender *(Lavandula latifolia)* has finally come into its own after twenty years or so of needed quality improvement. This oil, grown on large plantations in France and northern Africa, is perhaps the largest volume flower oil produced, and one of the largest volume of all natural fragrance materials. At one time lavandin was used to cut lavender oil, but is now a major raw material in its own right. It is reasonable in price, dependable in quality, and there is no problem with availability. It has supplanted lavender oil for most of its uses. Because of its fresh, clean character, it is a major component of soap fragrances.

Narcissus

The wild narcissus grows on slopes in the Maritime Alps above Grasse. The fragile yellow and white flowers have a delicate and unusual scent which makes narcissus much sought after as a perfume ingredient. Because of its rarity (annual production being less than 200 lbs.) only a few of the world's most expensive perfumes contain this fragrance. One of them is Worth's *Je Revien,* the marketers of which tell its story in a memorable advertisement which is captioned "The world's smallest harvest."

Oakmoss

Oakmoss has long been important as one of the "natural" perfume raw materials, but today is receiving even more attention because of the current emphasis on "mossy" notes in many perfume creations. There are many supplements and substitutes for oakmoss, but perfumers still demand the natural material because in addition to its persistent "woody" or "earthy" note, it is also one of the finest fixatives.

Oakmoss, a lichen, grows, as might be suspected from its name, on oak trees. It is believed to have originated on the island of Cyprus, but most of the supply today comes from Yugoslavia. Initially oakmoss has no noticeable fragrance, but after being dried and stored, it develops a musk-lavender note. The fragrance is extracted from the dried moss by treating with boiling solvent.

Oakmoss is indispensable in chypre bouquets, and is also an excellent base for fougere perfumes. In lavender, it is equally essential. It imparts an agreeable freshness to bergamot, lemon and citronella, and adds a special quality to the fragrances of jasmine, rose, cassie, and orange blossoms. It lends a certain "tone" to ylang ylang, coriander, sassafras and vanillin, and gives depth and stability to patchouli, vetiver, coumarin and musk.

Gums, Resins and Balsams

The items of importance in this category are benzoin Siam, galbanum, labdanum, myrrh, olibanum, opoponax, Peruvian balsam, styrax and tolu balsam. Most of these materials have been used since Biblical times. They are found on the Mediterranean islands and on the mainland in Africa and Asia as far east as Sumatra and Thailand. They are also found in South America. In some cases the natural raw material is used as such, and in other cases it is processed to produce essential oils, resinoids or tinctures. Since these materials generally have a high content of sesquiterpenes or non-volatile components, they possess fixative properties. They are processed in different ways to develop better solubility, fluidity, improved color and the like. In general the fragrance application is much the same regardless of the form of the raw material used. The nature of the finished product being made dictates whether the raw material will be incorporated as an oil, a resinoid of a tincture. The fragrance note achieved will be somewhat similar regardless of the form of the raw material.

Siam Benzoin

The finest Siam benzoin comes from Thailand. Like the other gum resins, it adds weight and body to perfumes, as well as its own delightful aromatic fragrance. It is also an excellent fixative. It is commonly used to scent face creams, nail polishes and hair pomades and is a favorite ingredient of incense. Siam benzoin is offered as a resinoid and a tincture, but not as an essential oil.

Galbanum

Galbanum has long been used as an incense, and also in medicine. The Levant variety is the best offered commercially. This resin is musky and balsamic. Like labdanum and myrrh, it is used primarily in oriental blends. The oil has a green leafy character, with a woody, balsamic note. It is powerful and penetrating. The oil and resinoid have many applications, and can be used for fougere, chypre, oriental and pine notes, etc.

Labdanum

Labdanum has a powerful but sweet odor suggestive of ambergris. It is used as a fixative in many verbena and violet-type perfumes, but is most frequently used in making artificial ambers, which are powerful fixing agents. With its fragrant odor but bitter taste, labdanum has a place in medicine as well as in perfumery, having been so used since ancient times. Labdanum is supplied as the oil, as a tincture, and as a resinoid. An absolute as well as a decolorized absolute are also offered.

Myrrh

Myrrh is another bitter, aromatic gum with a long history. It is probably the earliest aromatic gum of which we have any record. Myrrh was used in ancient times in primitive medicines, for fumigation, embalming and as an offering to the Gods, in addition to its use in perfumes. Today it is used in dental preparations and in soaps. It serves as a fixative and fragrance in perfumes of the oriental type. It is available as an oil, tincture and resinoid.

Olibanum

Olibanum is the "frankincense" of the Bible, the word meaning, in old French, "pure incense." It had an early use in religious rituals, and for medicinal purposes. This resin, which permeates the leaves and bark of the Bosmellia tree, as well as exuding as a milky juice from the flowers, is highly aromatic. In addition to the gum, an essential oil, a resinoid and a tincture are also offered. Olibanum is valued for its marvelous fixative properties, and is frequently used in perfumes of the heavy oriental type, often combined with musks or ambergris. It blends well with bergamot, cedarwood, iso-eugenol, nutmeg, neroli, patchouli, rose and vanillin.

Opoponax

Opoponax could well have been the "myrrh" of the Bible. It has a delightfully aromatic fragrance, and a tart flavor. The odor is similar to that of myrrh, somewhat floral in character. Like the other gums, opoponax is an excellent fixative. It has a richer, more balsamic odor than either myrrh or olibanum, and sometimes an animal-like sweetness. It is available as essential oil, tincture or resinoid.

Balsam Peru

This balsam doesn't come from Peru, in spite of its name, rather from El Salvador. It is useful as a cicatrizer, antiseptic, and as an ingredient in cough syrups and throat lozenges, as well as a fixative for perfumes and soaps, being a particularly good fixative for such perfumes as rose, heliotrope and lilac. Its fragrance resembles that of benzoin and vanillin. Balsam Peru imparts a characteristic "warmth" to face powder fragrances as well as to perfumes. It is used mainly in the form of a tincture or a resinoid, but an essential oil is also available, which is rare for such a product.

Styrax (Storax)

There are two varieties of styrax (Levant and American) which are expressed from the bark of different varieties of the Liquid Ambar tree. The odor of styrax, is very much like that of naphtha, since it contains styrene. When refined and diluted for use, however, it resembles a bouquet of hyacinth, tuberose and jonquil. The perfumer uses styrax sparingly, for very little is required to give the desired finishing touch to such perfumes as narcissus, lilac, tuberose and wisteria. It is also used successfully with jasmine, mimosa, muguet and hyacinth. When using this material, the perfumer and flavorist can choose between the essential oil and the resinoid.

Balsam Tolu

This tree, which is a variety of balsam Peru, grows in South America. The balsam exuded has a soft, hyacinth-like fragrance, and is a wonderful base for heavy perfumes. It is also excellent in soap perfumery because of its tenacity, and fine for incense and sachets. It finds use also in antiseptics and as a flavor for cough syrups. It is used as a tincture, a resinoid, or an essential oil.

Natural Musks

We move on now to the Animal Fixatives, which are an even more fascinating group than the gums and resins. Although synthetic musks have now largely replaced natural musk, the musk deer has long played a unique role in the chronicle of perfumery. This animal is small, about 20 inches high at the shoulder, with a coarse, brittle coat of dark brown hair, and a pair of long, canine tusks. The musk comes from the scent glands or pods on the deer's abdomen, presumably a sexual attractant as the musk scent is strongest during the mating season. The dried pods are not particularly odoriferous, but when they are moistened the fragrance becomes overpowering. The aroma of musk has been variously described. Some call it ambrosial; others categorize it as luxurious, rich or full. Many perfumers say that of all the animal products musk has the most characteristic "animal note."

There are many stories about the lasting quality of musk. A tiny bit suffices to scent a room for weeks. Musk is more associated with the Orient than any other fragrant substance. In the Koran, Mohammed wrote of musk, "For this let those pant who pant for bliss."

Musk tonquin is used as an absolute (concentrated extract) or as a tincture (dilute alcohol extract). These products are aged or fixed, as with all animal products. They give lift and body to perfumes and flavor bases. Care must be taken to use the correct concentrations, not only because of cost considerations but mostly to get the lift without the animal note.

Ambergris

Ambergris must be discussed only in retrospect, since it can no longer be imported or used in fragrances in the United States. Ambergris is a "calculus" ejected by the sperm whale which is occasionally found floating on the surface of the ocean or washed up on beaches. It is a gray or vari-colored and mottled substance which is light, greasy, fatty and porous. It does not have the odor intensity of other animal fixatives, and therefore interferes but little with any of the fragrance notes in a perfumer's composition; but it possesses amazing fixative properties, and is valued for its tremendous ability to hold a composition together and make it long lasting. It is the most delicate, yet persistent of the animal fixatives, superior to all the others when used in delicate florals. Its use is still permitted in all other countries except the United States.

Civet

The use of civet in perfumery dates back at least to the early part of the 16th century. There is an old reference which describes civet as a "thick liquid found in a pouch which is under the tail of a beast like a Spanish cat, but much more fierce and voracious." The civet cat, native to Abyssinia, is a short-legged, long-bodied, bushy-tailed feline, about a foot high and three feet long.

The liquid recovered from the animal's pouch is, in its natural state, overpowering and repulsive. When diluted and mixed with other fragrances, the effect is pleasingly floral. Some call the odor of civet superior even to that of musk. Civet is valued primarily as a fixative, and there are some perfume compounds for which civet is the perfect, even essential, fixative. Civet has the capacity to "round out" or "bouquet" a perfume. For instance, tincture of civet smooths down the rough edges and gives body and beauty to a finished jasmine compound.

Castoreum

The castoreum-producing beaver, the largest member of the rodent family, has a paddle-shaped tail, webbed feet, and a luxurious fur coat. Two pear-shaped membraneous sacs, found in both the male and female near the genitals, are removed from animals killed for their fur. If beavers were not trapped for fur, the cost of castoreum would be prohibitive. The sacs contain a creamy secretion which coats the fur and tail to protect it from the water. The sacs are dried in the sun, or smoke-dried over burning wood; then the secretion is taken out and rubbed either with orris powder or clean sand. After maceration for six months with 90% alcohol, the solution is filtered and the castoreum recovered.

As with civet, the aroma of castoreum is at first unpleasant, but when diluted and prepared for use it is much less intense and decidedly attractive. In fact, castoreum has a definite pungent fragrance described as "birch-like." Castoreum is also a prized fixative, and is used in minute amounts to give a spicy or oriental note to perfumes such as chypres, ambers and heavy floral bouquets. Castoreum tincture is similar to musk, and is often substituted for musk because it is less expensive. Quality is highly important with castoreum, and special treatments have been developed to insure a finished product with superior finesse and clarity.

Oleoresins

Earlier we discussed the spice oils, including nutmeg, cassia, clove, etc., which are more important for flavor applications than for their use in fragrances. Extracts of the raw spices from which the solvents have been totally removed are called oleoresins. A long list of spices is processed to make oleoresins, including: allspice, basil, cardamom, caraway, celery, cubeb, cinnamon, clove, coriander, cumin, dillseed, fennel, ginger, mace, marjoram, nutmeg, black and white peppers, sage Dalmatian, tarragon, thyme and origanum. In addition, paprika and turmeric are produced, being used to color meat and pickles, respectively. Oleoresins may be used in any products to replace ground spices, for example in luncheon meats, sausages, baked goods, snack foods, canned goods, meat sauces, soups, pizzas and others.

Fruit Concentrates and Essences

Other important natural flavor raw materials include the fruit juices, their concentrates and essences. As described in the Production chapter (see pp. 279-81), the fruits are squeezed, depectinized, and concentrated to a specific solids content. Blackberry and black raspberry as well as grape skin concentrate are items used to impart natural color to food products. Other concentrates are available, such as apple, apricot, cherry, black cherry, cherry with pit, wild cherry, grape, loganberry, peach, pineapple, plum, raspberry and strawberry.

Flavor Concentrates

Tinctures and extracts of such flavor items as coffee, cola nuts, cocoa and the like are also common raw materials. Almost any food item — fruit, vegetable, etc. — can be processed to produce a flavor concentrate which will have superior flavor intensity, better color, or improved solubility or utility as compared with the original article. Even fruits such as banana, papaya or mango, which have limited aqueous juice, can be pureed and concentrated by removing the cellulosic material, pulp, etc., to produce a usable flavor raw material.

Vanilla

We close this chapter with a discussion of perhaps the most widely used flavoring material, — vanilla. The vanilla bean is the seed pod of an orchid. Finest quality beans come from Madagascar which completely dominates the market for natural vanilla. Mexico, Java and Tahiti produce minor amounts. The choicest beans are hand pollinated, permitted to flower, then harvested while still unripe just at the point when they are starting to turn yellow at the ends. When picked, they are odorless; it is the curing process which brings out the vanilla fragrance.

The method of curing differs with the area of production, but the following is typical. The beans are first immersed in water at 63 degrees Centigrade, then put into wool-lined barrels. After about 24 hours in the barrels they are spread out on trays, covered with cloths, and exposed to the sun for six hours a day for six or seven days. Next they go to drying sheds for four to six weeks, until they become hard, wrinkled, and dark brown in color. After curing is completed, they are graded, tied into bundles, and packed into parchment-lined tins for shipment.

Vanilla is used extensively in flavoring, being used in an endless variety of food products from cake mixes to puddings, from candies to ice cream, from malt powders to chocolate syrups, and on and on.

Chapter VIII

Raw Materials—Synthetic

In the field of fragrance and flavor raw materials, synthetics have gained increasingly in importance over the past fifty years as compared with natural materials. The first synthetics were isolates, made from natural products which had been processed to develop desirable materials: linalool from bois de rose oil to linalyl acetate or butyrate; geraniol from citronella oil to produce geranyl acetate and butyrate; all materials occurring in nature, but which could be synthesized, readily and economically.

In a second stage of the development of synthetics, technology was directed toward nature-identical substances which were totally synthetic: i.e. ethyl acetate and ethyl butyrate; racemic and l-menthol from thymol or isopulegol; coumarin; benzaldehyde; and like substances. Finally, fragrance chemists succeeded in producing materials that had odor or flavor value although they had no nature-identical counterparts: such materials as tolyl aldehyde for cherry flavors; so called aldehyde C-16, known in the trade as "strawberry aldehyde," but actually an ester; so called aldehyde C-18, referred to as "coconut aldehyde," actually a lactone; hydroxycitronellal; *Lyral;* cyclamen aldehyde, etc.

For many applications, natural materials are still the products of choice. One of their great advantages is that we tend to think of their fragrance or flavor as "natural." Natural materials are conglomerates of many diverse, complex, mostly odorous substances, the aroma of which we recognize as a harmonious blend. The trace components in a natural product, impurities if you will, result from natural reactions such as plant metabolism, enzymatic reactions and the like, the same reactions which produced the natural product. These trace components are natural materials themselves, and normally have the same odor character as the product in which they are present as trace components.

When producing a synthetic, however, the production process, carried out in a factory, is normally designed to produce a single chemical structure. True, the synthetic also contains impurities, due to reagents, solvents, side reactions and the like, but these impurities seldom have any chemical relation to the desired end product, and often introduce a character unlike that of the material whose aroma we have selected.

It is for this reason to a major extent that we almost unconsciously accept the natural compound as the "real thing," while identifying the synthetic as an imitation. In spite of the tremendous strides in chemistry in recent years, the complex nature of all natural fragrance and flavor materials has made it next to impossible to reproduce them precisely in the factory, or even in the laboratory.

Unfortunately, there are many problems which attend the use of natural materials. Supply is always subject to interruption. Cold weather, drought, disease, typhoons and hurricanes, and other such natural, uncontrollable factors can wipe out, or at least seriously limit the size of, a crop. Disasters such as earthquakes or floods can cause a suspension in production that may last for months. Wars, military or economic, can cut consumers off from their normal sources of supply.

Other factors such as rising wage rates in production areas, or competition by other crops for available land, can raise the cost of producing essential oil crops. Fifty years ago, the natives in undeveloped nations, where most essential oils are produced, had little choice of where they would work, or at what wage. Hand labor in gathering flowers, leaves, berries, bark, etc. was often the only way a native family had to support itself; and the only alternative to making a few pennies an hour by working in the field was not to work at all. Today all that has changed. Wage rates in the developing nations have risen sharply.

Years ago, there was still an ample amount of unused crop land. Acres used for production of essential oil-bearing crops were of limited value, and there was little competition for them. That situation has also changed today. In many areas, land formerly employed for the production of essential oil crops has become too valuable for such use.

This change is well illustrated in the case of mint production in the United States. Winship A. Todd of the A.M. Todd Company observed in a talk at the 6th Essential Oil Congress, held in San Francisco in September, 1974, that "there is now strong competition among the various farm crops" for land that was formerly used almost exclusively for mint production. Twenty or thirty years ago, Mr. Todd commented, there was an excess of land available for mint production. But, as he pointed out:

"The situation in 1974 is quite different. Virtually all U.S. farm crops are now in great demand throughout the world. World requirements for food and fiber have caught up with the capacity of our farms to produce them. There is now strong competition among the various farm crops for available land. Mint crops are in the thick of this competition. The mint farmer now has a choice of many crops which offer a prospect of profit, such as corn, wheat, sugar beets, grass seeds and vegetable crops. Obviously, therefore, the price of mint has had to increase in order to compete for land."

A similar situation can be found with production of lemon and orange crops in California. There used to be hundreds of lemon plantations on the slopes of all the valleys around Los Angeles, while oranges covered most of the flat land on the valley floors. For the past fifteen or twenty years, however, there has been such a wave of building in the area that lemon and orange growers have been driven out. They have moved east into Arizona, or up the coast into Ventura or Santa Barbara counties, where at least up to this point the real estate pressure has not been so severe. The same thing has been happening in Grasse, France, for a number of years.

We see then that cultivators of essential oil crops in many parts of the world are being moved out of some of their best producing areas by population growth. This interference with essential oil production raises prices of natural materials and contributes to occasional undependability of supply. It has stimulated the already existing demand for production of more synthetics.

Production of synthetic aroma chemicals in the United States started back in the closing years of the 19th century. One of the earliest significant dates was 1889, when Alois von Isakovics started Synfleur Scientific Laboratories to manufacture synthetic aromatic raw materials for the perfume industry (see p. 183). This was just a year after Baur, while working on dyestuff research in Germany, had synthesized nitro musk compounds from which musk xylol was eventually developed.

Further developments in this emerging industry came along at a rapid rate. In 1890 Dr. Francis T. Dodge of Dodge & Olcott Co. succeeded in isolating citral and citronellal. Then, Tieman and Kruger, while attempting to make irone, the principal fragrance element in violet flowers and orris root, discovered ionone instead. By condensing citral with acetone, and cyclizing the resulting compound, they synthesized a chemical which, although not found in nature, closely resembled the fragrance of the violet. (See also p. 160). This was the beginning of the manufacture of ionones/and methyl ionones, which have since then become such important aroma chemicals. It marked the first time that an aroma chemical not known to occur in nature had been made for the fragrance industry.

Another important date in the development of the aroma chemical industry was 1905, when hydroxycitronellal (more correctly it should be called hydroxy dihydro citronellal) appeared on the market for the first time. It is made by the hydration of citronellal. So incomplete is the early history of the aroma chemical industry that the name of the discoverer of this highly important aroma chemical is not known, although we do know that in 1915 a soap perfumer, Dr. E. Emmet Reid, clarified the structure of hydroxycitronellal and developed processes for its manufacture.

The development of new aroma chemicals proceeded by several major routes. An early logical step was the separation of important aromatic constituents out of essential oils. Thus, just as citral had been isolated from lemongrass oil, so essential oil chemists soon succeeded in separating geraniol and citronellal from citronella oil, linalool from bois de rose oil, and terpineol, safrol, geraniol and linalool from camphor oil. Similarly santalol was isolated from sandalwood oil, and linalyl acetate from oil of lavender.

In addition to being useful aroma chemicals themselves, these isolates often served as raw materials for the manufacture of still other fragrance and flavor materials. As mentioned earlier, hydroxycitronellal was made from citronellal, while the processing of citral resulted in development of the ionone and methyl ionone fragrance families. Safrol was further processed to yield heliotropine. Other examples illustrating the productivity of this approach include the synthesis of carvone from limonene; geraniol, nerol and linalool from turpentine; l-menthol from dextro citronellal; and dl-menthol from thymol.

Developments in coal tar chemistry produced still another group of valuable aroma chemicals. Included in these useful fragrance materials and intermediates were benzyl acetate, benzyl alcohol and benzaldehyde, as well as coumarin, phenyl ethyl alcohol, and the nitro musks.

In some instances it became possible to synthesize chemicals which were almost complete synthetic duplicates of natural products. Such an achievement was the production of vanillin from eugenol and from guaiacol. Later these synthetic processes were made uneconomic and were superseded when a technique was developed for separating vanillin at low cost and in ample quantity from lignin, recovered from the waste liquors of paper factories. Patents were issued in 1936 for the production of vanillin from sulfite liquor to Sanborn, Salvesen and Howard, and independently in Canada to Tomlinson. At a later date, in a similar development, a method was developed for the manufacture of eugenol from catechol, but it was never commercialized.

Through the 20's and 30's, developments in the manufacture of aroma chemicals accelerated. Workers in the industry soon discovered that the same product could often be synthesized by totally different methods, and in some cases starting with different raw materials. It was also discovered that certain purely synthetic chemicals, not derived from natural raw materials, possessed pleasant and interesting aromas.

We mentioned earlier the discovery of ionones by Tieman and Kruger, and the discovery of the first nitrated musk by Baur, who was trying to synthesize dyes. Now, further developments in synthetic musks came along rapidly. L. Ruzicka of Firmenich & Cie. succeeded in making synthetically the naturally occurring large ring musk compounds which are found in the scent glands of various animals. Then, in the 30's, while working on synthesis of the polymer, nylon, for du Pont, Carruthers made a series of large ring musks, among which was the entirely synthetic ethylene brassylate, a strong musk-like material. Subsequently du Pont brought it to the fragrance market under the name, *Astratone.*

Developments in the chemistry of musk fragrances continued. In 1948, Dr. Blomquist synthesized a macrocylic musk. In 1954, Dr. Marion Carpenter of Givaudan Corp. discovered polyalkyltetralines, a new class of musk compounds. About 1955, workers at Givaudan and at Trubek Laboratories, working independently, both filed patents for Musk 36A, which Givaudan later marketed as *Versalide.* During this same period, a range of polycyclic musk compounds was developed by Dr. Ernst Theimer of IFF, Dr. Marion Carpenter of Givaudan and Ernest Polak of Polak's Frutal Works. When introduced to the market, they helped to establish new trends in perfumery.

These advances in the chemistry of musk materials came at a most opportune time. Natural musk was becoming more and more scarce, and thus extremely expensive. There was also increasing recognition of the fact that it was ecologically unacceptable to continue sacrificing animals merely to produce a fragrance material. Synthetic substitutes had to be found.

It was also important for the industry to have a wider range of musk fragrances from which the perfumer could make his selection. The nitro-musks alone were not enough. Some of them are weak in aroma, and all tend to cause discoloration. With the advent of such new products as *Tonalid, Versalide, Galaxolide* and others, the perfumer's palette has been enriched by the addition of a series of new synthetic musks with which to replace natural musk, and to supplement or blend with some of the less satisfactory earlier synthetic products.

The fragrance and flavor industry benefitted through this developmental period from parallel work going on in other industries. Development of new pharmaceutical chemicals contributed in large measure to the availability of new fragrance and flavor raw materials. Such basic chemicals as acetylene, isobutylene and methylvinylketone were processed to yield ketones, required in the manufacture of vitamins A and E. It was soon recognized that this same technology could be used to produce such aroma chemicals as citral, the ionones, methyl ionones, linalool and geraniol, as well as a host of new chemicals having fragrance possibilities.

New and useful raw materials for fragrance manufacture also began to come from the petroleum industry. Back in the days when the only building blocks the chemist had to work with were a few coal tar chemicals and whatever he could isolate from plant materials, progress was inevitably slow, and most aroma chemicals traced back to ingredients in various flowers and fruits. But with the advent of "stream chemicals," which started to issue in profusion from oil refineries and cracking plants, a multitude of new building blocks became available. They could be used to manufacture aroma chemicals which were not only *not* present in nature, but many of which didn't even smell like anything with which perfumers had previously been familiar.

One of these new products was para tertiary butyl cyclohexyl acetate. When para tertiary butyl phenol first became available, esters of the phenol and its hydrogenated cyclohexyl derivative were routinely made, and the saturated acetate was quickly recognized as having pleasant, woody properties, combined with great stability. In time it became a favorite fragrance product, for it could be used in a wide variety of formulations and for a great variety of end uses. A chemical not found in nature.

Another such aroma chemical was Givaudan's *Sandela* (IFF's *Indisan* is based on similar technology). This material had been made years before, and some similar work had been reported in the Russian literature. It was not recognized at first as a valuable fragrance material, and it was only after it had finally been perfected by Givaudan that perfumers came to appreciate that its odor was close to that of sandalwood. Although it is completely different chemically from santalol, the principal fragrance component of sandalwood oil, it does smell remarkably like santalol, although it has a greasy character which must be taken into account by the creative perfumer before it can be used. Another chemical like nature, but not in nature.

An extremely important development of this middle period was the discovery that useful fragrance materials could be produced from products of the pine forests. In 1940, Dr. Bain and co-workers with Glidden Co. developed a process for the manufacture of terpene alcohols from beta pinene. Readily available and cheap, turpentine, a by-product of the wood pulp industry, became an important raw material source for the manufacture of such synthetic aroma chemicals as geraniol, citral and linalool. No longer did aroma chemical manufacturers have to depend on imports of bois de rose, lemongrass and citronella oils, the price fluctuations of which had long plagued them. Study of turpentine-based raw materials also led to the discovery of a number of highly unusual and useful new aroma chemicals, some of which are not found in nature, but in this case were provided by nature.

The first of these, if not the most important, was IFF's pseudo linalyl acetate. This material had the general odor characteristics of linalyl acetate, although it contained little or none of this compound. It did contain a mixture of chemicals, some of which were acetates of terpene alcohols, one of the principal ones being myrcenol. Pseudo linalyl acetate is still sold today, although now in smaller volume than before its components became available in their pure forms. But, it served in its own "unnatural" way for quite a few years, a mixture unknown in nature, appearing as a reaction mixture, and so difficult to keep to specification that huge lots had to bulked like wines to even out the swings.

The first pure aroma chemical from pinene to make a major impact on the fragrance market was IFF's *Lyral.* This chemical has the jaw breaking chemical name, 4-(4-hydroxy-4-methyl pentyl)-3-cyclohexene-1 carboxaldehyde.

It was extremely difficult to make and purify, but was accepted quite rapidly after the manufacturing problems were solved. The odor of *Lyral* has the same lily-of-the-valley character which is exhibited by hydroxycitronellal; however *Lyral's* fragrance is much longer lasting. When hydroxycitronellal is blottered, it loses its true fragrance within a few hours, and after two or three days is completely gone. Not so with *Lyral,* which will give off its same delicious floweriness for a considerably longer period. Here we have another purely synthetic product, surpassing in stability and persistence the corresponding fragrance note from a natural botanical source.

There are other examples of "unnatural" aroma chemicals which have proved to be very useful in perfuming. For example, there are the fatty aldehydes, which are present to some extent in nature but which are used in perfumery in a fashion not intended by nature. These fatty aldehydes have straight carbon chains, and correspond to the fatty acids obtainable from animal and vegetable fats.

The perfumer often takes advantage of the "shock value" of the aldehyde note in developing his creations. This shock value can be a definite plus in perfumery. When cleverly blended with other aroma chemicals and naturals, the aldehyde note produces a piquant quality which often typifies "modern" perfumes. A strikingly successful example of such a perfume, as a matter of fact, is *Chanel 5,* the first perfume to use an "unnatural" note. Fatty aldehydes were used in formulating *Chanel 5* to give lift and life to the beautiful floweriness of jasmine which is the underlying note.

Another chemical not found in nature is one which has allowed marketers to offer all kinds of functional products with a lemon note. The aroma chemical of choice for this purpose has always been citral, whether purified or as it appears in citral-bearing oils such as lemongrass or lemon oil. However a major problem which accompanies use of citral is the fact that it is very unstable and degrades quickly in alkaline media, either by polymerization or by oxidation. Thus it cannot be used successfully in perfuming soaps or detergents, for it tends to break down and rapidly lose its fragrance.

IFF discovered a very useful aroma chemical, geranonitrile, which has a stable, lemon-like odor. They called it *Citralva*. It is a highly potent material, several times as strong as citral, stable in most media, thus very useful in the perfuming of soaps and detergents, which are normally strongly alkaline and thus quickly break down citral and other aldehydes. A disadvantage of geranonitrile is that it has a greasy and disagreeable quality. But skillful perfumers found ways to use the lemoniness of geranonitrile while suppressing its greasy character. Hard work, backed up by skill and creativity and years of experience in solving practical perfuming problems, are the factors that allow perfume compound houses to solve such perfuming problems.

Two other source materials for the manufacture of aroma chemicals which have come into use in recent years are isoprene and acetylene. In 1948, Drs. Cope and Kimel of Hoffman La Roche developed processes for the manufacture of linalool and beta ionone from acetylene, the Roche process, and in the 70's Rhodia began to make linalool and methyl ionones from isoprene.

Another research approach which led to the development of a number of new and highly useful aroma chemicals was the molecular modification of existing chemical structures. One example of this technique is the replacement of vanillin by ethyl vanillin for certain applications. Ethyl vanillin has substantially more flavor power. Another such modification is the substitution of the methyl group normally occurring in maltol by an ethyl group, creating ethyl maltol.

Improved analytical methods also made their contribution to the discovery of new aroma chemicals. When the perfection of the spectrophotometer, and of gas liquid chromatography and other refined analytical methods made it possible to identify aroma chemicals present only in trace amounts in various essential oils, the way was open to make them useable. Once identified, they could be synthesized. Products which originated through this technique include rose oxide, a minor but highly important constituent of otto of rose, and jasmone, which is present in only minor quantities in jasmine absolute, but which is one of the most important constituents of the jasmine aroma.

Discoveries of new synthetic aroma chemicals have generally resulted from close observation by trained experts, — what Henry Smythe at Princeton used to call "observations by the prepared mind." These researchers were not looking blindly for a needle in a haystack and stumbling on it by chance. Rather they were for the most part looking for something they suspected was there, and were capable of recognizing and appraising it when they found it. It wasn't enough for Carruthers, for instance, simply to have discovered ethylene brassylate. He or one of his associates had to possess sufficient background in aroma chemicals to be able to recognize that in this material they had an interesting, useful new aroma chemical.

History of U.S. Industry

But, perhaps we should switch for a moment from further discussion of products, to go back and record the names of some of the companies that were responsible for emergence of an aroma chemical industry in the United States. Dodge & Olcott started manufacturing such products as citral, citronellal and citronellol well before the end of the 19th century. Another pioneer in the manufacture of aroma chemicals was Synfleur Scientific Laboratories, started by Alois von Isakovics in 1889 in New York City.

Monsanto Chemical Company was one of the first U.S. chemical companies to experiment with the manufacture of aroma chemicals. They brought a Swiss chemist named

Gaston DuBois over to the United States in 1904. He worked for Monsanto for many years, made the first vanillin produced in the U.S. (from clove oil), and also manufactured salicylic acid, methyl salicylate (artificial oil of wintergreen), coumarin, and a number of other aromatics. Dow Chemical Company also manufactured synthetic aromatics, including methyl salicylate, methyl anthranilate, and phenyl ethyl alcohol.

An early manufacturer* of aroma chemicals for the trade was Morana, Inc., which started operations in 1909. Then, in 1912, Van Dyk & Co. started to manufacture aroma chemicals at a plant in Jersey City, N.J. Next, Edwin Kutroff founded Verona Chemical Co. in Newark, N.J., in 1914, and by that time Fries Brothers were operating a small aromatics plant in Bloomfield, N.J. They had New York offices at 92 Reade St. Dr. Harold Fries was one of the principals in the company. This company was later operated by Dr. Stockelbach, who patented a number of processes for the manufacture of aroma chemicals.

Fritzsche Brothers started to make aroma chemicals in 1914. Rhodia Chemical Co. was established in 1915. Florasynth Laboratories was organized in 1916 by Louis A. Rosett and Charles L. Senior. The latter had been a salesman for Van Dyk & Co. George G. and Robert G. Fries organized Fries and Fries in Cincinnati in 1919.

Dr. Joseph Felton established Felton Chemical Co. in Brooklyn in 1921. Morana sold its Elizabeth, N.J. plant to van Ameringen-Haebler, Inc. in 1929, and this company, which later changed its name to International Flavors & Fragrances, Inc. became the largest flavor and fragrance company of them all. In about 1930 another company bearing the Fries name started operations. This firm was Kay-Fries Chemicals, Inc., affiliated with American-British Chemical Supplies, Inc., and controlled by the British firm

*For our information on the start of the aromatic chemical business in the United States, we lean on coverage of the subject by Williams Haynes in his series of books on "The American Chemical Industry." (Vol, III, which covers the period 1912-1922).

of Chas. Tennant & Sons. They manufactured a number of basic aroma chemicals at a plant in West Haverstraw, N.Y. Kay-Fries Chemicals became part of the Dynamit Nobel Group a few years back, and offices are now located at Northvale, N.J. However, except for producing diethyl phthalate, Kay-Fries is no longer active in the production of aroma chemicals.

Although Kay-Fries acquired the right to use the Fries name back in 1930, there was apparently no member of the Fries family connected with Kay-Fries Chemicals. According to information given to us by H. Kent Vanderhoef, now president of Kay-Fries Chemicals, Inc., principals representing America-British Chemical Supplies, Inc. and Kay Laboratories bought Fries & Fries from the receiver when the Cincinnati, Ohio, company which had been incorporated in 1920, was in receivership during the depression days of 1930. Later the same year, Dr. George G. Fries and Robert G. Fries bought back the flavor and extract business from Kay-Fries, including the perfume oil department.

There was still another Fries engaged in the manufacture of aroma chemicals, George Fries, who was a partner of Dr. Harold Simons. The two men operated affiliated plants in Brooklyn. We figured it could not be just coincidence that there were so many individuals bearing the same name identified with the aroma chemical business, and suspect a family connection. But our efforts to clarify the Fries story have been unsuccessful. Fries Brothers was absorbed by Stepan Chemical Company a number of years ago. That would seem to leave only Fries & Fries still active in the business today, and they are now a division of Mallinckrodt Inc.

Givaudan Corp. became an important factor in the aroma chemical business after they ended their relationship with Burton T. Bush in 1926, started to manufacture their own materials (See p. 200), and in time became the largest aroma chemical producer in the U.S.

Robert Trubek started to manufacture aroma chemicals in the early 30's, and became a supplier to the trade rather than selling direct to users of fragrance materials. (See p. 208). As his company grew, he absorbed a number of smaller manufacturers of aroma chemicals, including Clemenson Chemical, Truland Chemical, Harold Simons and George Fries. Trubek became a major factor in the industry, and many developments in new aroma chemicals came from the Trubek plant.

Norda Essential Oil & Chemical Co., which started operations in 1924, became a major factor in the aroma chemical business after setting up their first plant for the manufacture of aroma chemicals in Hanover, N.J. in 1941. (See p. 198).

The list of manufacturers of aroma chemicals now numbers approximately 20 or 25 companies. Among the major producers, other than those already mentioned above, are Stepan Chemical Co., Chem-Fleur, Inc., PFW, Inc., Union Camp Corp., UOP Naarden, Firmenich, King Aromatics, Monsanto Flavor/Essence, Elan Chemical Co., Haarmann & Reimer Corp. and the Glidden-Durkee Div. of SCM Corp. Production and sale of aroma chemicals grows steadily larger year by year.

PRODUCTS

But perhaps we should switch our attention back again from companies and individuals to products, singling out for brief review some of the most important of the aroma chemicals. We offer in the following pages capsule summaries of the discovery, history, methods of manufacture, characteristics of, and principal uses of some of the more important aroma chemicals. Those interested in pursuing the subject further should refer to more detailed studies of the subject such as are contained in books like "Perfumery and Flavoring Synthetics" by Bedoukian, "Perfume and Flavoring Chemicals" by Arctander, and others. (See Chapter XVI).

Alcohol (Ethanol)

Actually, the material with which we start our discussion, ethyl alcohol, is not an aroma chemical at all, but it is a very important raw material for users of aroma chemicals. Ethyl alcohol (95.5% Vol.) is widely used *per se* in flavor compounds, and when denatured to prevent its consumption as a beverage, is one of the most important materials used by the fragrance industry.

There are two types of ethyl alcohol in common use in fragrances, designated as SDA (specially denatured alcohol) #39C and #40. The denaturant used in #39C is diethyl phthalate (1 gal. per 100 gals. ethanol). SDA #39C is widely used in the formulation of perfumes, and for some aerosols. The denaturant used in SDA #40 is brucine sulfate (3 ozs. per 100 gals. ethanol). This denaturant is a very bitter tasting, odorless material. SDA #40 is used primarily in the manufacture of colognes and toilet waters.

Most marketers of perfumes and colognes still buy perfume oils and alcohol separately, and since they may want to make other additions they prepare their own dilutions of fragrances. However, in recent years it has become more common for essential oil houses to make up colognes as finished products, since they have alcohol permits and are fully familiar with required procedures. They are then in position to ship ready-to-bottle mixtures to their marketer customers, or to their custom packagers.

Almost all the alcohol used in the fragrance industry is now made synthetically from stream gas ethylene, where formerly it was made by the fermentation of sugar or grain.

Ethanol, in varying degrees of aqueous dilution, is used to make tinctures such as: ambergris; musk tonquin; civet, kola nuts, tonka bean, etc.; essences such as cocoa and coffee; and extracts such as vanilla, almond, lemon, etc. Ethanol is also used in the processing of floral absolutes, some resinoids, and for solubilizing and dewaxing citrus oils, sesquiterpeneless oils, and concentrates of lemon, orange, grapefruit, mandarin, etc. These are of course natural products, but in many cases imitations are also produced which are combinations of ethanol and synthetics.

A major and highly important use of ethanol is as a reagent in the manufacture of ethyl esters and other derivatives. Since ethanol is present in many natural products, due to fermentation of sugar or starches, or develops as an artifact during processing of these products, alcohol is thus a common component of most natural flavor items such as juices, etc. And ethyl esters, which develop from the ethanol as metabolic or chemical artifacts, are thus also common components of flavor products. Ethyl esters are used in large volume in flavor compounding, the total consumption of ethyl acetate and ethyl butyrate for this purpose running to several million pounds a year.

Ethyl acetate is present in almost all fruit flavors such as grape, cherry, etc., while ethyl butyrate, a more specific ester, finds wide application in fruit flavors, and particularly in orange and grapefruit flavors. Other esters and derivatives which depend on ethanol for their manufacture are ethyl formate (rum flavor), ethyl caproate, ethyl caprylate and ethyl caprate (cognac or peach flavors), ethyl propionate (cherry and strawberry flavors), and ethyl isovalerate (apple and cheese flavors).

Amyl and Hexyl Cinnamic Aldehydes — Development of amyl cinnamic aldehyde, a purely synthetic F/F material with no known natural counterpart, is reported to have resulted from an attempt back in the 1930's to find a use for heptaldehyde, a by-product of the cracking of castor oil. A logical classical chemical reaction would be condensation with other aldehydes or ketones. This strong floral jasmine note resulted, and attempts were made initially to disguise the nature of the product by use of a tradename *(Florsal).* The material is now sold under many different trade names. Volume of production in the United States is estimated at around 300,000 lbs. annually. The method of production has not changed to any extent, and is still the condensation of benzaldehyde with heptaldehyde, carefully controlled to prevent the self condensation of either aldehyde.

Hexyl cinnamic aldehyde is a homolog of the amyl material made with octylaldehyde and benzaldehyde. Production of this material is estimated at about 200,000 lbs. yearly, and increasing. The odors of these two aldehydes are green, floral and jasmine-like, with the hexyl homolog being slightly less fatty in spite of its longer carbon side chain. Both items are FDA GRAS approved, but only small amounts are used in flavors. Both have excellent lasting power and intensity for use in perfuming of soaps, detergents and almost any green floral fragrances.

Minor amounts (less than 20,000 lbs.) of the dimethyl and diethyl acetals are manufactured. They are used where particular stability is required in alkaline bases.

Amyl Salicylate — This material was formerly made by the reaction of amyl alcohol (derived from fusel oil) and salicylic acid. Now, however, almost all the commercial amyl salicylate (more accurately iso amyl salicylate) is made from synthetic iso pentyl alcohols and salicylic acid. Amyl salicylate has a characteristic, slightly fatty, but flowery aroma. It is used in many floral applications, and is particularly useful in soap fragrances, since it is highly stable in that medium.

Benzyl Chloride — One of the earlier materials made from "coal tar" was benzyl chloride, made by reacting toluene with chlorine. It is the starting point for the manufacture of all the benzyl derivatives.

Benzyl Alcohol — is essentially odorless when freshly made, but gradually takes on a mild, almondy character as small amounts of benzaldehyde are formed by oxidation on exposure to the air. Benzyl alcohol may be made by reacting benzyl chloride with soda ash.

Benzyl Acetate — is a very flowery, jasmine-like chemical, made either by treating benzyl chloride with anhydrous sodium acetate, or by reacting benzyl alcohol with acetic acid or acetic anhydride. It finds wide use in perfumery, both for fine perfumes and in soap and detergent applications.

Benzyl Benzoate — which occurs naturally in balsam tolu and balsam Peru, as well as in some cinnamon species, has a faint, balsamic note. It is prized mostly for its fixative properties. Benzyl benzoate is made by reacting benzyl chloride and sodium benzoate, or from benzyl alcohol and benzoic acid.

Benzyl Salicylate — has somewhat more odor value than benzyl benzoate, and finds similar applications. It is normally made from methyl salicylate and benzyl alcohol by ester exchange, but can also be made from benzyl chloride and sodium salicylate.

Cedarwood Derivatives — Among this important group of aroma chemicals, one of those enjoying widest use is cedryl acetate, of which quite a few qualities are offered commercially. They differ significantly in purity, although all exhibit the same basic note, an intense, very pleasant odor of cedarwood. Usage is diverse, wherever a woody note is desired. In the manufacture of one of the types of cedryl acetate, cedarwood oil (either the *Virginiana* or the *Mexicana*) is separated into fractions by distilling off the lower boiling sesquiterpene components. The alcohols, of which there are several, remain behind as a pot residue. They may be distilled to lighten the color, and are then separated by crystallization.

True cedrol is removed as the crystal, leaving behind a variety of alcohols such as pseudo cedrol, cedrenol and others. If there is an active demand for cedrol, this product is removed as described, and used either as such or converted to cedryl acetate (sometimes called cedrenyl acetate). The other alcohols, when acetylated with acetic acid or acetic anhydride, produce a mixture of acetates, also useful for similar perfuming applications.

Another type of cedryl acetate is produced by acetylating the whole oil and topping the reaction mixture for odor quality. This product is sometimes called cedarwood acetate. Further fractionation increases the ester content, improves the color, and yields by-products for further processing.

Cedryl acetate as commonly sold varies widely in ester content, and thus in quality. Often the difference cannot be readily discerned by simple olfactive examination, but a change in character may result when the material is mixed with other aroma chemicals in making a perfume. The source of the alcohol used also has an effect on the quality of the final product.

Another group of cedar wood derivatives is made by the acylation of cedarwood oil sesquiterpenes. There are two basic types. The one made from *Juniperus Virginiana* contains a little more acetyl cedrene, while the one made from *Juniperus Mexicana* contains a larger amount of so called "acetyl thujopsene." The latter material is an aroma chemical pioneered by IFF under the name, *Vertofix,* and now made in various grades by several other aroma chemical manufacturers. It has an intense fragrance of the precious wood* type, with an underlying muskiness. It is widely used in fine perfumery, and its stability and moderate price also permit its use in perfuming soaps and detergents. It is the "acetyl thujopsene" which is believed to yield the rich, full, woody odor of the compound, whereas products made from the less expensive *Virginiana oil* have not nearly so much of the precious wood character, do not seem to persist as long, and are now manufactured only infrequently.

Cinnamic Alcohol — This is a pleasant, hyacinth-like material, used mostly in fragrance formulations. It is found in balsam Peru and in gum styrax. In the latter it occurs principally in combination with cinnamic acid. Some years

*The precious woods include sandalwood, rosewood, cedar, etc.

ago, cinnamic alcohol was produced by saponifying cinnamyl cinnamate. Currently, cinnamic alcohol is made by cautious reduction of cinnamic aldehyde. It is useful in many perfume applications to take advantage of its floweriness and balsamic properties.

Cinnamic Aldehyde — This chemical, most widely used in flavors, contributes to a formulation the intense odor of cinnamon, especially when it is used along with small amounts of eugenol, a constituent of cinnamon leaf oil. Cinnamic aldehyde is prepared by reacting benzaldehyde and acetaldehyde. Before it was made by synthesis, cinnamic aldehyde normally went into a perfume formulation in the form of cassia oil (which consists predominantly of cinnamic aldehyde, with a few other aromatic materials such as cinnamyl acetate, salicylic aldehyde and coumarin present in small quantities). Because of its unique flavor, the price of cassia oil always remained high in comparison with that of cinnamic aldehyde, although cassia oil contains about 70% cinnamic aldehyde.

There is very limited use for cinnamic aldehyde in perfumery, but it is a favorite product for flavoring tooth paste, mouth wash, and chewing gum.

Citral — This important aroma chemical has an intense fragrance, much like that of lemon peel. Until fairly recently it was derived from lemongrass oil, *Litsea cubeba,* and other plant materials in which it occurred in concentrations of 75% or higher. More recently, most of the citral used in flavoring and perfumery, and virtually all that used in synthesis, comes either from vitamin manufacturers who make it as an intermediate, or is made from beta pinene or isoprene.

A number of aroma chemicals are now made from beta pinene. These include geraniol, citronellal and linalool, in addition to citral. Beta pinene is a constituent of turpentine, used for many years as a diluent and extender by paint manufacturers. When paint manufacturers changed many of their formulations to synthetics, demand for turpentine dropped, the price of turpentine declined, and it became more attractive as a raw material for the manufacture of aroma chemicals. Producers of aroma chemicals welcomed an alternative raw material which would free them from dependence on the imported oils, lemongrass and citronella. The naval stores industry as well as aroma chemical manufacturers did considerable development work, and patented many products and processes. Shortly a number of aroma chemicals were being made successfully and economically from beta pinene.

Citronellol and Geraniol — These alcohols were formerly made by separation from the naturally occurring mixture of citronellol and geraniol in citronella oil, Java type. Some natural citronellol is present in Citronella Java, (35% citronellol, 65% geraniol), but was rarely separated in processing. The usual method was to distill the citronellal out carefully, saponifying the remainder, which contained the geraniol/citronellol, as a pot residue. The geraniol/citronellol was then distilled, and if citronellol was desired it could be made by hydrogenating the geraniol/citronellol mixture.

Other plant sources also yield geraniol, notably palmarosa oil. Citronella oil, Ceylon type, also yields small amounts of geraniol/citronellol, but because of the relatively low content is not processed for geraniol. Geraniol made from palmarosa oil, which contains no citronellol, is quite different in odor, but is useful for special compounds. Ceylon citronella oil contains low concentrations of citronella (about 10%), but higher concentrations of methyl eugenol, methyl iso eugenol and elemol. It is thus uneconomic as a source for citronellal and geraniol. Geraniol has a pronounced rosy odor, and is often used in conjunction with citronellol, linalool and phenyl ethyl alcohol in rose notes, both in fine perfumery as well as in the perfuming of soaps and detergents.

In recent years, most of the geraniol used has been produced from beta pinene by synthesis. The mix of the synthetic product results in a 60/40 ratio of geraniol and nerol, rather than the natural 65/35 ratio of geraniol and citronellol which results when using citronella oil as the starting raw material. Geraniols now offered commercially contain more geraniol/nerol, and less geraniol/citronellol. With less and less citronella oil now being processed, production of synthetic geraniol from beta pinene has resulted in availability of good quality nerol at a moderate price. Since there is no great demand for nerol, however, most of it is hydrogenated to produce citronellol.

Development of processes for making geraniol and citral synthetically has been of benefit to aroma chemical manufacturers, since it freed them from their former dependence on lemongrass and citronella oils, which often were subject to wild price gyrations. Availability of the synthetic material has now made it possible to maintain supplies and price on a more level basis.

Coumarin — This material, which exhibits the heady sweetness of new mown hay, was formerly made by extraction from tonka beans, deer tongue, or similar type plants. It is now made from ortho cresol, or directly from salicylic acid by reacting the acid with sodium acetate and acetic anhydride, or by trans alkylation methods.

Until the mid 50's, coumarin was used as a flavorant, to add punch to vanilla flavors, particularly in baked goods. It had a good record behind it, and had been used for 70 or 80 years with no question of its safety to humans ever being raised. However, in the wave of suspicion directed at so many materials, coumarin came under attack. Tests on laboratory animals raised suspicion as to its safety, and even though the tests were inconclusive they resulted in issuance of a ban on use of the chemical in foodstuffs. Being ruled out for use as a flavorant, the sole use of coumarin is now in perfumery. It is an extremely useful aroma chemical, employed widely in soap perfuming. Many soap fragrances include a generous portion of coumarin.

Diethyl Phthalate — This is not actually a perfuming material, as the best quality diethyl phthalate has no discernible odor. Nevertheless DEP is used to a considerable extent in the fragrance field, in several different ways. First, it has some fixative properties and may thus be used for this purpose. It is a fairly large molecule which seems to have a binding effect on other fragrance materials, preventing the lower boiling ones from disappearing too quickly from fragrance mixtures.

Diethyl phthalate is frequently used as a diluent for powerful aroma chemicals. When a perfumer intends to incorporate a very small amount of a strong material such as a fatty aldehyde in his formula, he may dilute the powerful material ten or even 100 to one with diethyl phthalate so that the quantity of this ingredient going into the formula will be comparable to the quantities of other materials being used. This makes it possible for the compounder to weigh with acceptable accuracy the very small amounts of super strong materials being used. The compounder thus need not be so concerned about the limitations of his scale.

Finally, diethyl phthalate is often used simply to dilute fragrances so that the price per pound of the finished fragrance will come within the limits the buyer has arbitrarily set. Such addition is not considered an unethical practice, because the diethyl phthalate has definite value. However, in general the buyer's interests are best served if he buys his fragrance in terms of dollars per 100 lbs. of perfumed product rather than on the basis of dollars per pound of fragrance compound.

Eugenol — This is the chemical which gives cloves their odor. There is a large amount of eugenol in a number of plant materials, including clove buds, clove leaves, cinnamon leaves, and clove stems. Besides its use in flavors and fragrances, eugenol (or the oils containing it) is also used to make iso-eugenol, which has a carnation-like character. In earlier years, eugenol was the starting material for the manufacture of vanillin. (See vanillin).

Eugenol is isolated from the oils mentioned above by distillation, and sometimes by extraction, using sodium hydroxide. Eugenol does not need to be purified for the manufacture of iso eugenol. Iso eugenol has a tendency to become sweeter with age, due perhaps to the formation of a small amount of vanillin, lending the qualities of that chemical to those of iso eugenol.

Heliotropine — This very sweet chemical is reminiscent of the heliotrope flower from which it gets its name. It is made by the oxidation of iso safrol, which in turn is made from safrol in a manner similar to that in which iso eugenol is made from eugenol. Safrol is present in the oil distilled from the camphor tree, as well as in the roots of the sassafras tree. It is also present in oil *Ocotea cymbarum,* produced in Brazil and first brought to the United States as oil sassafras artificial.* *Ocotea cymbarum* is now the principal source for the safrol required for the manufacture of heliotropine, as well as that used for manufacture of the insecticide synergist, piperonyl butoxide.

In earlier days, heliotropine was made by the chemical oxidation of iso safrol by methods analogous to the preparation of vanillin from iso eugenol. These chemical methods are necessarily accompanied by the problem of having to dispose of troublesome by-product wastes. Accordingly, the chemical methods have to a great extent been superseded by direct air oxidation, using catalysts, or by use of ozone, which under correctly controlled conditions is the most favorable method of all. Japanese aroma chemical manufacturers have perfected the method of ozonolysis, and as a result are now the world's major suppliers of heliotropine.

Hydroxycitronellal — This is one of the truly beautiful aroma chemicals, smelling more like a perfume than a

*When the first shipments of *Ocotea cymbarum* reached the U.S., labeled as oil sassafrass artificial, the oil was denied entry by the U.S. Customs. Eventually the origin of the oil was clarified, and entry was allowed as oil *Ocotea cymbarum.*

chemical, and possessing the soft, flowery qualities of linden blossoms, lilies and lilacs. Hydroxycitronellal is used very widely in perfumery, but has stability characteristics which limit its use in soaps, detergents, or powders. For those applications, the dimethyl acetal is used, or IFF's *Lyral,* a material which smells very much like hydroxycitronellal, but is much more stable.

Hydroxycitronellal is made from citronellal by hydration, after protecting the aldehyde group either with bisulfite or sulfite. The hydration is carried out with acids, after which the hydroxycitronellal is liberated from its addition compound with an alkali.

Ionones and Methyl Ionones — As previously mentioned, ionones occupy a particularly important slot in the aroma chemical family because they were the first fragrance materials made in the laboratory which were not known at the time to have a counterpart in nature, although certain ionones have since then been detected in raspberry and black tea essences. Both ionones and methyl ionones are made from citral, ionones by condensation with acetone, and methyl ionones by condensation with methyl ethyl ketone. Both types have pronounced irone-like odors, and both products are of major importance in perfumery. There are two types of ionones: the alpha, which is the preferred isomer for use in perfumery, and the beta, which has more limited use in perfumery, and was at one time used primarily to make Vitamin A.

There are many isomers of methyl ionone, some of which are important in the F/F industry. For a full discussion see Chapter XII (pp. 311-313).

Iso Bornyl Acetate — This is an inexpensive aroma chemical made by acetylation of camphene, a turpentine derivative. It finds wide use because it supplies an excellent pine-like odor at a low cost.

Linalool — This extremely important aroma chemical, with its lavender character, is another terpene alcohol, as are geraniol and nerol. It is found uncombined in rosewood oil (from Brazil or Cayenne), as well as in linaloe oil from Mexico. Linalool is present as the acetate in a wide variety of plant and flower materials, including lavender, lavandin, neroli, rose and clary sage oils. For many years, linalool was made commercially by distillation and isolation from Brazilian bois de rose oil. In fact small amounts are still made in this way for special accounts who demand natural linalool. However, about twenty-five years ago, Hoffman La Roche found that by making a slight variation in the process they were using to produce Vitamin E, they could produce a precursor to linalool much more economically than producers who made the product from bois de rose oil.

There was a problem initially because of the high degree of purity of their product. Perfumers were not used to the quality of the pure linalool, and refused at first to accept it. A way was found to adjust it by adding natural trace components. With these modifications, the value of the new product was in time recognized, and in fact it became the standard. It readily developed a market in new compositions, having the immediate advantage that potential buyers already knew how to use it effectively. Perfumers found that if they mixed the chemicals occurring naturally in bois de rose oil with the pure linalool, in the correct proportions, they got an acceptable approximation of naturally derived linalool. From this point on, synthetic linalool took over the market.

Still later, synthetic linalool made from beta pinene made its appearance, and also linalool made from isoprene. Linalool and its esters, principally linalyl acetate, are widely used in all sorts of perfumes, from fine fragrances to detergent perfumes. Linalool represents a good value from an olfactory standpoint, and is quite stable except in acid media. The three synthetic processes for manufacture of linalool — from acetylene, beta pinene and isoprene — remain in close competition for the market.

Methyl Anthranilate is a synthetic grape flavorant which has been in common use for a number of years. It is found naturally in Concord grape juice. Production of the synthetic is of medium volume (about 200,000 lbs. annually). It is used primarily in flavors, but also to some extent in fragrances (jasmine and neroli).

Methyl anthranilate was formerly produced by the reaction of anthranilic acid with methanol, but more recently by the methanolysis of isatoic anhydride. It has a very strong, sweet, fruity odor, and is normally employed sparingly in compounds because of its intensity. The N-methyl homolog, (N-methyl ortho amino benzoic acid methyl ester), called dimethyl anthranilate, is more rounded and fruity in odor character. It too is used as an F/F raw material, but not in as great a quantity as methyl anthranilate.

Musks — This is such a large subject that we shall divide it into four sections, first taking up natural musk, which has in recent years become a material of almost secondary importance in perfumery. The days when the industry depended upon musk obtained from the musk deer are long past, although some small quantities are still available from this source. Supply is extremely limited, however, and price is high.

Our second classification would be the synthetic macrocylic materials related to the naturally occurring macrocyclic musks. These synthetics were first made by Ruzicka in the 20's. The best known of the early representatives of this type are cyclo pentadecanone *(Exaltone),* and pentadecanolide *(Exaltolide).* Another very important member of this class developed at a later date is ethylene brassylate *(Astratone).*

The third class are the nitro musks, the most important representatives of this class being musk ambrette, musk ketone, and musk xylol. The latter is the least expensive and the most widely used. It is made by nitrating tertiary butyl xylene. Musk xylol has a rather weak odor, but works well in soap in which its low cost permits its use. Musk ambrette is the strongest of this family of musks, and has a cedar note in addition to its musk fragrance. It is made from meta cresol. Musk ketone is the finest of the nitro musks, having a fine, powdery note. Unlike musk xylol, which is a tri-nitro compound, musk ambrette and musk ketone are dinitro compounds.

The fourth class of musk chemicals is the one which has assumed the greatest importance to perfumery in recent years. These musks are neither macrocylics nor nitro compounds, but have as a common feature a benzene ring connected to at least one hydro-aromatic ring of 5 or 6 carbons, so that they are either tetralin or indane derivatives. Because of their relatively low price' their general stability in most media, and the fact that they do not discolor soap, in addition to their olfactive properties, these new musks now enjoy wide use. Among the list are Givaudan's *Versalide,* Polaks Frutal Works' *Tonalid,* and IFF's *Galaxolide* and *Celestolide.*

It must always be borne in mind that it is difficult to get perfumers to make substitutions in existing, successful formulae. This is a major factor which makes it a problem to get acceptance for even an excellent new perfuming material. No one, whether it be the marketer of a successful product, his fragrance supplier, or a prospective new supplier, wants to risk changing an accepted product. The marketer may be willing to consider a straight substitution of one perfuming ingredient for a material already being used if the materials are essentially interchangeable; but any change which goes beyond a straight replacement must be explored very carefully.

There are problems of strength of aroma chemicals which must be studied closely before any change can be considered. A change in the strength of one ingredient may completely unbalance a formula. If a material of higher strength is to be substituted, and a smaller amount of one aroma chemical will replace a larger quantity of another, the relationships of all the other ingredients in the formulation may change. It is partly for this reason that the newer musks have not replaced nitro musks as completely as they might have.

Also musk ketone, and particularly musk ambrette, have odor qualities other than their muskiness which contribute to their value; and musk xylol will continue to be used in large volume as long as it remains the cheapest member of the musk family on a dollar per pound basis (or perhaps we should now say on a Swiss franc per kilo basis). The new musks tend to be used in new formulas. In the long run, as they attain still wider use, perhaps the importance of the nitro musks will eventually decline.

Phenyl Ethyl Alcohol — This is a product with a soft and appealing rose note. It is used extensively in combination with other rose note chemicals such as geraniol and diphenyl oxide. Phenyl ethyl alcohol is highly stable in soaps and detergents, and can be used in everything from fine perfumes to odor masks. In addition to its rose character, phenyl ethyl alcohol has a remarkable ability to blend with almost any type of fragrance, and for this reason is used in most perfumes. It can be made from styrene oxide by hydrogenation, however at present the process most widely used commercially is the interaction of ethylene oxide with benzene in the presence of anhydrous aluminum chloride.

Numerous esters of phenyl ethyl alcohol are used in perfumery, including phenyl ethyl acetate, phenyl ethyl propionate, phenyl ethyl butyrate, phenyl ethyl formate and phenyl ethyl phenyl acetate. All of these are variants on the rosy background, with different nuances.

Terpineol and Terpinyl Acetate — Terpineol is made from turpentine, through terpin hydrate. Terpinyl acetate is made from terpineol by acetylation. Both terpineol and its acetate occur naturally in numerous essential oils, including bois de rose (see linalool), bergamot, petitgrain and neroli. Of the isomers of terpineol, the so-called alpha, beta and gamma find use in perfumery, the latter to only a very limited extent. The alpha isomer is used in by far the largest volume, but the beta is preferred for some special applications. The fragrance of the terpineols is in the rosy, woody category. All these materials perform well in most media, being highly stable in soaps and detergents. If it has been poorly fractionated from other turpentine components, alpha terpineol may have a rather disagreeable and medicinal, piney character.

Vanillin — This widely used flavor and fragrance material is the principal component of the odor of vanilla, without the resinous character associated with vanilla beans. It is, incidentally, one of the few F/F materials which has appreciable water solubility. It was initially isolated by alcohol washing of vanilla beans. The first synthetic vanillin was produced by oxidizing iso eugenol with chromic acid. However, the commercial process which later evolved employed nitrobenzene as the oxidizing agent. The nitrobenzene does not destroy as much of the iso eugenol as does chromic acid. Vanillin was later made in large quantity from guaiacol.

The development which at still a later date ended the manufacture of vanillin from eugenol was the discovery that vanillin could be separated from waste liquors from paper manufacture. This is the source of almost all the vanillin sold today.

Unusual and Exotic Aroma Chemicals

Many new materials have been introduced over the past few years, by the development laboratories of some of the major manufacturers of aroma chemicals. The Swiss fragrance house, Firmenich & Cie., has been responsible for a number of interesting new products.

One new material introduced by Firmenich which quickly built an active market for itself is *The*, (sold as *The 17.617*) which reproduces closely the odor of Ceylon tea. It creates an interesting effect in a large variety of fragrance types, even when used in very small quantities. It is said to be stable in most personal products, including soaps, deodorants, antiperspirants, etc.

Another interesting new aroma chemical introduced by Firmenich is *Fixateur 404*, a synthetic fixative which allows the perfumer to achieve special fixative effects. It was discovered by Firmenich in the course of research on the odoriferous constituents of ambergris, although in use it has a different effect than ambergris. It evaporates very slowly and is unusually persistent. It has an amber note, slightly woody, very dry, fine, yet powerful.

IFF makes a chemical similar to *Fixateur 404* which they market under the name, *Grisalva*. Both these products have exceptional strength, and combine the best features of ambergris and tobacco.

Through the years, various aroma chemists have devised loosely bound chemicals called Schiff bases, made from anthranilates, especially methyl anthranilate, and various aldehydes. One of the most widely used of the Schiff bases is a combination of methyl anthranilate and hydroxycitronellal. It is sold by different manufacturers under a variety of trade names, including *Aurantiol* (Givaudan), and *Auriol* and *Auralva* (IFF). These products have the flowery beauty of hydroxycitronellal combined with the appealing grape note of methyl anthranilate. They are highly prized, and much used in fine perfumery.

Another class of "addition products" is the acetal and hemi-acetal group, made by adding various alcohols, including methyl and benzyl alcohol, to the fairly unstable aldehydes. When the addition is not accompanied by treatment to remove the water which is formed, the products are hemi-acetals. The hemi-acetals readily revert to the original components. For example, benzyl alcohol added to phenyl acetaldehyde will loosely bind the latter, with no significant change in fragrance, slowly liberating phenyl acetaldehyde to the perfume mixtures in which the material is incorporated.

If methyl alcohol is added to phenyl acetaldehyde, however, and the water of formation driven off, a quite stable new chemical is produced, with an odor much different from that of the original material. Instead of the pungent hyacinth note of phenyl acetaldehyde, a green note is produced. The end product of this process is the chemical, phenyl acetaldehyde dimethyl acetal, which comes to market under the name *Viridine* (Givaudan Corp.)

Similarly, a new and very useful aroma chemical results from the interaction of citral and methyl alcohol. This product is citral dimethyl acetal. In this case the odor is modified, rather than being completely changed. The resulting fragrance material has a more stable, milder lemon aroma than does citral.

Another way to make an aroma chemical with the fragrance of lemon, much more stable than citral, was devised with the discovery of geranonitrile. (For details on geranonitrile see p. 145).

The Future for Synthetics

As indicated earlier, synthetics have a number of built-in advantages. Supply is more dependable, since production is not affected by weather or natural occurrences. Synthetics can be made domestically, rather than thousands of miles away, so there is less hazard of being cut off from sources of supply. And synthetics have at least in the recent past been more stable in price, since they are produced from raw materials which have not fluctuated widely in cost. There is some question now as to whether they will continue to enjoy this advantage, for most synthetic aroma chemicals are based on petrochemicals, and the price of crude oil is now subject to arbitrary decisions by the OPEC nations.

Because of the economic advantage of synthetics, assurance of supply, and close control over physical and chemical characteristics, demand has grown rapidly over recent years and there has been a tremendous increase in output. Today the pound and dollar volume of aroma chemicals used is far greater than comparable totals for natural fragrance and flavor materials. More naturals are being produced today than ever before, but it would be physically impossible to satisfy the demand for aroma chemicals from natural sources alone. The percentage of the market naturals have is declining, simply because the overall market continues to expand at such a rapid rate, and the only way this demand can be filled is by increased production of synthetics. There is no problem today in selling every pound of natural raw material produced, particularly in the flavor category, but it is a practical impossibility for production of sufficient natural raw materials to keep pace with the rapidly expanding fragrance and flavor raw materials market. It is not a processing problem, but an agricultural one.

It is fact, however, that materials of natural origin still hold a preferred place in perfuming and flavoring. Most perfumers and flavorists insist that, to impart life, lustre, body and brilliance to their compositions, and to imitate nature with a high degree of verisimilitude, the formula for a fragrance or flavor must contain at least a minimum amount of "naturals." Formulations based exclusively on synthetics can never duplicate the character and appeal of those which contain an appreciable percentage of "naturals." So, no matter what developments lie ahead in synthetics, it is probably safe to say that marketers of finished products using fragrances or flavors will always be able to absorb every pound of natural fragrance and flavor raw material the world can produce, except for perhaps lemongrass and citronella oils, where the synthetics have supplanted these natural oils almost completely.

We might pause to note that Steffen Arctander, well known writer and consultant in the fragrance field, is among those who believe that synthetics will not replace natural materials, in the foreseeable future. He points out that "only about a dozen synthetic essential oils have been compounded to such perfection that they meet the demands and expectations of the knowledgeable and discriminating perfumer or flavorist."

In his opinion, successful perfumery almost always requires a blend of synthetics with a certain amount of a natural fragrance material. He does suggest, however, that, perhaps in time, when there is a completely new generation, relatively unfamiliar with the fragrance of "naturals," fragrances based exclusively on synthetics could have a considerably wider appeal.

If there is a cloud on the horizon so far as synthetic aroma chemicals are concerned, it lies in the attitude toward them of government agencies and consumer protection groups. Over recent years advances in science and technology have made it possible to detect the presence of minute qualities of potentially toxic substances in all materials, whether natural or synthetic. A public, lacking the training to read and interpret labeling correctly, has developed fear of all "chemicals," at any rate of exposure. This anxiety and suspicion has been stimulated by some extremist consumer groups.

Meanwhile the Federal Government has piled one regulatory agency and law on top of another. New food and drug legislation has been proposed, new cosmetic regulations, and now we have the Toxic Substances Control Act. Attitudes have become polarized, and a "natural vs. synthetic" confrontation has developed. Unfortunately, neither those who make the laws nor those who argue for. or against them are doing anything to educate the public in how to read labels, or to understand that any material, whether natural or synthetic, can be toxic in at least some degree, depending on dosage and conditions of exposure.

It is the earnest desire of all those in the fragrance and flavor industry to remove from commerce any materials which are sensitizing, irritating or allergenic. All the major fragrance components and additives, both natural and synthetic, have already been tested by the Research Institute for Fragrance Materials, and up to this point only a very limited number have been found to pose any problem.

Indictment or exoneration of the thousands of other less used materials presents a mammoth testing job for the future.

Meanwhile, it must be anticipated that, in the days ahead, producers of aroma chemicals may be less willing to fund the substantial research budgets necessary to develop new aroma chemicals. They may hesitate to spend the substantial sums required to test the safety of new raw materials which may never reach the market. Instead, they may well be fully occupied with problems of detecting, isolating, identifying and synthesizing the trace components of natural products. And, such new product development as they can afford to continue will probably be concentrated in research on "natural identicals" and away from the discovery of new synthetics.

(For a further discussion of the impact of government regulation on the aroma chemical industry the reader is referred to Chapter XIV).

Chapter IX

History of The U.S. Industry

Fifty years ago the essential oil and aromatic chemical industry of the United States consisted of about seventy firms, three-quarters of which are now no longer in existence. Over fifty of these companies were located in New York City, most of them in a small area of two or three square blocks on the east side of lower Manhattan, close to the docks where the fragrant materials in which they traded arrived on ships from such distant ports as Marseilles, Messina, Penang, Colombo or Surabaya. The heart of the American essential oil industry was this tight little nucleus of firms situated on Pearl, Cliff and Water streets as the north and south axes, and Beekman, John, Platt and Fulton streets as the cross streets.

If you wanted to talk over a deal with someone else in the business, you usually just put on your hat (people still wore hats in those days) and stepped around the corner to see him, rather than bothering with the telephone. Everyone knew everyone else, there was an air of informality, you seldom bothered to make appointments, and the people you called on usually saw you with a minimum of delay. Many of them lunched together at

Office of Roure-Bertrand Fils in lower Manhattan, about 1905

White's on John Street, at Sweet's on Fulton Street, at Angelo's on Pearl Street, or perhaps at the Drug & Chemical Club. Many a time this reporter called on six or eight firms in a morning, picking up news items, or checking the essential oil markets. In many establishments you made your entrance through the shipping department, stepping around crates and over packages as you made your way to the office of Joe Magnus, Peter Dreyer or A. L. van Ameringen.

The accompanying list of perfuming materials houses is taken from several sources: the 1924 directory of the American Manufacturers of Toilet Articles, the first edition of the "Soap Blue Book," published in 1929, the membership list of the Essential Oil Dealers Association, and the memories of the writer and his friends. The particular assistance of Grant A. Dorland, publisher of "Soap/Cosmetics/Chemical Specialties" is acknowledged. The only names omitted from the list are those of brokers like George Uhe & Co., whose office was located at 11 Cliff St., Charles Huisking & Co., on Platt St., and the two chemical companies, Monsanto Chemical Co., who produced coumarin and vanillin, and Dow Chemical Co., who manufactured methyl anthranilate and artificial oil of wintergreen. While these two companies were obviously manufacturers of aroma chemicals, this was a relatively insignificant part of their business, and they did not offer a general line of aroma chemicals as did firms like Givaudan, Norda, van Ameringen, Synfleur, Van Dyk, Verona Chemical and other such firms.

Suppliers of Essential Oils and Aroma Chemicals — 1927

M. L. Barrett Co., Chicago
Edward T. Beiser Co., Riverside, Conn.
Belgian Trading Co., 44 Whitehall St., New York
O. A. Brown Co., 246 Pearl St., New York
W. J. Bush & Co., 370 7th Ave., New York

Antoine Chiris & Co., 147 Waverly Place, New York
Coignet Chemical Products Co., 17 State St., New York

Compagnie Duval, 32 Cliff St., New York
Compagnie Parento, Croton-on-Hudson, N.Y.
Coupey Fils, 7-11 Water St., New York
Cox, Aspden & Fletcher, 26 Cortlandt St., New York

Delphi Products Co., 95 Beekman St., New York
Dodge & Olcott Co., 87 Fulton St., New York
P. R. Dreyer Inc., 109 Beekman St., New York
A. C. Drury & Co., 420 Rush St., Chicago

Elson & Brewer (on Gold or Platt St.) New York
C. G. Euler Co., Platt or Pearl St., New York

Felton Chemical Co., 65 Taffe Pl., Brooklyn
Florasynth Laboratories, Olmstead & Starling Aves., Bronx, N.Y.
Benj. French, Inc., 160 Fifth Ave., New York
Fries Brothers, 92 Reade St., New York
Fries & Fries, 242 Pearl St., New York
(plant and main office at 1500 W. 6th St., Cincinnati)
Fritzsche Brothers, Inc., 82 Beekman St., New York

Givaudan-Delawanna, Inc., 101 5th Ave., New York
Geo. V. Gross Co., 30 Old Slip, New York

Heine & Co., 54 Cliff St., New York
James B. Horner, Inc., 3 Platt St., New York
D. W. Hutchinson & Co., 94 John St., New York
Hymes Brothers Co., 233 Pearl St., New York

Imperial Export Co., 15 Whitehall St., New York
Industrial Aromatic Co., 4020 Tenth Ave., New York
C. E. Ising Corp., Flushing, L.I., N.Y.

E. M. Laning Co., 273 Water St., New York
Lautier Fils, Beekman St., New York
Leghorn Trading Co., 59 Pearl St., New York
E. Lelong, 30 Pearl St., New York
Pierre Lemoine et Cie., 294 Pearl St., New York
George Lueders & Co., 427 Washington St., New York
Julian W. Lyon & Co., 35 Fulton St., New York

Magnus, Mabee & Reynard, Inc., 257 Pearl St., New York
J. Manheimer, 271 Pearl St., New York
Morana, Inc., 118 East 27th St., New York
Clarence Morgan & Co., 355 W. Ontario St., Chicago

Natural Products Co., 175 Fifth Ave., New York
Neumann-Buslee & Wolfe, 224 West Huron St., Chicago
Norda Essential Oil & Chemical Co., 121 E. 24th St., New York
Northwestern Chemical Co., Wauwatosa, Wis.

Orbis Products Trading Co., 215 Pearl St., New York

Pfaltz & Bauer, 300 Pearl St., New York
Polak's Frutal Works, 180 Lafayette St., New York

Rhodia Chemical Co., 89 Fulton St., New York
Riviera Products Co., 510 N. Dearborn St., Chicago
Roure Bertrand-Justin Dupont, 18 Cedar St., New York
H. C. Ryland Co., 52 Laight St., New York

E. J. Schmidt & Co., Fulton St., New York
C. A. Seguin Co., 500 N. Dearborn St., Chicago
George Silver Import Co., 461 4th Ave., New York
R. Walton Smith & Co., 60 East 43rd St., New York
Chas. V. Sparhawk Co., 278 Pearl St., New York
A. A. Stilwell & Co., 28 Cliff St., New York
Synfleur Scientific Laboratories, Monticello, N.Y.

A. M. Todd Co., Kalamazoo, Mich.

Ungerer & Co., 124 West 19th St., New York
United Laboratories, 15 S. William St., New York

van Ameringen, Inc., 15 Irving Pl., New York
Van Dyk & Co., 6 Platt St., New York
L. A. Van Dyk & Co., 110 West 14th St., New York
Verona Chemical Co., Newark, N.J.
Albert Verley & Co., 11 E. Austin Avenue, Chicago
Vietor & Hosken, 160 Pearl St., New York

Wangler-Budd Co., 205 West 14th St., New York

The principal firms surviving from those early days, and still doing business today under somewhat the same names are:

Felton International, Inc.
Florasynth, Inc.
Fritzsche/Dodge & Olcott, Inc.
Givaudan Corporation
D. W. Hutchinson & Co.
Hymes Aromatic & Import
J. Manheimer, Inc.
Neumann-Buslee & Wolfe, Inc.
Norda Inc.
Compagnie Parento, Inc.
Polak's Frutal Works, Inc.
Rhodia, Inc.
Roure Bertrand Dupont, Inc.
Synfleur Scientific Laboratories
A. M. Todd Company
Ungerer & Company
Van Dyk & Company, Inc.
Albert Verley & Company

There are other firms that have undergone mergers and name changes: van Ameringen, which first became van Ameringen-Haebler, Inc., and then International Flavors & Fragrances; Lueders, which is now a subsidiary of Monsanto; Fries & Fries, which is a division of Mallinckrodt, Inc.; and Magnus, Mabee & Reynard, Inc., which was taken over by Gentry in 1966, and has since then been incorporated into Crompton & Knowles Corp. Belgian Trading Co. survives as Danco In'tl. But many other firms, such as Pierre Lemoine, Morana, Fries Brothers, Benj. French, Heine & Co., O. A. Brown, Schimmel & Co., Trubek Laboratories and James B. Horner, once important names in the essential oil business, are gone with the wind. On the other hand, major factors in today's perfuming materials industry such as Dragoco Inc., Firmenich, Inc., Haarmann & Reimer Corp., Hoffman La Roche, Inc., Perry Bros., Inc., Robertet, Inc., Stepan Chemical Co., UOP Fragrances and Union Camp Corp. had not yet appeared on the scene, fifty years ago.

Representation of Foreign Firms

Many of the American essential oil firms of fifty years ago represented European producers on their lines of perfume specialties, in addition to importing natural oils. Ungerer & Co. had several of the best foreign accounts: including Charabot, Chuit Naef (later to become Firmenich), Stafford Allen on their English lavender, and Botu Pappazoglou & Co. on otto of rose. Morana represented Bruno Court of Grasse, Haarmann & Reimer of Holzminden, Germany, and Christo Christoff, another Bulgarian producer of otto of rose. Dodge & Olcott had the prestigious DeLaire account. Julian W. Lyon represented Polak & Schwarz of the Netherlands, and an Italian citrus oil firm, La Zagara. Lyon later lost the P & S account, for Polak & Schwarz sent a young Hollander, A.L. van Ameringen, to the United States, and in time transferred their account to him.

Benj. French represented Descollonges Freres, a French firm. Rhodia had another French account, Societe Chimique des usines du Rhone. P.R.Dreyer, who had learned the essential oil trade as a salesman for Ungerer and later managed Rockhill & Vietor's essential oil department, set up his own business in the 20's and got U.S. representation for Bertrand Freres, another French account. He also represented Paolo Vilardi, Italy, and Naarden Chemical Works, the Netherlands. O. A. Brown had a Swiss account, Chemical Works Flora. Coupey Fils were agents for Coupey Fils & Dehais, Paris. Lueders represented Camilli, Albert and Laloue, Grasse, and Botzuroff, Bulgarian producer of otto of rose. Lueders also had the Givaudan account for a time. Givaudan & Cie., a major Swiss producer of perfuming materials, later switched their account to Burton T. Bush, Inc., and eventually opened their own branch in the United States. Belgian Trading Co. had two French accounts, Nadal, Desparmet et Cie and Hortus Nerolium. Coignet represented Societe Francaise de Produits Aromatiques. Elson & Brewer were agents for Societe Usine Chemie Rhone, known usually by their initials, SUCR. E.M. Laning Co. had two French accounts, H. Eiziere & Co. and Societe des Produits de Synthese. W.J. Bush & Co., an English house, imported sandalwood from India and distilled the oil at their New Jersey plant. Cox, Aspden & Fletcher represented the Mysore Government on sandalwood oil, while Imperial Export Co. were representatives for Plaimar, Ltd., Perth, West Australia, another Australian sandalwood oil producer.

European suppliers were not always happy with their U.S. representatives. The suspicion occasionally arose that the U.S. agents might be substituting their own products for the higher priced imported materials, thereby widening their profit margins and decreasing the commissions paid to their suppliers. Thus it was not at all unusual for a foreign producer of perfume specialties to change his U.S. representative from time to time, or occasionally to decide to solve the problem permanently by opening a company-owned branch in the U.S.

U.S. importers of natural oils also had their problems. In those days before chemical controls had been perfected, and when the gas liquid chromatograph had not yet been developed to give a speedy and precise read on composition and quality of the material under test, it was not at all unusual to find adulteration in imported oils.

Bergamot oil was readily susceptible to sophistication, and complaints on quality were common. Ceylon citronella oil was widely adulterated by the addition of kerosene or other petroleum fractions. Lemongrass oil often tested low on citral content. Lead and rosin were frequently found on examination of cassia oil shipments. Civet, a costly material of animal origin, was frequently adulterated with honey, mashed bananas or petroleum jelly. The buyer had to be constantly on the alert. And when, occasionally, adulterated oils were received, the buyer was never sure whom to blame. It could be the supplier, or it might be some third party intent on taking advantage of both of them. But no one recalls any such fraud as that recently perpetrated on essential oil buyers all over the world by Indonesian shippers who in 1976 sent huge quantities of watered merchandise instead of the oils that had been ordered. This is a hazard against which it is difficult to defend, for the custom still prevails of accepting advance samples as an indication of merchandise to be shipped, and paying for material ordered before shipment.

A satisfactory long-term arrangement between producer and seller obviously demands full good faith and trust between the contracting parties. The personal honesty of both seller and buyer are the "priceless ingredients" in the products being sold. From the long continued and apparently profitable relationships between many foreign producers and their American agents, it seems apparent that at least some of these arrangements for representation are soundly based, for they have continued to be successful over a long period of years.

In reviewing the early history of the fragrance and flavor industry in the United States, we find that the first company to start importing essential oils, and one which later moved into the manufacture of aroma chemicals was Dodge & Olcott Co. It dates back to the tail end of the 18th Century.

Dodge & Olcott

A young New York City druggist named Robert Bach established a drug and chemical importing business at 128 Pearl St., New York, in 1798. The materials he imported included essential oils, perfumes and soaps. By 1821 he had acquired a partner, and the business was renamed Bach & Bradish. By 1840, there were two new partners, the business had another name change to Dodge, Cuming &

Co., and the firm had opened an office in London. By 1850, there was still another partner, another name, Dodge & Colvill, and a second foreign office, in Paris.

The first Olcott joined the firm in 1859, reflected in the change of name to Dodge, Colvill & Olcott. It was in this year that the company went into the manufacture of aromatics. Colvill withdrew from the firm in 1861, and after his departure there was a name change to Dodge & Olcott, by which this very early American essential oil house was known for so many years.

The manufacturing operation moved ahead swiftly. By 1862, D&O was making extracts of caraway, sandalwood, nutmeg, clove and bay oils. By 1880, the distillation business had expanded so much that a factory was built in Brooklyn, followed by another plant in Bayonne, N.J., in 1904.

In the early 1900's D&O offices were located in a building on William St. just south of Platt St. Both Mr. Dodge and Mr. Olcott were blind in later years, but still came regularly to the office, according to Charles Huisking ("Herbs to Hormones" page 238). The company later had its own building at 87 Fulton St. until the move was made to 180 Varick St. in 1930.

It was a prophetic day for D&O when Dr. Francis T. Dodge joined the firm, perhaps in the early 1880's. His accomplishments in synthesizing a number of major aromatic chemicals helped Dodge and Olcott attain a prominent position in its industry. He was the first chemist to isolate citral, and also figured prominently in the development of citronellal, citronellol, and various aldehydes.

Dodge & Olcott Co. was a charter member of the Essential Oil Dealers Association. Herman Weicker of D&O was president of EOA in 1933-34, and another top D&O executive, Val H. Fischer, headed EOA in 1942-43.

Dodge & Olcott got into the insecticide business during World War II. They supplied the pyrethrum extract for millions of "bug bombs," which protected U.S. troops in the South Pacific from insect-borne diseases (see also p.

Fritzsche Brothers factory in Garfield, N.J., about 1893

236). The company followed up this initial success with development of the leading insecticide synergist, piperonyl butoxide.

During this period Dodge & Olcott was taken over by National Distillers. However, on a long term basis, a perfuming materials subsidiary proved not to fit in with National Distillers' corporate planning, and in 1952 Dodge and Olcott was sold to Fritzsche Brothers, Inc. D&O continued to operate independently until 1969, at which time the two companies were merged.

Fritzsche Brothers

Fritzsche Brothers dates back to 1871, when Paul, Herman and Ernest Fritzsche formed a partnership with Schimmel & Co. of Leipzig, Germany, and established an essential oils business in the United States. Fritzsche had a manufacturing plant in Hoboken, N.J., and an office in New York City. By 1892 the Hoboken plant had become inadequate, so they closed it and built a new plant and research headquarters in Garfield, N.J. At about the same time the New York office was moved to 34 Barclay St.

Frederick H. Leonhardt Sr. joined Fritzsche Brothers during this period, initially as director of laboratories. In time he became active head of the company after the Fritzsche brothers had withdrawn, leaving Fritzsche purely an American operation. The Garfield plant was sold in 1900, and the company decided at that time to let Dr. Clemens Kleber handle manufacturing as well as operation of their laboratory. Dr. Kleber, who had previously directed the Fritzsche laboratory, opened his own independent laboratory in Clifton, N.J., for this purpose. In 1907, the New York office was relocated in larger quarters at 82 Beekman St.

In 1914, Fritzsche was cut off from its German source of supply, the Schimmel plant in Leipzig, by the outbreak of World War I. Thus the U.S. company had to start making its own aromatics, which Dr. Kleber was able to handle for them. In the post World War I period, Fritzsche bought its own essential oil production facilities in Southern

Battery of stills in Fritzsche Garfield factory

France, Parfumeries de Seillans, and thereafter was more independent of commercial suppliers of flower oils. About 1935, Fritzsche also bought the Clifton factory of Dr. Kleber. In the same year the Beekman St. office was wiped out in a fire. At that time the company moved to its present location at 76 Ninth Ave. Fritzsche Brothers acquired Dodge & Olcott in 1952. In 1969 the two companies were merged into Fritzsche Dodge & Olcott. Production for the merged companies is now centralized in a plant on a 65-acre site in East Hanover, N.J., and a naturals plant in Clifton, N.J.

Over the 50 years of EOA's history, three top executives of Fritzsche Brothers have served as president of the Essential Oil Association: A.D. Armstrong in 1937, John Montgomery in 1944-45, and H.P. Wesemann in 1949-50.

Synfleur Scientific Laboratories

This firm, the first American manufacturer of compounded perfuming materials, was started by Alois von Isakovics in 1889. This young Austrian chemist, born in Prague, 1870, was a graduate in chemistry of the University of Vienna, and came to the United States in 1886. He saw a future in the manufacture of aromatic chemicals and built a factory for producing them at 1750 Park Ave. in Harlem (New York City). He was a brilliant chemist and was greatly aided in building his new company by his wife, Mary Upshur von Isakovics, a graduate of Cooper Union.

As he was in frail health, von Isakovics transferred the site of his operations to Sullivan County, in the heart of the Catskill Mountains, in 1901, building a plant in Monticello, N.Y. In 1909 Luis de Hoyos, a native of Spain, joined the company, and soon was made general manager. He also shortly became the von Isakovics' son-in-law, as he married Marie von Isakovics, their eldest daughter. Marie had a brother, Alois Jr., who joined the company in 1925 after completing his education.

Mr. von Isakovics Sr. died at a relatively early age — in 1917 — and his widow took over direction of the company, assisted by Mr. de Hoyos. In addition to running the

Office of Ungerer & Co., about 1905. Fred Ungerer is seated at desk in right foreground, with brother Will standing at the desk behind him

business, de Hoyos became active in local politics and served as Mayor of Monticello for many years.

Like his father, Alois Jr. died at an early age — in 1935 at the age of 41. However Luis de Hoyos and Marie von Isakovics produced a son, Luis de Hoyos Jr., to carry on the company. Luis Jr. was born in 1921. After graduating from Colgate University in 1942, he came into the business, and by the time his father died in 1951 he was able to step into his place.

Control of Synfleur was sold to The Nestle Company in August, 1975. Mr. de Hoyos will remain as president of the company for several years by contractual arrangement.

Ungerer & Company

Ungerer & Company trace their origin back to 1865, when William P. Ungerer, in partnership with a brother, Henry Ungerer, started Ungerer Brothers in Rochester, N.Y., for the manufacture of specialties. This firm became dormant after a time, and in 1872 William Ungerer became chief perfumer for Colgate & Co., soap manufacturers. He stayed with Colgate for over twenty years, then left them in 1893 to start a perfuming materials supply business, having obtained the American agency for a leading European supplier of perfuming specialties. In a sense he was merely reactivating the business that he and his brother had started years before back in Rochester.

His son, William G., succeeded him as perfumer at Colgate, but about 1895 decided to join his father in the essential oil business. They incorporated the firm in 1901, and by that time Will's brother, Fred, was also active in Ungerer & Co. The first office was at 35 S. William St., from 1893 to 1896. They were then located at 18 Cedar St. for eight years, at 15 Platt St. for three more years, and at 273 Pearl St. for ten years, before moving "uptown" to 124 West 19th St. in 1917. In 1930 came another move to 13/15 West 20th St., where they stayed until 1942. Since then they have been at their present location at 161 Sixth Ave. (Avenue of the Americas). The Ungerer factory at 650 Union Blvd., Totowa, N.J., was established in 1946.

The Ungerer name has long been a highly respected one in the perfuming materials business. Through the 20's Ungerer represented some of the most prestigious European suppliers of perfuming specialties. However, for many more recent years, Ungerer's emphasis has been on products designed through their own creative abilities. The company no longer represents any foreign producers, but creates original flavor and fragrance formulations in its own laboratories.

In 1935 Ungerer & Co. took over Budd Aromatics, a company started by Ivon H. Budd, who had previously been sales manager for Julian W. Lyon & Co. and later a partner in Wangler-Budd Co. Mr. Budd later became sales manager of Ungerer & Co. and remained with the company until his death in 1969.

The Ungerer business has remained under the control of the Ungerer family for over a hundred years, and has seen four generations of Ungerer descendants pass through its doors. Kenneth Voorhees, son-in-law of Frederick Ungerer, was made president in 1949 and held this post for twenty-seven years. He was succeeded early in 1977 by his son, Gary, who joined Ungerer in 1957.

Magnus, Mabee & Reynard, Inc.

The predecessor company to Magnus, Mabee & Reynard, Inc., was McKenzie Brothers & Hill, organized in 1895 with offices at 52 Water St., New York City. Percy C. Magnus Sr., who had started his business career in an Atlanta, Ga., drug firm, came to New York City and took a job as sales manager for McKenzie Bros. & Hill. He soon became a partner. Then, in 1898, he and Monroe W. Lauer took over the firm and changed the name to Magnus & Lauer. Their office was originally at 4 Cedar St., but was shortly moved to 92 Pearl St.

Magnus and Lauer did a considerable business in essential oils, vanilla beans, drugs and chemicals, and represented a number of foreign producers including Morana, a Swiss manufacturer of aroma chemicals, Spurway & Cie., Grasse, France, G.E. Peirce, Messina, Italy, and the Swiss-American Chemical Co., Basle, Switz., producers of vanillin and coumarin. M&L later took on the products of E. Sachsse & Co., Leipzig, Germany, who manufactured artificial flavors for blending gin, cordials, rum, kummel, wines, etc.

In 1905 Lauer died, and shortly thereafter Magnus accepted two new partners, D.W. Mabee and G.C. Reynard. Then, in 1907, Magnus & Lauer was merged with the National Essential Oils Distilling Company of Chester, Conn., and a new corporation, Magnus, Mabee & Reynard, Inc. was formed. Mabee and Reynard were only nominally active in the company, and in 1910 Mr. Magnus bought out their interests, although retaining their names in the corporate title.

Meanwhile Percy C. Magnus, Jr. had joined the company. When his father died in 1916, he succeeded him as president, although he was only 22 years old. His younger brothers, Joseph Baird Magnus and Robert Burke Magnus, also joined the company about this time.

In addition to members of the Magnus family, several other individuals who gained considerable prominence in the essential oil industry worked for MM&R. One was Bill Fisher, sales manager of MM&R for many years; another was George McGlynn. McGlynn served as president of the Essential Oil Association of U.S.A. in 1952.

Offices had been moved in 1905, to 257 Pearl St. In 1919, MM&R bought the building at 32 Cliff St., immediately in the rear of their Pearl St. quarters and both buildings were enlarged. From that point on they used the Cliff St. address. The plant in Chester, Conn., was sold to Middlesex Chemical Co. In 1938 MM&R moved out of the lower New York City area, to their own building at 16 Desbrosses St.

In 1966 MM&R was sold and became the Fragrance and Flavor Division of Gentry International. Then, in 1975, came another ownership change, with Crompton & Knowles, Fair Lawn, N.J., buying the division from Gentry. MM&R now operates as the Flavor and Fragrance Division of Crompton & Knowles Corp.

The Chiris aroma chemical plant in Delawanna, N.J. in 1918

Antoine Chiris Company

This company, the U.S. branch of a famous French essential oil firm, was an important factor in the U.S. market up until World War II, but thereafter lost ground and was eventually purchased by UOP when Universal Oil Products became interested in the perfuming materials field.

The U.S. branch of Antoine Chiris Co. was established in 1896 by Georges Chiris, son of Leon Chiris, head of the French firm at that time. Georges was the great-great-great-great grandson of Antoine Chiris, who founded Chiris in Grasse in 1768. Some time prior to World War I, Burton T. Bush joined Chiris in New York, and by 1913 had been made general manager. He persuaded Leon Chiris that Chiris should go into manufacturing in the United States, and they started to build a factory in Delawanna, N.J. It was planned to distill essential oils of U.S. origin, and to manufacture synthetic aromatics and pharmaceuticals. Their first products were vanillin and terpineol.

The timing on the project was extremely unfortunate, for shortly after it was under way the outbreak of World War I made it impossible to get capital out of France to finance construction. Bush arranged alternate financing in the U.S., and it was reported that Henry Colgate of Colgate-Palmolive-Peet Co. personally supplied funds for the project. Colgate would have had a natural interest in helping to develop a U.S. source for the perfuming materials they might soon no longer be able to obtain from France. Bush ended up with a considerable personal investment in the Chiris plant.

Mr. Bush left Chiris in 1924, having persuaded Givaudan & Cie., Geneva, Switzerland, to name him as their agent. When he left Chiris, Bush bought out their interest in the plant. He was succeeded as general manager at Chiris by Charles Swan, who had been with the company for many years. The Chiris New York office was located at 147 Waverly Place until about 1930, then was moved to 23rd St. When Swan left to join Aromatic Products, Inc. in 1935,

Chiris sent Louis Rapin from France to head the U.S. operation.

Not only was Chiris a leading company in the U.S. fragrance industry from a sales standpoint for many years, but many individuals who later became top figures in the industry at other companies got much of their early training at Chiris. George Cadgene, who retired in 1970 as v-p aroma chemical sales for IFF, told us that George Warren, for many years Givaudan's assistant sales manager for aroma chemicals, told him that in the early 50's he attended an informal reunion of former Chiris employees who were then working elsewhere in the trade. There were over 50 there. Chiris was serving unintentionally as a training school for the essential oil industry.

The Chiris organization continued operating until the late 60's, when they were taken over by Universal Oil Products, who had earlier bought Trubek Laboratories and Fleuroma Laboratories as the nucleus of a new fragrance operation. In 1976 UOP decided to dispose of its fragrance unit to Naarden; thus the old Chiris operation is now part of this Dutch perfuming materials firm.

Hymes Aromatics and Import

Another essential oil company that dates back to the 19th century is the firm which now operates as Hymes Aromatics and Import, but which when organized in 1891 was known as Hymes Brothers Company. The firm was founded by Henry M. Hymes, Edward I. Lowell and Jacob Hymes. They owned their own 2-story building, which was located at the corner of John and Pearl Streets. They dealt in essential oils, vanilla beans, gums, olive oil and spices and also manufactured flavors and synthetic fragrances. The business prospered and Hymes Brothers continued active until 1932, when Henry Hymes retired and the original firm was discontinued. Leonard J. Hymes, son of Henry Hymes, then started another company under his own name. In 1952 the name was changed to Hymes Aromatics and Import. The firm is now located at 43 West 61st St., New York.

D. W. Hutchinson & Co.

One of the earliest U.S. firms in the essential oil business was D. W. Hutchinson & Co., which started operations in 1896 in New York City on South Street in lower Manhattan. South Street was known as the "Street of Ships," dating back to the days when sailing ships used to stretch their bowsprits across the cobblestones with which the street was paved.

With the passing of years, as the essential oil industry moved away from lower Manhattan, Hutchinson moved its base of operations to Mt. Vernon, N.Y. They are now located there at 700 South Columbus Ave. D. W. Hutchinson has a unique record of having kept its business in the hands of four generations of the same family for over eighty years.

Roure Bertrand Dupont, Inc.

The French firm of Roure Bertrand Fils was founded in 1820 by Claude Roure. In 1898 Jean Amic, who had married one of the Roure descendants, became president of the company, which he ran with the assistance of two brothers-in-law, Jean and Louis Roure.

In 1900 a decision was made to add the manufacture of aroma chemicals to the production of essential oils, but as there was at the time a rather strong prejudice against synthetic fragrance materials, a separate company was set up to handle aroma chemical production. Operating as S.A. des Ets. Justin Dupont, it was housed in a small factory in Argenteuil, near Paris.

When Jean's son, Louis Amic, joined the firm in 1928, he felt it was uneconomic to run two separate perfume laboratories, so the businesses were merged under the name S.A. des Ets. Roure, Bertrand Fils and Justin Dupont, with two centers of production, one in Grasse and the other in Paris. Louis Amic ran the Paris operation which served the domestic market as well as England, the United States and other English-speaking countries, while his brother, Francois, was in control of operations in Germany, South America, and the Middle and Far East.

Roure Bertrand had established its first office in the United States around 1900, with a Mr. Burr in charge. Justin Dupont was at that time not yet a part of the company. After Justin Dupont was organized, both companies for a time had separate offices in New York City, the Roure Bertrand office for a time being located at 18 Cedar St. About 1922, George Silver Import Company, 461 4th Ave., was named to represent both Roure Bertrand and Justin Dupont. Albert Delavigne was sent over from France to join the George Silver staff, with his specific responsibilities being to look after the Roure-Dupont interests. When George Silver, Sr., retired, about 1930, Roure-Dupont took over his company, Mr. Silver Jr. remaining with the firm.

In 1969 a decision was made to make a major expansion in the U.S. operation, adding manufacturing and full laboratory facilities. Alvin F. Lindsay joined the U.S. subsidiary of Roure Bertrand Dupont as president. Construction was started shortly thereafter on a building in Teaneck, N.J., to house offices, plant and laboratory. The move to the new headquarters was made in September, 1972.

Van Dyk & Company

Van Dyk & Company, Inc. was formed by Samuel Isermann and L.A. Van Dyk in 1904 to manufacture and import perfuming materials. Imports were handled through August Maschmeyer, Jr., with offices in Amsterdam and Paris. Van Dyk & Co. was first located at 96-98 Maiden Lane, New York City, but moved to 4-6 Platt St. in 1911. That year L. A. Van Dyk, who was president of the company, resigned to form his own company, L. A. Van Dyk. He was succeeded as president of Van Dyk & Co. by Samuel Isermann.

Max Isermann, Sam's brother, had joined the company in 1906. He was named vice-president in 1918, then president in 1928, serving in this post until his retirement in 1932. At that point, Samuel Isermann resumed the presidency and held the office until his death in 1949. Max died in 1947.

Howard Isermann, son of Max Isermann, a graduate in chemical engineering of Rensselaer Polytechnic Institute, joined the firm as plant engineer in 1947. Shortly after Samuel Isermann's death, his nephew, Isidore Hollenberg, a chemist and lawyer who had been with the company for many years, was named president. He left Van Dyk & Co. after a few years to organize his own company, and was succeeded by Mrs. Samuel Isermann, widow of the founder. In 1970, Mrs. Isermann became chairman of the board and Howard Isermann was elected president.

Van Dyk acquired a factory on Wilkinson Ave., Jersey City, N.J., in 1912, retaining its office in New York City until 1933 when the office was moved to the plant. Ten years later offices and plant were moved to their present location at 11 William St., Belleville, N.J.

Starting in the late 1930's, Van Dyk & Co. diversified its business by adding ultraviolet absorbers and cosmetic chemical specialties to its original line of perfumery raw materials. About 1950, production of aroma chemicals was discontinued, as it was found more economical to buy these materials from larger manufacturers.

Morana, Inc.

One of the major perfuming raw materials firms of fifty years ago that failed to survive the 20's was Morana, Inc., a manufacturer and importer of aroma chemicals affiliated with a Swiss firm of the same name. Morana was one of the first U. S. companies that emphasized manufacture of synthetics rather than importation of natural materials.

The U.S. Morana company was organized by Carl Schaetzer, Warren E. Burns and Willard A. Walsh about 1909. They had a New York office at 115 East 27th St., and a large factory in Elizabeth, N.J. In addition to manufacturing their own aroma chemicals, Morana represented three major European producers of perfuming materials in the United States: Bruno Court S.A. of Grasse, France; Haarman & Reimer of Holzminden, Germany; and Christo

Christoff of Kazanlik, Bulgaria, producers of otto of rose. Burns and Walsh handled sales to their major accounts, and they also had an aggressive young salesman working on their staff, Hermann Kohl, who was in time to start his own company, Norda, Inc. Walter Mueller was secretary of Morana, Inc. He later joined Van Dyk & Co. Burton T. Bush also worked for Morana for a time, so there were some top line personnel on the Morana staff.

In spite of what was apparently a big and highly successful operation, the company was apparently not profitable, and the partners had to look outside for working capital. They found their man in Theodore Haebler, who made a substantial investment in Morana, and assigned one of his two sons, William T. Haebler, to keep an eye on the investment. With the infusion of new capital, Morana was able to continue its operations and in September, 1925, moved to their own building at 61-63 Vandam St., New York City.

Still all was not well. Whether the original three partners wanted to get out, or whether the Haebler interest was dissatisfied with the way the partners were operating the business is not clear, but the Haeblers started looking around for new partners. They found a man who was definitely interested, A. L. van Ameringen, who had an excellent background in the industry, and who was preparing to expand his business. Morana ceased operations in 1929, and from the ashes rose van Ameringen-Haebler, Inc., a union of the Haebler holdings in Morana and the business which van Ameringen had been conducting as van Ameringen, Inc.

It was the bad fortune of the original principals in Morana to disappear from the scene just as the aroma chemicals industry was poised to take off on a major period of growth.

Polaks Frutal Works

Polaks Frutal Works was started in Amersfoort, Holland, in 1914 by Jacob and Henri Polak, who had withdrawn from the firm of Polak & Schwarz. Jacob had two sons, Bernard and Frits, and Henri had one son, Ernest. In time Frits and Ernest took over operation of the company.

Eric Vles and John Ter Veer, who joined the company when it was founded in 1914, established a branch office in the United States in 1923 on Washington St. The office was apparently later moved to 180 Lafayette St., for Polaks Frutal was listed at this address in a 1929 trade directory. They moved in 1930 to 350 West 31st St., where the office remained until late in the 30's. Vles and Ter Veer continued to handle the business through those years, with Ter Veer in charge of administration, and Vles responsible for sales.

In 1938 Polaks Frutal started a major expansion of its U.S. operations by occupying a building in Long Island City at 3614 Astoria Blvd. Later, after Polaks Frutal moved out, this building was taken over by Duveen Soap Co. In 1947 Polaks Frutal made the decision to leave the metropolitan area. They moved their manufacturing operations to Middletown, N.Y., but for several years kept the office and laboratory in Long Island City. Finally, in 1950. the entire operation was moved to Middletown.

For over fifty years Polaks Frutal was primarily a flavor house. However about 1969 or 1970 increased emphasis was placed on fragrance development, and now the company's interest is about evenly divided between flavors and fragrances. Dr. James Davis is president of the company, Marvin Preiser is in charge of operations, D. R. Levering of research and development, Dr. Jan Stofberg of standards, John Cook of finance, Maurice Raviol of marketing, and Gerard Goupy, at the Paris office, is director of fragrance creation.

Polaks Frutal Works was acquired by Hercules, Inc. in 1973, and now operates as a subsidiary of Hercules.

International Flavors & Fragrances, Inc.

Here is one of those great success stories that could have happened only in America. A. L. van Ameringen was an immigrant from the Netherlands who was sent to the United States in 1917 by the Dutch firm, Polak & Schwarz, to work for their U.S. agent, Julian W. Lyon & Co. Van Ameringen later opened his own small office (500 sq. ft.) at 13 Gold St., New York City, with two employees. The business grew, and by the mid-20's Mr. van Ameringen was manufacturing aromatic chemicals and flavors at a plant in Orange, N.J. By this time he had moved his New York office to 15 Irving Place.

In 1929, he and Dr. William T. Haebler organized van Ameringen-Haebler, Inc., acquiring the aromatic chemical plant that Morana, Inc. had been operating in Elizabeth, N.J., in which the Haebler family owned an interest. Soon after the new operation was under way, the Morana building on Vandam St., which they had acquired, was sold, and van Ameringen-Haebler offices were moved to 315 4th Ave.

Mr. van Ameringen was president of the new company, Dr. Haebler was treasurer, and Sidney Friend, who had worked for a number of years with Mr. van Ameringen, was secretary. Russell Stoddard, who had previously been sales manager for Ungerer & Co., was engaged to take charge of the Essential Oil Department, and S.A. DeVries was hired to have responsibility for flavor sales. DeVries had formerly been with Polak's Frutal Works, but had operated his own business for about a year prior to joining van Ameringen-Haebler.

The new company grew rapidly, in spite of the problems of operating through the "depression" years. By 1958 they were ready for the big step of going international by merging with the worldwide operations of Polak & Schwarz, the same firm that had sent a young Hollander to the United States forty years earlier to work for their agents. Prior to the consolidation, van Ameringen-Haebler's business had been mainly in the United States, and about 90% of their volume was in fragrances. The Polak & Schwarz operation was international in scope, and P&S did as much business in flavors as in fragrances. Thus, as a result of the merger, the new firm enjoyed a pre-eminent position in the flavor field, and also became a major factor in the international market for both flavors and fragrances.

In 1961, the company made its next important decision — to "go public", with a change in name to International Flavors & Fragrances, Inc. The infusion of new capital made it possible to expand in most any direction management selected, and also provided ample funds for research, as well as for equipping plants and laboratories with the sophisticated apparatus required for production and quality control of aroma chemicals under modern conditions. Sales expanded rapidly, reaching $57 million in 1956, $103 million in 1970, and $275 million by 1976.

In 1956 Mr. van Ameringen became board chairman and was succeeded as president by Charles P. Walker, Jr., who served until April, 1962. Mr. van Ameringen then resumed the presidency, until December 31, 1962, when Henry G. Walter Jr. was named president. Mr. van Ameringen remained as chairman of the board until his death in January, 1966.

Under Mr. Walter's direction, IFF's growth accelerated. The company now employs approximately 3,100 persons worldwide, with about 1,100 of this number in the United States, 600 in Holland, 200 in France, 300 in England, and 900 elsewhere. Sales now break down into approximately 71% fragrances and 29% flavors. In the five years ended December, 1976, IFF's sales of fragrance products accounted for approximately 72%, 71%, 71%, 68% and 71% respectively of total sales, with the balance represented by flavors.

For the year ended December 31, 1976, 35% of IFF's sales were made in North America, 37% in Western Europe, and 28% in the rest of the world. During 1976, the company's twenty largest customers accounted for about 53% of total sales, with the four largest customers accounting for 11%, 8%, 7% and 6%, respectively.

IFF spent a little over $16 million on research and development in 1976, and expected to spend about $19 million in 1977. The company had 527 persons engaged in research in 1976. IFF maintains a research center at Union Beach, N.J. The work of IFF's perfumers and flavorists is conducted there, and at 21 fragrance and flavor laboratories in 20 countries.

Norda, Inc.

This major U.S. manufacturer of aroma chemicals, originally called Norda Essential Oil & Chemical Company, was started by Hermann J. Kohl in 1924. Kohl, who had come to the United States from Germany as a boy, gained his early experience in the perfuming materials industry as an in-house chemist and salesman for Morana, Inc., in the early 20's. Then, deciding to branch out on his own account in the same business, he started Norda Essential Oil & Chemical Co., with three employees: Herbert Schwarz, Nick Mulfese, and George Fortunato, and very limited capital. The firm grew rapidly.

The first office was at 121 East 24th St., in Manhattan, but as the business expanded and more space was required, they moved some time prior to 1936 to 601 West 26th St. The first Norda plant was established in East Hanover, N.J., in 1941. A year later a second plant started operating in Boonton, N.J.

One way in which the new company grew was by buying up competitors. Companies that experienced financial problems often found Dr. Kohl willing to lend them assistance in return for a part interest. Thus, beginning in the 30's, Norda became part owner of such firms as P.R. Dreyer Inc. (acquired in 1933), Orbis Products Co. (1948), Heine & Co. (1955), Schimmel & Co. (1961), and others. In some cases these companies operated in their own offices, apparently still independent companies. In others, they were absorbed at once into the Norda organization: L. A. Van Dyk & Co. (1929), A. A. Stillwell & Co. (1932), James B. Horner, Inc. (1948), R. D. Webb Co. (1961), and F. Huber & Co. (1965). If you wanted to stir up a friend in the essential oil business in those days, you simply asked him if it was true, as you claimed to have heard, that Doc Kohl had bought a piece of his business. Or you suggested that, one of these days, his company might end up too as one more desk in the long row of acquired companies in Norda's big office.

But while he bought up companies, Dr. Kohl also helped to start new ones. He helped, with advice and financial assistance, to start a number of new companies in the fragrance industry which turned out, in time, to become substantial factors in the trade. Some were competitors of Norda's, others were customers or suppliers.

Dr. Kohl also supported and financed numerous base operations for the production of essential oil crops, thus ensuring a continuity of supply of raw materials, not only for his own companies, but for the industry generally.

Norda moved in 1964 to offices at 475 Tenth Ave., New York City. By the time of Dr. Kohl's death, in 1971, the firm had over 500 employees and its operations were international in scope. In 1977, headquarters were moved to East Hanover, N.J., where they are now located in a complex at 140 Route #10. Edward E. Benedict succeeded Dr. Kohl as president of Norda, then in 1973 was named chairman of the board. At that time Louis Amaducci was elected president.

Felton International, Inc.

Felton International Inc. was established in 1922 as Felton Chemical Co. by Dr. and Mrs. Joseph Felton. The original headquarters, at Taffe Place in Brooklyn, consisted of a sales office, laboratory and chemical manufacturing plant. The product line concentrated on aroma chemicals and essential oils. By 1927 these facilities had been outgrown, and a new factory was built in the industrial complex in Ridgewood, Brooklyn.

The first building was a simple two-story structure with manufacturing and warehousing on the first floor and laboratory and sales office above. In the fifty years since then, additional construction has produced a complex of five factory buildings which now contain over 200,000 sq. ft. of laboratory, office and manufacturing space.

Following World War II, Felton began to broaden its horizons. In 1947, the company's first European plant was opened in Versailles, France, followed by factories in Milan, Italy, and Northampton, U.K. In North America, factories were opened in Montreal and then in Toronto,

Canada, as well as in Mexico City. Far Eastern operations were expanded from a factory in Hong Kong to include facilities in Tokyo, Japan, and Sydney, Australia.

Felton now employs more than 450 people in its worldwide operations, and generates in excess of $30 million in total sales. The product line has expanded to include a complete line of flavors and fragrance materials as well as essential oils and aroma chemicals.

Ira B. Kapp, son-in-law of Dr. and Mrs. Felton, is chairman and chief executive officer of the company. Philip Rosner is president and chief operating officer. Mrs. Sophie Felton is treasurer.

Givaudan Corporation

Givaudan Corporation is the United States affiliate of L. Givaudan & Cie. S.A., Vernier, Geneva, Switzerland. The American company has been in operation since 1924 when it acquired Burton T. Bush, Inc., a firm which had represented Givaudan in the United States for several years prior to that time. Earlier, Givaudan, Geneva, had been represented at different times by George Lueders & Co., Morana, Inc. and by Lum & Lemmermeyer, a partnership that operated in the midwest and whose principals later joined Givaudan in marketing positions.

Burton T. Bush, Inc. continued to operate for a number of years as the manufacturing arm of Givaudan. The firm, Givaudan-Delawanna, Inc., which Mr. Bush organized, carried on Givaudan's marketing activities for aroma chemicals, perfume compositions and specialties, while flavor compositions were marketed by Givaudan Flavors, Inc. Industrial odorants, germicides, fungicides and related chemicals were distributed by an affiliated company, Sindar Corporation. In 1965, the separate corporation set-up was ended and all activities were united under Givaudan Corporation.

The first president of Givaudan was Dr. Eric C. Kunz under whose guidance the company broadened its activities and sales approach. One of his early steps was to acquire the essential oil dealing firm of George V. Gross & Co. which was merged into the Givaudan organization in 1930. It was during this time that the slogan "Sell by Smell" was coined and Givaudan began a program aimed at the deodorization and reodorization of industrial products such as textiles, rubber goods, printing inks, plastics, etc. Also during this period, Givaudan's interest in germicidal and fungicidal agents resulted in the development of hexachlorophene (G-11®), a widely used germicide, and dichlorophene (G-4®), a fungicide which was adopted by the military.

In 1953, Dr. Kunz was succeeded as president of Givaudan by Ernest R. Durrer. He was followed by Salvadore Sanz de Acedo in 1967 and by Dr. Francois Petitpierre in 1971. The present chief executive officer is Norman Greif, who took over the post in 1974.

When Givaudan-Delawanna, Inc. was first organized, the Clifton factory consisted of an administration-and-laboratory building and several manufacturing structures, for a total of 20 units. Executive and sales offices were in New York City, and the firm employed about 75 people. Today, Givaudan produces over 500 different perfume and flavor chemicals. The company occupies more than 52 acres of land with over 100 buildings and employs about 750.

In 1971, Givaudan constructed the six-story Xavier Givaudan Building in Clifton, which houses executive, sales and administrative offices as well as the flavor and fragrance creative and applied research laboratories.

Since its inception, Givaudan Corp. has been research oriented and has devoted a great deal of its time to the development of new fragrance and flavor chemicals. The company has been responsible for many important developments in perfume and flavor chemistry, and the numerous patents and scientific publications issued in the Givaudan name attest to some of the major contributions the company has made to the fragrance and flavor industry. An obvious example is the work done in the area of synthetic musk development and production.

Offices of George Lueders & Co. on Pearl St. in lower Manhattan about 1900

Givaudan is widely known throughout the industry for its publication, the "Givaudan Index," Several years ago, control of Givaudan Corp. was acquired by Hoffman-La Roche, Inc.

Monsanto Flavor/Essence, Inc.

The predecessor company to Monsanto Flavor/Essence was George Lueders & Company, Inc., whose original essential oil importing business was started in quarters at 13 Gold St., New York City, in 1885. When Mr. Lueders founded the company, with Henry J. Heister as sales manager and Ferdinand Weber as treasurer, he was a representative of Lautier Fils on assignment in the United States. Six years after the company opened its doors, Edward V. Killeen, a young chap who was later to play an important role in Lueders' growth, joined the company as a stenographer.

From its inception, George Lueders & Company was a manufacturer as well as an importer of basic raw materials for the fragrance and flavor trade. In the early 1900's Lueders represented such European fragrance firms as Givaudan, Batzouroff, and Camilli, Albert & Laloue. They were also one of the few companies distilling essential oils other than mint and citrus oils in the United States. For a time Lueders had a factory in Brooklyn and were one of the first U.S. firms to distill sandalwood oil from logs as well as oil of rosewood. They eventually moved this manufacturing operation to a large building in Freeport, Long Island.

The Lueders business expanded rapidly. In 1887 they moved to 213 Pearl St., then in succeeding years took over several other buildings on Pearl St. until they had perhaps the largest quarters of any U.S. essential oil dealer. When the industry later started to move uptown, Lueders was one of the first firms to leave the old essential oil district, and in 1927 moved office and plant to 427 Washington St.

Mr. Killeen was active in organizing the Essential Oil Dealers Association in 1927, and was the second president of EOA, in 1928-29. Another Lueders executive, Ferdinand Weber, became president of EOA in 1935-36.

Reviewing the history of top management of the company, Mr. Killeen became president in 1930. He later moved up to chairman of the board, holding this post until 1952. He died in 1956. Frederick J. Lueders, son of the founder, succeeded Mr. Killeen as president in 1939, then became chairman of the board in 1958. He retired in 1968, and died in 1976. "Fritz" Lueders worked for the company in both the San Francisco and Chicago branches, acquiring friends all across the country. He was an ardent and expert golfer, able to play the pros without a handicap, and had many tight matches with Johnny Farrell.

Frederick G. Buehler was elected president of Lueders in 1958, and held this post until the company was acquired by Monsanto Company in February, 1968. Dr. Lee A. Miller became president of the reorganized company and continued through the name change in 1971 to Monsanto Flavor/Essence, Inc. In 1974, John C. Livengood, took over. He was succeeded in 1977 by Thomas H. Lafferre.

Lueders' emphasis over recent years has been on compounding rather than the importation of essential oils, and the dealership function, which gave the company its start. When Lueders became a subsidiary of Monsanto, it was able to take advantage of the larger company's strong position in basic research and its ample supply of capital. The new operation combines the creativity of the relatively small, privately-owned flavor and fragrance specialty house with the technological strength of a big chemical company. Its new world headquarters and modern, fully equipped laboratories at Montvale, N.J., are a far cry from the Gold St., Pearl St., or Washington St. offices it occupied earlier.

Louis M. Allstadt, senior vice-president, joined Lueders in 1935. He is the second generation of his family with the company, as his uncle, William Allstadt, was head of the Lueders Flavor Department from 1929 to 1969. Following in the tradition of Lueders executives of taking an active role in the direction of the Essential Oil Association, Louis Allstadt served as EOA president in 1969.

Gerard J. Danco International Corporation

The predecessor company to the present Danco firm was Belgian Trading Co., started by Pierre Danco, a native of Belgium, in 1914. Danco, who had previously operated an essential oil and petroleum oil business in Belgium as far back as 1897, was in the United States when World War I broke out. When Belgium was overrun by the German Army, he elected to stay in the United States. His son, Leon, joined him in the business in 1919, after coming out of the Belgian Army. Then in 1922, a second son, Gerard, joined the firm. Offices were originally located at 44 Whitehall St. The company specialized in essential and flower oils, soluble gums, aroma chemicals, animal fixatives, and certain dyes, and also manufactured aromatic compounds.

Later, Leon left Danco to go with McKesson & Robbins, and Gerry took over the business. He changed the name of the company to Gerard J. Danco Inc. in 1932, then later to Gerard J. Danco International Corp. Offices were moved to several locations around New York City, then in July 1966 to their present location at 17 Elm St., Morristown, N.J.

Gerry Danco was prominent in the activities of the Essential Oil Association of U.S.A. for many years and served as president in 1951. Pierre Danco died in September 1946, and brother Leon, in June 1976. Gerry celebrated his 78th birthday in August 1977, and remains as the active head of his company. The firm specializes in the importation of lavender, lavandin and refined grape seed oil, and also handles sale of a concentrated witch hazel made for them under a patented process.

Rhodia, Inc.

Societe des Usine du Rhone established Rhodia Chemical Co. in the United States in 1915. Rhodia was one of the first American manufacturers of aroma chemicals. The Rhodia office was first located at the site of the present plant in New Brunswick, N.J. Later an office was opened in New York City at 89 Fulton St. General manager of the company was Dr. Meuller, a Swiss organic chemist, who initially supervised the importation of aromatics from

Europe and later their domestic manufacture at the Rhodia plant. They manufactured coumarin, hydroquinone and rhodinol.

In 1929, Societe des Usines de Rhone (predecessors to Rhone Poulenc) became disenchanted with the American market and sold Rhodia Chemical to Newport Industries. About a year later E.I. du Pont acquired Newport Industries and for a time Rhodia became the Aromatics Division of du Pont. Du Pont continued use of the Rhodia name for a time, but then dropped it.

Through the 20's and early 30's Rhodia was represented by a salesman named Charles Kelly, who was very well known in the cosmetic, chemical and perfuming materials industry. The presence of a second Charlie Kelly in the industry posed something of an identity problem. The second Kelly was sales manager for Hagerty Bros., suppliers of glass bottles. To keep the two individuals more readily identifiable, the Rhodia representative of the Kelly clan was generally known, to the initiated, as "Rhodia" Kelly, or "Baldy" Kelly, while the Hagerty Kelly was always "Bottles" Kelly.

The Rhodia name disappeared from the perfuming materials industry for a time, but in the 40's Raymond J. Picard arrived from France and set up Rhodia Inc. He later bought back the Aromatics Division from du Pont and has continued to operate Rhodia Inc. since then. Ed Bush was sales manager of Rhodia in the 50's. He left the company after a time, then returned to work in the Compound Department. He retired about 1972, and is now living on Cape Cod.

Lautier Fils, Inc.

Lautier Fils was incorporated into Rhodia in 1974. Lautier had been represented in the U.S. by George Lueders & Co. from well back in the 19th Century until shortly after World War I. In 1920, Lautier Fils opened their own office in New York on Beekman St. under the direction of Camille Bourguet. The office was later moved to 321 Fifth Avenue, where it remained until 1969. In that year Lautier occupied a plant in Parsippany, N.J. and moved their offices there. The plant was closed down in Sept., 1976 and Lautier is now a division of Rhodia, Inc.

E. M. Laning Company

One small essential oil firm that prospered in the 20's, but failed to last to present times was E. M. Laning Co. The story of how it was started is an interesting one. In 1919, "The American Perfumer" announced that a major U.S. toilet goods manufacturer was offering a $2,000 award to the U.S. perfumer who developed the best original new perfume compound.

Dozens of leading perfumers competed, and the winner was the comparatively obscure Elmer Laning, who was chemist/perfumer for La France Soap and Perfume Company, Long Island City, New York. Instead of taking his family on a vacation, or buying two new cars, Laning promptly used the money to start his own business. He did very well until the grip of the Depression tightened in the middle 30's. Then, in a move to reduce expenses he moved the operation out to quarters adjoining his home in Irvington, N.J. He was operating there as recently as fifteen or twenty years ago, but the business was apparently discontinued sometime in the 50's or early 60's.

Aromatic Products, Inc.

Here was another manufacturer of aroma chemicals that operated successfully for almost thirty years, was for all this period a major factor in the U.S. market, then was absorbed by a larger firm and lost its individual identity. Aromatic Products, Inc. was largely the creation of Michael Lemmermeyer, who resigned as sales manager of Givaudan-Delawanna, Inc. in 1935, to join Arthur W. Mudge and Edwin T. Booth in starting Aromatic Products.

Mudge had been in the sales department of Antoine Chiris Co. in the 20's, then had been called back to France by the parent company and given an opportunity to study operations at the Chiris plant at Grasse. After he left Chiris, he gained additional experience with the aromatic chemical divisions of Newport Chemical Co. and du Pont, before organizing his own company, Organic Products, Inc., about 1933, to manufacture aroma chemicals and import essential oils. In 1935, Lemmermeyer and Booth teamed up with him to start Aromatic Products, Inc. Organic Products, Inc. was later bought by Elan Chemical Co.

The third member of the team, Edwin Booth, was a member of a family of famous perfumers. After gaining early experience with the Lundberg Co., manufacturers of perfumes and cosmetics, he became an officer and director of Pacquin, Inc. in 1920. Then, in 1924, he became perfumer for the newly organized Givaudan-Delawanna, Inc., where he was a close associate of Lemmermeyer's. The two men left Givaudan to start Aromatic Products, Inc.

Charles Swan also joined Aromatic Products, although not as one of the owners. Swan had spent many years with Antoine Chiris Co., and succeeded B.T. Bush as U.S. manager for Chiris when Bush withdrew from the company in 1924.

Aromatic Products, Inc. continued to operate successfully until about 1964, when it was taken over by Stepan Chemical Co.

Trubek Laboratories

Robert Trubek started Trubek Laboratories in East Rutherford, N.J., in 1932. He had previously worked for his father who, under the name, Franco-American, manufactured such products as amyl and ethyl acetate for the lacquer trade at a plant in Carlstadt, N.J. Young Trubek decided to manufacture flavor grades of these chemicals, and early in his career made the important decision to sell exclusively through dealers rather than to consumers.

He hired Dr. Joseph Ziv as chemist. Ziv had received his original training in Europe, then, after coming to the U.S., worked briefly for Felton Chemical Co. Herbert Halpern joined Trubek in 1938, initially as laboratory assistant. He later became plant manager and vice-president.

One of Trubek's first products was amyl acetate. The first batch was made in a gas-fired boiler that had seen service in a tailor shop. The run just filled a 5-gallon jug. Trubek put it in his car and took it out to show samples to a few potential buyers. Percy Magnus of MM&R liked what he saw, and ordered five gallons. Trubek made the delivery, but was out of stock until he could get back to the plant and fire up the boiler again.

Trubek also made vanillin from clove oil. This was a major product for the company until about 1933 when paper companies developed a process to make vanillin from the waste liquors of lignin. This closed Trubek out of the vanillin market, but in a way it may have helped them for it was then the firm started to make a wider range of aroma chemicals.

With the start of World War II, imports of aroma chemicals from Europe were shut off. Essential oil dealers who had previously imported their chemicals had to look for other sources of supply. At the same time U.S. manufacturers of synthetics had to use so much of their output in their own captive operations that they were unable to take care of the needs of dealer customers. The dealers were delighted to find a new supplier in Trubek.

At this time, Trubek started an affiliation with the Dutch firm, Polak & Schwarz, whose owners had been driven out of Holland by the Germans. Trubek acquired the right to use some of the P&S processes, and was soon making a number of the specialty aroma chemicals that had been developed by P&S.

After the war, Trubek went through a major expansion period, buying up and merging several other manufacturers of aroma chemicals. Two firms bought were Harold Simons, Inc., and George Fries, who had affiliated plants in Long Island City where they made benzyl alcohol, benzyl acetate, cinnamic aldehyde and similar products. Like Trubek, they also sold through dealers. During the post-war depression, Simons sold his business to Trubek, which got Trubek started making these additional products. Harry Hachen, plant manager for Simon, joined Trubek.

Another product Trubek started to make at this time was phenyl acetic acid, a precursor for penicillin. This not only moved Trubek into the pharmaceutical field, but also many esters of phenyl acetic acid later proved to be useful products in the aroma chemical field.

Trubek also bought Clemenson Chemical Co., Newark, N.Y., who had been making coumarin. Trubek made this chemical for a time, but when the government placed restrictions on the use of coumarin in foods Trubek discontinued sale for food uses. Trubek then started to test all their aroma chemicals for toxicity, using the Food Materials Laboratory run by Bernard Oser for this work.

During this expansion period, Trubek also bought the business of Truland Chemical, Inc., Union, N.J., the partners in which were Dr. Ed Trueger and Dr. Don Friedland.

Rob Trubek was unquestionably one of the most highly respected and best liked figures in the aroma chemical industry. He was especially esteemed by his employees because of his progressive policies — employee committees were organized to facilitate solution of plant problems, and management committees (sales, research, etc.) were created to widen the leadership of the company beyond Trubek himself.

The company's role of being a supplier to essential oil dealers rather than their competitor built its standing with this segment of the industry. Incidentally, Rob Trubek helped train a number of top people who went on to make major careers for themselves elsewhere in the industry after Trubek Laboratories was sold. Otto Linke, now aroma sales manager for Givaudan, first worked for Trubek, as did Arthur Edwards, vice-president of aroma sales for Fritzsche D&O. Bob Engel worked for Trubek. George Cadgene moved from Givaudan to Trubek, then left to wind up his career with IFF. Mickey Harris worked for Trubek, then later returned to Norda as a production consultant. Herbert Halpern, who spent many years with Trubek, organized Elan Chemical Co., Newark, N.J., after he left Trubek. It was Halpern who gave us much of the history of this unique man and the equally unusual company he built. Hyman Kaplan, who was associated with Halpern in starting Elan, is another Trubek alumnus, having once been chief engineer of the Trubek operation.

It was tragic that Robert Trubek didn't survive to see the company he built go on to bigger things. Instead, he sold the company to Universal Oil Products in 1959, then died shortly thereafter, in 1964.

Firmenich & Co.

The history of Firmenich begins in Geneva, Switzerland, in 1895, when Philippe Chuit and Martin Naef founded the firm of Chuit & Naef. The firm name was changed to M. Naef & Co. in 1910 and continued under this style until 1934 when the company started operating as Firmenich & Cie.

Chuit Naef was represented in the United States for many years by Ungerer & Co. However in 1936 Firmenich established its own branch office in the United States, naming Rupert Watson as managing director. Watson was succeeded in 1947 by Mr. Charles C. Bryan, who took the post of president of the U.S. subsidiary, Firmenich & Co. Upon Mr. Bryan's retirement in 1975, the presidency was assumed by Mr. George A. Trigaux. Mr. Trigaux resigned on June 3, 1977, and Mr. Henri G. van Mameren was appointed as his successor.

The Firmenich offices were located in New York City for many years. In July, 1975, headquarters were moved to Plainsboro (Princeton), N.J. There, administrative offices, fragrance and flavor laboratories are housed in Firmenich's own building.

J. Manheimer, Inc.

This company is an outgrowth of Leo Bernard & Co., which was conducting a business in essential oils and perfume and soap raw materials at 228 Pearl St., New York

City, as early as 1860. Mr. Bernard was a native of France. In addition to selling floral oils made by one of his brothers in a plant at Grasse, he was also agent for such major European fragrance suppliers as G. DeLaire & Co., A. Ducros, D. Pappazoglou, Hugues Aine and Haarmann & Reimer.

A young fellow named Jake Manheimer took a job with Bernard in 1876 as office boy. He was 12 years old at the time, the son of a New York butcher, born on 43rd St., when that was still "way up-town." His salary was $2 a week, and for this lavish sum he worked ten hours a day, six days a week as office boy, delivery boy and general factotum. When he got to be 16 years old, Bernard gave him a chance as a salesman, and Jake was an instant success. Three years later Bernard sent him on a trip to France where he had an opportunity to learn something about essential oil production at the Bernard plant in Grasse.

It wasn't long after he returned to New York that Mr. Bernard told Jake he had decided to close up shop and go back to France. Manheimer rounded up some money and bought the business, including the plant in France. Jake's brother, Isaac, went in with him and they moved their office to 1 Cedar St. The firm continued to operate under the name of Leo Bernard & Co. until at least 1895. Old records indicate that what was apparently an affiliated company, The French & American Trading Co., with Jos. Bernard as manager, was doing business at 58 Pine St. in 1896 and had a warehouse at 28 Gold St.

In 1898 the French & American Trading Co. ceased operating their drug and essential oil department and turned it over to a new firm of Manheimer & Eben, whose office was at 28 Gold St. In 1903 Manheimer & Eben were advertising themselves as a branch of Warrick Freres, Paris and London.

Eben retired from the firm in 1904 and the name was changed to J. Manheimer. The company was incorporated in 1963. Manheimer moved the site of their operations to Long Island City in 1970, consolidating offices, processing department, warehouse and laboratory there. They added a second warehouse in 1974. J. Manheimer, Inc. has continued over the years to be one of the leading dealers in essential oils.

The present head of the company is Stephen Manheimer, a grand nephew of Jake Manheimer, and the son of the late Edwin J. Manheimer. In addition, Paul H. Manheimer remains as chairman, with Bruce, Alan and Arnold Manheimer rounding out the management team. Steve Manheimer served as president of the Essential Oil Association of U.S.A., Inc. in 1973-74.

Seeley & Company

Seeley & Company distributed vanillin, originally supplied to them by Sterwin Chemical Company. During and after World War II, when Sterwin would not supply material needed, the Seeley Company turned to Monsanto Chemical Company as their source. Seeley made flavors, and was one of the first manufacturers of hydroxycitronellal. They also made several other aroma chemicals.

The company was owned and operated by John Beach. Beach had a son who is still active in the fragrance and flavor business in California. He operates under the name John Beach Company.

One of the key men at Seeley & Company was Dr. Arthur Wendt, who is still acting as a consultant in the flavor field. Another Seeley employee was Jack Wennis, who after leaving Seeley finished his career in the flavor and fragrance business by working for Norda Inc. Wennis died a number of years ago.

The Fragrance Industry in the 30's

But, perhaps we should swing back for a moment from the history of specific companies to recall what was happening to the essential oil industry in general as the 20's drew to a close and the country launched into the dreary, depression-hobbled 30's. Business was off, economy was the order of the day, and companies were all scrambling for sales in an effort to stay solvent. It was this atmosphere that spawned two major frauds that cost essential oil importers and producers of aroma chemicals thousands of dollars.

The first of these unfortunate incidents was the Euro-American Corp. raid on gullible credit managers. This concern opened offices at 52 Beekman St. one day in 1929, did some advertising in the trade press, and began to buy and sell all types of aromatic products. At first Euro-American met their commitments promptly, their suppliers were lulled into a feeling of false security, happy to find an active new account. Once a credit line had been established, they began to buy more heavily but slowed down on payment. It wasn't long before payments stopped completely, and when the company shortly went broke, there were very few assets.

The stocks of aromatics they had bought had been sold as fast as they were delivered, often at cut prices to move them rapidly. The principals vanished, leaving a figurehead holding the bag. All that was left were desks, telephones, and an empty warehouse. The suppliers, including a number of essential oil houses, suffered substantial losses. Euro-American Corp. entered a voluntary petition in bankruptcy in June, 1931, reporting 141 unsecured claims aggregating $103 thousand. In August, 1932, twelve of the principals were indicted by a Federal Grand jury on charges of concealing $75 thousand in assets. ("Oil, Paint & Drug Reporter," June 22, 1931, pg. 52, Aug. 8, 1932, pg. 13).

Six years later, another much larger fraud came to a head. Three brothers, whose real name was Musica, but who had been masquerading under other names, gained control of the respectable drug firm of McKesson & Robbins. Philip Musica, representing himself as Donald Coster, succeeded in having himself elected president of McK & R. Another brother was purchasing agent, and a third, Arthur Musica, who had adopted the name of George Vernard, was head of the company's operations in Canada.

In addition to handling crude drugs and pharmaceuticals, the company, under Coster, opened an essential oil department, and sought in 1929 to join the Essential Oil Dealers Association. However by the time the fraudulent operations were exposed, in 1938, McKesson & Robbins were no longer dealing in essential oils, as they had sold their essential oil department to Dodge & Olcott in the mid-30's.

The Musicas operated with two sets of books, and succeeded in deceiving a highly respected firm of auditors for several years. Many of their "assets", represented by warehouse receipts for non-existent stocks, existed only as ledger entries. The physical inventories conducted by the auditors must have been something less than adequate. The principals in the fraud were indicted by a Grand Jury, and as the authorities began to close in, "Donald Coster" committed suicide. When the smoke cleared, it became apparent that most of the firm's assets had vanished, leaving claims of about $3 million. ("Oil, Paint & Drug Reporter," Dec. 18, 1938, pg. 5).

McKesson & Robbins had a firm enough base and a good enough reputation so they survived the episode. Under new management they won back their respected place in the wholesale drug business.

Perfuming Materials and Prohibition

During the early 30's, Prohibition was still the law of the land. Because of their heavy legitimate use of ethyl alcohol, suppliers of perfuming compounds, as well as manufacturers of perfumes and cosmetics became the target of bootleggers looking for the alcohol they needed to make gin. For those who do not go far enough back to recall the days when the 18th Amendment was still part of our Constitution, perhaps a note of explanation is in order.

Making gin was no trick at all, if you had a source of potable alcohol. You could make your own gin, and many did, right in the comfort of their own homes in any handy vessel of adequate size. This gave birth to the term "bath tub gin". You mixed one part of alky with two of distilled water, added a little glycerin to smooth out the mixture, and a few drops of gin essence, which consisted for the most part of oil of juniper. (One essential oil house did its part to ease the problems of the thirsty by selling small bottles of "gin essence.") After mixing with the alcohol and aging for about three minutes, the resulting concoction made a delightful mix for the manufacture of a Tom Collins, Gin Sling, Orange Blossom, or you name it. If you wanted a more potent beverage, you simply adjusted the proportions a bit — two parts of alky to three of water.

The bootleggers learned early in the game that they could buy alcohol-based preparations, give them a quick pass through a still, and end up with a drinkable alcohol. It was not at all amazing, then, that in the closing days of the Prohibition era, alcohol-based theater sprays became big sellers. There had been a small market for water-based theater sprays for several years, but suddenly, in the late 20's, a heavy demand developed for alcohol-based sprays. Sales zoomed, although many of the buyers had no theaters to spray. Somewhat the same thing happened with colognes. A number of new "barber shop supply houses" suddenly materialized and began to buy cologne in multiple drum lots, obviously because they were planning to cook off the denaturant and recover the alcohol.

It didn't take long for the Volstead Act sleuths to become suspicious. (The Volstead Act was the act passed by Congress to enforce the 18th Amendment. It forbade the manufacture, or transportation, of alcoholic beverages, but not their consumption). In 1930, Federal authorities began refusing to approve formulas for many toilet preparations on the grounds that "on simple distillation and manipulation they yield potable alcohol". They were so right, but a little late in trying to lock the barn door.

Later that same year, in a last final gasp at enforcing the Volstead Act, the Feds issued a long series of indictments in which they accused several alcohol manufacturers, a dozen or so essential oil houses, and anyone else they could think of, of conspiring to violate the Volstead Act. These blanket indictments accused them of "knowingly selling their products to firms and individuals who have in turn used them in some manner contrary to the law".

In this ridiculous action the government was trying to make the sellers of legitimate merchandise equally guilty with those who might use it for illegal purposes. The indictments were a one-day sensation in the newspapers, and members of the Women's Christian Temperance Union were easily persuaded that the Hoover administration was making a valiant effort to enforce Prohibition. Then, all was forgotten and the whole thing vanished down the drain, that is it did for all but one luckless seller of theater spray who had made the mistake of pleading guilty.

The Perfumers' Dinner

The big social event of the year in the perfuming materials business was always the "Perfumers' Dinner". Although the Manufacturing Perfumers Association had changed its name to the American Manufacturers of Toilet Articles in 1918, the association banquet continued for years to be called the "Perfumers' Dinner". Perhaps the high point in these parties came along in the early 30's, when several of the dinners were held aboard the French liner *Ile de France.* We are indebted to Grant Dorland of "Soap/ Cosmetics/Chemical Specialties" for a first hand description of the 1930 banquet.

"In those days when essential oil dealers traveled to France regularly to visit with their principals, and when the French and Swiss aromatic chemical and essential oil producers came to the United States to visit their representatives, it was considered obligatory to travel on the French Lines' *Ile de France.* In 1930 the AMTA had their annual banquet aboard this great ocean liner, which was docked at a Hudson River pier across town from the convention hotel. It was my good fortune to have been invited to sit at the van Ameringen-Haebler table with Van, Bill and their wives. Among other guests were Dave and Nedra Bennett of Albert Verley & Co., Chicago, mutual friends, who helped make the party a perfect evening.

"After the cocktail parties at the hotel, conventioneers were bussed across town to the beautiful *Ile de France.* As I stood at the top of the wide staircase, leading down to the lavishly decorated dining saloon, it was the most magnificent sight I had ever seen. The food was something I had never been exposed to anywhere, let alone at an annual dinner where you normally expect cold chicken or tough beef. The evening was made memorable by the dance music of the 'Ipana Troubadors', radio's leading orchestra at the time. There was also a plentiful supply of authentic liquid refreshments, which was a unique experience to a group of Americans who had lived for more than twelve years under the yoke of Prohibition, and whose palates had been conditioned to look forward to nothing better than muscatel wine, bath tub gin or homemade whisky."

The Industry Moves Uptown

During the late 20's and early 30's the essential oil industry was starting to move away from Cliff, Pearl, Gold, Platt, John and Fulton Streets. Ungerer & Co. moved up to 124 West 19th St., as early as 1917. A. L. van Ameringen moved to Irving Place, just above 14th., in the mid-20's. George Lueders & Co. moved to 427 Washington St. in 1927. Pierre Lemoine moved from Pearl St. up to 200 Varick St. about 1927-28. Dodge & Olcott moved in 1930 from 87 Fulton St. to 180 Varick St. After fire destroyed their offices at 78 Beekman St. in 1935, Fritzsche Brothers moved to the New York Port of Authority Building at 76 Ninth Ave. In 1938, after the death of P. R. Dreyer, his firm, of which Fred Theile had become president, moved up to 12 East 12th St., merging with Charles Fischbeck & Co. Magnus, Mabee & Reynard moved to 16 Desbrosses St. in 1938. Orbis Products moved from Pearl St. to 76 Ninth Ave. Roure Bertrand-Justin Dupont, who had originally been located at 18 Cedar St., moved uptown to 461 4th Ave., where they shared quarters with their U.S. agent, George Silver Import Co.

In addition to firms moving out of the lower Manhattan "aromatic circle", many firms were also going out of business in the depression-laden days of the 30's. The list of casualties included Pierre Lemoine, Imperial Export Co., Morana, Pfaltz & Bauer, George V. Gross, James Horner, United Laboratories, Elson & Brewer, O. A. Brown, A. A. Stillwell, L. A. Van Dyk, Vietor & Hosken, Coignet, Coupey Fils, and Delphi Products, to name just a few.

Some were absorbed; others just faded away. As a result, it wasn't long before the half/mile circle around the corner of John and Cliff streets no longer marked the geographical center of the American essential oil industry.

In the general exodus from lower Manhattan, many of the firms settled in the area around 14th St. As they moved, they necessarily found new eating places. The top brass at van Ameringen usually ate most of their lunches at "Luchow's". The Fritzsche staff discovered an excellent Italian restaurant called "Casa Johnny" on West 15th St. Will Ungerer haunted "The Little Venice" on West 12th St., and his Rolls-Royce was often parked in front of this establishment from noon until well into the evening. If you really wanted to find Will, it was often a good idea to look for him there, rather than to seek him at his office.

So many essential oil people frequented "Felix's" bistro, near the corner of 13th St. and 7th Ave., that the owner had a mural painted on the wall in which several of them are immortalized, including Al. Pillet, chief perfumer for DuPont's aromatics department. Rupe Watson, who ran the Firmenich office in the 40's, spent considerable time at the Gramercy Park Hotel, and gave a big birthday party there every year. Another place where you could usually find a group of essential oil people lunching was Charles French Restaurant, on 6th Ave., around 10th St. It was a favorite spot for Rob Trubek, the Magnus brothers and others.

Philip Chaleyer introduced the writer to what later became a favorite retreat in the upper eastside fifties. Chaleyer had been a perfumer with Givaudan in Europe. He came to the United States in the late 20's, and soon gained prominence by winning a competition for developing a fragrance for a new perfume that McKesson & Robbins was introducing. Chaleyer detested the American idea of "grabbing a sandwich", and insisted that the only civilized way to eat lunch was to dine quietly, at a deliberate pace, in a fine restaurant, accompanying the meal with a bottle of acceptable wine. He took me to the Hapsburg House, which featured excellent Viennese cooking.

Someone told me later that this agreeable place had been the private speakeasy of Walter Chrysler during Prohibition days. Its walls were decorated with original drawings by Ludwig Bemelmans, unmovable and irreplaceable. Unfortunately, like so many other attractive places, the Hapsburg House too succumbed to progress and has disappeared.

With this introduction of personalities into our historical study of the essential oil industry, perhaps it might be a good time to continue along this line, describing some of the individuals who gave the industry such a stimulating, engaging character back in the 20's and 30's.

A. L. van Ameringen

Certainly the first "character" we should describe is Arnold van Ameringen. Someone may once have called him "Arnold", but to his hundreds of friends he was always "Van". Occasionally "Mr. van American" to those who stumbled over his Dutch name, or often "All American Van" to his sparring partners around the card table, where he was a consistent and always agreeable contributor.

Van was an extremely friendly and cheerful individual, who had a fine sense of humor and enjoyed a joke on himself to a refreshing degree. At the same time he had the stubbornness and determination universally attributed to the Dutch, and when there was an actual or a potential customer in the vicinity, Van was all business.

Starting out as a small importer of essential oils from Holland, van Ameringen had the drive, the vision and the ability to combine his talents with those of Bill Haebler and his other associates, and eventually succeeded in building International Flavors & Fragrances, Inc., the largest and most profitable firm in the U.S. manufacturing flavors, fragrances and aroma chemicals. He derived considerable satisfaction in later years out of the fact that he eventually bought the company, Polak & Schwarz, which had sent him to the U.S. to represent them, then changed their minds and fired him.

But business, and his instinct for going after and getting the big accounts through intelligent employment of his personal charm, as well as his business acumen, was only one side of the man. He was also an avid golfer, an amateur magician, an eager card player, a gourmet, and an enthusiastic teacher. In 1947, Mr. van Ameringen was invited by Columbia University to conduct a series of seminars on perfumery, which were highly successful. Incidentally, Van was a pretty fair perfumer himself.

There was still another side to this man of which not even all of his friends were aware. Mr. van Ameringen was a philanthropist who aided Fountain House in New York City with generous gifts. This institution helps the mentally ill to bridge the gap between confinement and their eventual return to life in the community. He made a similar contribution to the cause of mental health in Holland by funding the Pandora Foundation.

Mike Lemmermeyer

The present dean of the essential oil and aroma chemical industry has to be Michael Lemmermeyer, who first entered the business as an office boy for Geo. Lueders & Co. back in 1907, and who is now living in retirement in Sarasota, Fla. at the age of 86. Mike recalls that his starting salary at Lueders was the princely sum of six dollars a week. He held various clerical positions in the New York office, then went to France with the 77th Division during World War I. He served with the 304th Field Artillery, and took part in a number of major actions. Because of his drawing ability, he was made a scout.

Mike returned to Lueders after the war, and in 1921 was made manager of the Lueders Chicago office. There he worked closely with Dudley Lum, a Lueders salesman for the mid-west. At the time Lueders represented Givaudan & Cie, but in 1924 they lost the account. Mike and Dudley Lum could see an inevitable drop in commissions, and decided to go after the Givaudan account themselves. When they were successful in persuading the new agent, Burton T. Bush, Inc., to let them handle mid-west sales, they organized their own company, Lum & Lemmermeyer.

This firm operated successfully for three years, then was taken over by Givaudan-Delawanna, Inc. Lum was appointed Chicago manager for G-D, and Lemmermeyer moved to the New York office as sales manager. After holding this position for eleven years, he resigned in 1935 to form Aromatic Products, Inc. along with Arthur Mudge and Edwin Booth.

Almost thirty years later, when Aromatic Products, Inc. was taken over by Stepan Chemical Co., Mike became secretary-treasurer of the Aromatic Products Division of Stepan Chemical. He retired in 1967.

During his active business life, Mike was a long time member of the Essential Oil Association, and served for many years on EOA's Aromatic Chemical Committee. He attended most of the meetings of the major trade associations in his industry, including CSMA and TGA. He served on the CSMA board of governors, 1959-1961. Mike was very popular personally, a tough opponent around the poker table, and his considerable talent as a painter made him known to many people in his industry. For over fifteen years he regularly painted a portrait of each retiring president of CSMA.

Dr. Joseph Felton

The founder of Felton International Inc., Dr. Joseph Felton, recently celebrated his 86th birthday. After receiving his Ph.D. from the University of Breslau, where he studied chemistry under Professor Semmler, he did extensive post-doctorate work with Professor Dutoit at Lausanne University, Lausanne, Switzerland.

Dr. Felton came to the United States in 1919. After working in the industry for three years, at least for part of this time with Van Dyk & Co., he founded his own company in 1922. As a research chemist he developed manufacturing processes for numerous aromatic chemicals which were then being imported from Europe. Aided by his wife, Sophie, herself a chemist, Dr. Felton built Felton International into one of the leading aromatic chemical companies in the world. The Feltons are now living in retirement at Cokesbury, in Hunterdon County. N.J.

Burton T. Bush

Certainly one of the most dynamic and energetic figures in the U.S. essential oil business in the early years was Burton T. Bush, who first came on the local perfuming materials scene as the general manager of Antoine Chiris Co. He had sufficient vision to see that the big profits in the aromatic business were going to be made by manufacturing one's own products in the United States, rather than by just selling imported materials as the agent for a European producer.

Thus he interested Georges Chiris in building a manufacturing plant at Delawanna, N.J., and as described earlier in this chapter, ended up by owning it himself. When he started Burton T. Bush, Inc. and became U.S. representative for Givaudan & Cie., Geneva, in 1924, Bush sold a one-third interest in the plant to them. This arrangement did not last for long. Givaudan-Delawanna, Inc. was organized as a sales company, which Bush headed. Again, this arrangement proved to be only temporary, for in 1926 Givaudan parted company with Bush. Bush resigned as vice-president of Givaudan-Delawanna, Dr. Eric Kunz took over as president, and Givaudan-Delawanna bought out Mr. Bush's interest in the plant.

"B.T.'s" next move took him out of the fragrance field into publishing. He became editor of a new trade magazine, "The Perfumers Journal", published by Trade Publications, Inc. Again the timing for this project was bad. When the "Depression" hit in 1929, magazine advertising slumped. Mr. Bush moved on and shortly thereafter the publishing firm went bankrupt. Most of its properties, including the perfume magazine, were sold to Hearst Publications who renamed the magazine "Aromatics" and tried for a couple more years to put it over. However, even with all the backing of the Hearst organization, with Phil Van Italie as editor and Ed Flynn as advertising manager, the magazine couldn't get off the ground. Lou Levy and

"The American Perfumer" proved too tough a nut to crack.

Next, Mr. Bush was instrumental in persuading Naugatuck Chemical Co., a division of U.S. Rubber Co., to enter the aroma chemical business. Naugatuck was making rubber chemicals, one of which was an accelerator made by cracking castor oil. It proved to be a useful intermediate for the manufacture of amyl cinnamic aldehyde, an important perfuming material used in substantial volume by soap manufacturers. Bush sold Naugatuck on the idea of manufacturing amyl cinnamic aldehyde, and they became the first U.S. manufacturer of this product.

This single product was apparently not sufficient to stimulate any permanent interest in the aromatic chemical business on their part, and they soon left manufacture of the product to other firms which made full lines of aromatic chemicals. Van Ameringen-Haebler Inc. was one company which learned about the process and began to use it. Their first production of the item, a fairly sizeable quantity to be supplied to Lever Brothers, was produced, we were told, in 36 liter flasks, an exceptionally small batch even in an industry where small batch production is commonplace.

Bush joined Newport Chemical Works in March, 1930. Newport had acquired Rhodia Chemical Co. and was U.S. agent for Societe des Usines Chimiques Rhone-Poulenc. However, again the timing was bad, because in August, 1931, Newport was taken over by E. I. du Pont, and Mr. Bush did not fit into du Pont's plans. Thus we find him that year organizing still another firm, Bushfield, Inc., 67 Lexington Ave., New York. Bushfield failed, and next "B.T." tied on with Albert Verley in Chicago.

In 1938 he started Bush Aromatics, Inc. and Bush Pan America, Ltd. These concerns were quite successful during World War II, importing essential oils produced in South and Central America. His two sons joined him in the business, Ed in 1941, and Tom after he came out of the U.S. Navy during World War II.

Bush Aromatics was taken over by Dow Chemical Co. in 1951. Dow had been producing coumarin, methyl salicylate and phenyl ethyl alcohol for some time, and sought to expand sale of these products as well as the aromatic specialties developed by the Bush organization. The venture failed to "take off" and Dow lost interest. Ed left to take a job with Rhodia. Tom worked a while longer in Midland, Mich., then in 1952 moved to Australia. He is still there, teaching business administration. Dow later sold the aromatic division. B. T. Bush retired in 1955, and lived with his son, Ed, in Boonton, N.J. until his death in July, 1975, at the age of 94.

In his series of books on "The American Chemical Industry", Williams Haynes makes highly complimentary reference to Burton T. Bush, describing him as "a man of enormous energy and enthusiasm". Certainly he must have been. At various stages of his career he was right in there scheming and planning with the top figures in the perfuming materials industry. But, somehow when the time came to pick up the chips, "B.T." was always on his way to some other arena. He didn't stay put long enough to be around for the payoff.

"Uncle Jake" Manheimer

The man who spent the most years in the business and who was widely and affectionately known wherever essential oils were traded, was "Uncle Jake" Manheimer. Jake started as office boy for the firm of Leo Bernard & Co. back in 1876 at the tender age of 12.

In the course of his long career, Jake made literally hundreds of trips to all parts of the world. He travelled on each ocean and to every continent to visit the principals who produced the perfuming materials he sold. He had one of the most dependable "noses" in the business for assessing quality and character of perfume raw materials, and was the final judge in many differences of opinion as to whether or not some lot of musk, civet or ambergris was or was not the true, legitimate article. Bill Schilling of Norda was quoted as saying "Maybe Uncle Jake doesn't know one specific gravity from another, but I'll put my money on any nose test he makes".

In 1946, when a large group of his friends got together to help him celebrate his 70th anniversary in business, Jake gave a short talk on the factors to which he attributed his success. His guiding principle, he said, had always been to "play it straight", never taking advantage of some threatened emergency to "load up" a buyer. Little did those friends think at the time that he would outlive many of them. He did, remaining active in his beloved essential oil business up until his death in 1964.

The Magnus Brothers

There were a number of brother combinations in the essential oil business, starting with the three Magnus brothers, Percy, Jr., Joe and Bob. Percy, the eldest, took over the presidency of Magnus, Mabee & Reynard, Inc. when he was only 22 years old, after his father died in 1916. Percy Jr. was one of the organizers of the Essential Oil Dealers Association, and a prominent figure in the industry for many years.

He later graduated to larger fields, and was president of the New York Board of Trade and became a director of several banks. He eventually went into the political arena, and at one point was prominently mentioned as a possible "business man" candidate for the post of Mayor of New York City. This venture into politics was probably a mistake, for in the process "P.C." attracted too much attention and made powerful political enemies among members of the liberal contingent around President Franklin Roosevelt. Percy is now (1977) 84 years old, and lives in New York City.

Second brother Joe was in the U.S. Navy during World War I. When he rejoined MM&R after being discharged, he took over responsibility for running the Flavor Department. He served successively as treasurer, vice-president and finally executive vice-president of the company. When the Essential Oil Dealers Association was organized, he was MM&R's representative at the meetings for years, and served as the second secretary of the EOA. Joe also regularly attended the annual meetings of the CSMA, among whose members Magnus had many customers. Joe is now 81 years old (1977) and lives with his wife, Norma, at Putnam Park, Greenwich, Conn. Much of his spare time is given to his activity with New York State's Veteran Corps of Artillery, of which he was named commandant in 1959. He has also held office in a number of patriotic societies.

Youngest of the three Magnus brothers, Robert, was born in 1900. At MM&R he had charge of the Pharmaceutical Department. He followed Joe as perennial secretary of the EOA. Bob was the youngest of the brothers, but the first to die. He passed away in February, 1976. He was the only one of the Magnus brothers to have a son to enter the industry. Robert Jr. is sales manager of Norda, Inc.

The Ungerer Brothers

Then there were the Ungerer brothers, William G. and Frederick H. Seldom were two brothers less alike. Born in Rochester, N.Y., in 1868, Will was educated in Paris and was said to have been the first American permitted to serve as an apprentice in a French perfuming materials plant. It was there that he got his excellent background in the essential oils business. It was not long before he succeeded his father as chief perfumer for Colgate & Co.

In 1893, he left Colgate to join his father who had started Ungerer & Co. two years earlier. Having progressive ideas, Will made the company's price list into an informative publication which he called "Ungerer's Bulletin" and which in time became "The American Perfumer".

Will was as colorful and flamboyant as brother Fred was steady and reliable. Having spent considerable time in France, Will had understandably developed a strong pro-French bias. During World War I he helped raise funds, and made a substantial contribution himself, to supplying ambulance units to the American Ambulance Field Service for use on the French battlefields. He also collected funds for the relief of a colony of French war orphans at Grasse.

After the war he was honored by being elected a Chevalier of the French Legion of Honor.

A ready clue to Will's personality is apparent in his reaction to a divorce in which he figured in the year 1922, or thereabouts. He and his young wife had parted after a relatively short marriage, and Will probably ended up with a substantial commitment for alimony. He gave a big party to celebrate the event, and published a book titled "What I Know About Women". It proved, when inspected, to consist simply of blank pages. Ira MacNair, in covering the story for "Drug & Chemical Markets", of which he was then editor, wrote a solemn, tongue-in-cheek review in which he complimented Mr. Ungerer on his extensive knowledge of famous women through the ages, from Eve and Cleopatra down to Mary Baker Eddy and Carrie Nation. Will got a bigger kick out of the review than he had out of publishing the book.

Fred was born in Jersey City in 1878. He went to Stevens Institute and Packard Business College, then joined Ungerer & Co. in 1896. He was made secretary-treasurer when the firm was incorporated in 1901. When Will died in 1930, Fred succeeded him as president.

Fred lived originally in Westfield, N.J., but later moved to Essex Fells. He headed Ungerer & Co. until his death in 1966. His daughter, the late Adele Ungerer, became the wife of Kenneth Voorhees in 1931, and Ken continued the family identification with Ungerer & Co. He was made general manager in 1936 and president in 1949.

The Voorhees had four children, including twin daughters. Their second child, a son, Gary, now 37 years old, joined Ungerer & Co. in 1957. He served as vice-president in charge of sales until early 1977, when he succeeded his father as president of the company, the fourth generation in the Ungerer family to take an active role in the essential oil business.

Charles Fischbeck

While reviewing the history of the Ungerers, we should pause to trace the record of a man who worked the first 25 years of his life for Ungerer & Co., then left to launch an independent career in the essential oil business. Charles Fischbeck was born in 1892 in New York City. He joined Ungerer & Co. in 1910. A junior employee, as Charlie was at the time, didn't figure to have much contact with the boss, but Charlie remembers very well that, in 1912, Will Ungerer gave him two tickets to the annual dinner of the Manufacturing Perfumers Association being held at the Astor Hotel in New York City. Charlie took his future wife, Viola, to the party.

Fischbeck was in the trading department at Ungerer, and had become executive vice-president by the time he left, in 1935. Teaming up with Wallace Bush, Charlie started his own company. The Charles Fischbeck Company operated for only three years, then was consolidated with P.R. Dreyer Inc. in 1938. Mr. Fischbeck was a vice-president of P.R. Dreyer Inc. for six years, and while with Dreyer served (1940-1941) as president of The Essential Oil Association. He left Dreyer in 1944 to become vice-president and manager of the Essential Oil Department of George Uhe Co. He stayed with Uhe until 1957, when he retired at age 65.

Charlie now lives, with Viola, in Short Hills, N.J., and is 84 years old. Their son, Lloyd, worked with his father at Charles Fischbeck Co., moved with him to Dreyer, and then after serving in the U.S. Army, joined George Uhe Co., where he is now a vice-president.

Peter R. Dreyer

Mention of the firm of P.R. Dreyer Inc. takes us back to its founder, Peter Dreyer, a kindly native of Alsace-Lorraine with a twinkling smile, a "Foxy Grandpa" type beard, and the general mien of Santa Claus. After selling for Ungerer for a time, Dreyer became manager of the essential oil department at Rockhill & Vietor. When this concern was dissolved, about 1922, Pete Dreyer continued the business under his own name. His assistant was Fred Theile, whom Mr. Dreyer always addressed as "Teel".

Fred Theile

Fred Theile, Sr., a graduate of the New Jersey College of Pharmacy (later part of Rutgers University), served his apprenticeship in the essential oil business with the National Aniline Company; then for a time was with Charles V. Sparhawk Company. He moved from Sparhawk to P.R. Dreyer when Pete Dreyer started his own firm about 1922, and was vice-president and Mr. Dreyer's assistant for many years.

Theile's keen nose and warm personality made him well known and liked in the fragrance industry across the country. While he was with Dreyer, Fred for a time had a small business of his own creating and manufacturing finished toiletries. Operating as the Vargish Company, he marketed a line of perfumes and cosmetics. He later took a partner, Charles Friche, and they branched out into the manufacture of flavors. In time Friche bought out Theile and later sold the business to Polak & Schwarz (U.S.) which was still later merged with van Ameringen-Haebler, Inc.

At Mr. Dreyer's death, Fred Theile succeeded him as president of P.R. Dreyer Inc. and directed the company until his untimely death in 1944.

Fred's son, also Fred C. Theile, followed his father into the essential oil business. Born in 1922, he worked summers for Dreyer during the late 30's; then went with the company full time after graduation from Rutgers and a stint in the U.S. Navy during World War II.

When Norda took over Dreyer, Fred was transferred to another Norda-controlled firm, Orbis Products, where he was plant manager. In 1951 he joined Shulton, Inc. as production manager of their Fine Chemicals Division. He advanced through several positions including research administrator, assistant to the president, and marketing manager for Shulton's Jacqueline Cochran Division. Then in 1970 Fred was named executive vice-president and general manager of Lautier Fils, Inc., which later became the Fragrances Division of Rhodia Inc. He left Rhodia in December, 1976, and was until recently in charge of the Fragrance Division of Dragoco, Inc.

Fred's son, a third Fred Theile, born in 1950, also stayed close to the fragrance field. He was with Revlon and now works with Avon Products Inc. in cosmetic marketing.

Albert Delavigne

As might have been anticipated, there were many individuals of French origin in the U.S. essential oil importing business, some of them having specific backgrounds in growing or processing of flower crops, and others selected largely because of their ability to handle French correspondence, or their familiarity with French business methods and French ways of thinking. Common in the industry were such names as Couderchet, Pillet, Bourguet, Chaleyer, Coutin, Rapin, Delavigne, et al.

Let's take Albert Delavigne as a typical case. President of Roure-Dupont, Inc., he was born in France, and educated at St. Cyr, the West Point of the French Cavalry. In World War I, Delavigne was a Captain in the French Aviation arm, and was awarded the Croix de Guerre and the Legion of Honor. He came to the United States in 1925 on the staff of George Silver Import Co., who at that time were U.S. agents for Roure Bertrand Fils-Justin Dupont.

I remember being impressed, when I first called on Mons. Delavigne, by his casual attitude toward business and his determination to avoid completely the stereotype of the typical American businessman. Mr. Delavigne told me that he had been reluctant to come to the United States, since he was not interested in filling any position that demanded a great deal of work. He had finally agreed to undertake management of the Roure-Dupont interests in New York City, he insisted, only after his French principals assured him that there was really not much work involved in the job, and specifically that there was very little mail to be answered.

I was trying to sell him some advertising space, but he successfully shied away from giving me an order with the explanation that the last thing he wanted to do was to encourage inquiries from prospective customers that he would just have to answer. I realize now that he must have been pulling my leg a bit, but at the time his pose shook me up considerably, and must have undermined my assurance for weeks.

Dr. Ernest Guenther

Dr. Ernest Guenther, recently retired executive vice-president and director of Fritzsche Dodge & Olcott, Inc., was born in Munich, Germany, May 22, 1895. He attended the Technische Hochschule and the University of Munich, and served in World War I as a volunteer in a Bavarian cavalry regiment. He received his Ph.D. in chemistry from the University of Zurich in 1920.

After graduation, he became assistant to Prof. Albert Hesse, leading world authority on natural flower oils. After the latter's death, Dr. Guenther joined a soap, cosmetic and perfume house in Berlin and was soon traveling through Europe setting up manufacturing branches for the firm.

In 1924 Dr. Guenther emigrated to the United States and joined Fritzsche Brothers, Inc., New York. His background made him well suited to carry out on-the-spot investigations into sources of essential oils, which assisted Fritzsche in establishing standards of quality. In the late 1920's Dr. Guenther began his worldwide travels, investigating the production of essential oils and securing samples of absolute purity. By the outbreak of World War II he had visited every essential oil producing region in the world. During World War II he made many trips to Central and South America to aid in developing new sources of essential oils to replace those on which supplies had been shut off by the war.

Over the years Dr. Guenther has published more than 150 articles reporting on his work. Motion pictures he took were developed into six full length films: Essential Oils of the Mediterranean; South America; North and Central America; Africa; the Far East; and the West Indies.

Dr. Guenther's greatest contribution to the essential oil industry is his six volume treatise, "The Essential Oils," which fifteen years after publication of the 6th volume is still accepted world-wide as the definitive work on essential oils.

Dr. Hermann J. Kohl

Dr. Kohl, who founded Norda Essential Oil & Chemical Co., was without doubt one of the most interesting and complex characters in the essential oil business. A psychiatrist would have had a field day in psychoanalyzing him. He was the typical American prototype of the first generation immigrant who made good — the successful, self-made business man who built a company through his own single-handed efforts, and in which he continued to hold all the strings of authority in his own hands. Yet, in his compassion and his eagerness to help others, and in the wide variety of his interests, he was a lot more complicated a man than the foregoing would suggest. Before he died, Dr. Kohl became a skilled chemist, an agronomist, a metaphysicist, an ecologist and a scholar.

Kohl came to the United States from Germany in 1902 at age 15. His first job in the U.S. was with Liggett Drug Stores, and he later worked for Morana, Inc., essential oil dealers and manufacturers of aroma chemicals. Dr. Kohl married Hertha Trapp in 1913. He didn't have time, nor the money either, for a college education as a young man, but he was an avid reader, traveled widely, had an amazing memory, boundless energy and ambition, and in time educated himself in the sciences, economics, philosophy, history, the classics, art and religion. Many years later he received an honorary doctorate in recognition of his work in the field of food chemistry.

After serving an apprenticeship in the perfuming materials business as a chemist-salesman for Morana, Dr. Kohl was determined to stay in the fragrance/flavor industry. So he started his own company, Norda Essential Oil & Chemical Co., in 1924. The company prospered and expanded. By the time Dr. Kohl died, Norda was a world-wide factor in its industry and employed over 500 people.

Dr. Kohl served not only as chief executive of the company, but as sales manager as well. He was aware of essentially everything that went on in the company. His capacity for work was legendary. His top executives had to get accustomed to doing as much work week-ends as on normal work days. Dr. Kohl took a great deal of pride in the management and production teams he selected and trained to carry out the policies he had shaped. He was vitally concerned with every phase of Norda operation and growth until his final illness and death in 1971.

This unusual man demanded tremendous devotion and loyalty from his people, and extended the same measure to them. He and his wife were childless, but they adopted a foster daughter, Elena Duke Benedict (nee Amaducci). Elena's husband, Edward D. Benedict, who went to work for Norda in 1939, has been a worthy successor to Dr. Kohl as the head of the company. He is now chairman, while Louis Amaducci, Elena's brother, has been president of Norda since 1973.

After Dr. Kohl's death, the Benedict family established "The Benedict Foundation in Memory of Hermann J. Kohl" to assist educational and charitable organizations with which he had been concerned throughout his career.

Charles Sparhawk

An unusual character in the essential oil business I remember, pungently, after all the intervening years, is Charles Sparhawk. I don't know just how early Charlie started dealing in essential oils, but in 1924 he had an office down in the heart of the essential oil district at 278 Pearl St. At one time Fred Theile, Sr. worked for him, before moving on to Dreyer.

By the time I got to know Charlie, he had moved his base of operations up to Sparkill, N.Y., and seemed interested almost exclusively in developing a perfume fixative from the glands of the skunk. The idea appears logical, since the gland of the civet cat, which gives off an equally horrendous odor, has proved extremely valuable as a perfume fixative, and brings a high price. For reasons I have never understood, Charlie's project didn't work out, but that didn't stop him from pursuing it. He was a steady caller at the office of the magazine where I worked and always asked the same question — whether we had found anything new in the literature on skunks or "essential oil of skunk". The answer was always the same — no one but he was doing any work in this field.

We always knew when Charlie came in the door. The aroma identifiable with his particular occupation accompanied him. Accompanied him? Rather I should say it preceded him.

Louis Spencer Levy

One of the leading figures connected with the essential oil industry in the 20's, not directly in the industry but certainly on the edge of it, was Louis Levy, publisher of "The American Perfumer". In the 20's, the "Perfumer" or "The American Presumer", as its competitors often dubbed it, was a standard fixture in the office of every concern where perfuming materials were bought or sold. It was readily recognizable with its characteristic bright yellow, embossed front cover. Its monthly issues were crammed with advertising, for Lou really had his advertisers trained to go all out in publicizing their products. In one issue published in 1912, Synfleur Scientific Laboratories carried a 48-page insert. Eight and sixteen page inserts were not at all uncommon, and four page inserts were run-of-the-mill for the bigger aromatic houses.

Friends were buttered up regularly. The editor ran page after page of news items, accompanied by pictures, in the oval shape made famous in college yearbooks of the nineties. Every move a big advertiser made was an event. He never just took a vacation. In "Levy language" it was always a "well-earned vacation". Lou made it all pay off. He built a tremendous magazine which none of the publishers who followed him could keep within a fraction of the size to which he expanded it. He convinced the perfuming materials industry to wage aggressive sales and advertising campaigns, which in the long run benefitted them immensely by helping them to expand their markets.

Russell Stoddard

There were two other figures who made important places for themselves in the perfuming materials business who worked at least briefly in the trade paper field. In the early 20's, Russell Stoddard, a laconic, dour, New Englander was working for Ungerer & Co. as a salesman. Not being satisfied with his progress up the ladder of success, Russ left Ungerer to become editor of a trade magazine called "Soda Fountain", published by the firm of David O. and Williams Haynes, whose other magazines included "Drug & Chemical Markets". It wasn't long, however, before Will Ungerer coaxed Stoddard back to Ungerer & Co., this time as sales manager.

When A. L. van Ameringen was putting together his staff for van Ameringen-Haebler, Inc. in 1929, he took Stoddard from Ungerer to head his essential oil department, and lived to regret it. Stoddard guessed wrong on the essential oil markets in the early years of the depression, prices on some of the materials on which the new company had taken a "long" position collapsed, and Va-H found its capital tied up in inventory which could be moved only at a loss. Stoddard had to walk the plank, and was out of the industry for a few years.

In time, however, he made a big come-back. In the late 30's he joined the staff of R. J. Prentiss & Co., dealers in crude drugs. In that post Stoddard, who had been a brilliant chemist as well as a very smart business man, learned of a new process for manufacturing a highly refined pyrethrum extract. When Prentiss showed no particular interest in promoting the product, Stoddard took the idea to Dodge & Olcott. The refined extract perfected at D & O made possible the aerosol "bug bomb" of World War II, made D & O a major factor in the insecticide raw materials business and guaranteed Stoddard a job for life.

Rupert C. Watson

Watson was another graduate of the Haynes "School of Journalism", as its numerous alumni jokingly called it.

Watson was working on the Haynes brothers magazine, "Drug & Chemical Markets", in the early 30's, when he had a chance to go with Ungerer & Co. as a cub salesman, an opportunity which he promptly accepted. Among the products he was in time given to handle were the specialties of Chuit Naef, French perfuming materials house which later became Firmenich & Co.

When Firmenich decided to open its own office in the U.S., it was only logical for them to name Watson to head the operation. Shortly thereafter, World War II broke out, and after the fall of France most of the perfuming materials that came out of Europe moved through Switzerland, a substantial volume through Firmenich. Watson and his technical director, Dr. A. T. Frascati, had to fight the customers off with baseball bats.

In 1947 Watson left Firmenich and took a job as sales manager for Fleuroma, Inc. Unfortunately, his health gave way and he died in 1953, at a relatively early age. Dr. Frascati is reported to be living in Florida, having retired a number of years ago.

George Uhe

George Uhe, who became one of the essential oil industry's leading brokers retired just last year (1976), after 55 years in the business. Born in 1900, George first worked for Julian W. Lyon in 1916-1917 as an office boy. He went into the Army in World War I, and when he came out in 1919 got a job in the Essential Oil Department of National Aniline Co. He later worked for Pfaltz and Bauer and A. L. van Ameringen. When he left van Ameringen, on excellent terms, "Van" assured him that he would never amount to a great deal, and George in turn predicted that van Ameringen's role in the perfuming materials business would always remain a minor one. They remained good friends for years, and often joked about their uncomplimentary and highly inaccurate assessments of one another's ability.

George established his own firm in 1921, specializing in menthol and camphor as well as essential oils. The company has grown over the years to become a leading broker.

Charles L. Huisking

Another firm comparable to Uhe's in the 20's was Charles Huisking & Co., located on Platt St. They acted as purchasing agents for a number of out of town wholesale druggists. In addition to dealing in essential oils, they took on the American sales agency for "Conti" castile soap and "Devold" cod liver oil. Charles Huisking headed the company, and his brother, Joe, worked with him. Joe eventually left Huisking to go with Fritzsche as assistant to the president. He is now retired, and lives in Boca Raton.

After he retired, Charles Huisking wrote an interesting account of his career in the brokerage business; the book "Herbs to Hormones", was published in 1968. It provides highly interested recollections of some of the many interesting personalities in the drug, chemical and essential oil field, and was one of our source books for this volume.

David Bennett

Dave Bennett founded Albert Verley & Co. He met Dr. Albert Verley in Paris in 1926 and arranged to represent Verley in the United States. Ed Strobl had been selling the Verley perfume specialties in Chicago, and continued as part of the U.S. Verley operation. There was an office in Chicago on Illinois Street, just north of the "Loop," and also a New York office on Beekman St. in the same building with Heine & Co., then a prominent essential oil house run by the brothers Schulz-Berge.

Bennett was an outstanding personality and a top salesman. He built up a prosperous business, with some major key accounts as a strong base. He received a great deal of help from his charming wife, Nedra, who was not only an exceptionally beautiful woman but also a hard worker. The Bennetts lived in an apartment on the near North Side of

Chicago overlooking Lake Michigan, and did considerable entertaining. Dave kept a sizeable yacht on the Lake, which he regularly took to Florida for the winter season.

Edward Strobl

Ed worked as a boy in a drug store where perfume was sold in bulk. When the druggist told him that is was on perfume that "they made their money," Ed determined to get into some phase of the perfuming materials business. He represented Albert Verley during 1924 and '25, then in 1926 when David Bennett obtained Verley representation in the U.S., Strobl became part of the deal. He was in charge of creation and compounding of fragrances, while Bennett was in charge of sales. After Mr. Bennett's death in 1952, Strobl became president of the company. He retired in 1960 and now lives in Huntington Station, N.Y. Ed's son, Raymond G. Strobl, was with Albert Verley & Co. from 1949 to 1962. He left in 1962 to join Charabot & Co. as vice-president. In 1972 he was named president of this Englewood, N.J. based company which represents Charabot & Cie of Grasse in the United States.

Dr. Eric C. Kunz

Dr. Kunz was one of several Swiss trained scientists who contributed in a major way to scientific advances in the perfuming materials business. After obtaining his doctorate from the Polytechnicum in Zurich, he worked for a time for L. Givaudan & Cie. in Geneva. He came to the United States in 1915, and first worked for United Piece & Dye Works in Lodi, N.J., a firm in the silk dyeing and printing business that had started to make their own dyestuffs when they were cut off from German sources of supply during World War I. Dr. Kunz was the first to make phosgene in the U.S., initially as a dye intermediate, and later as a poison gas for use by the U.S. Chemical Warfare Service.

When the war ended, United Piece & Dye Works lost interest in the dye business and Dr. Kunz negotiated sale of the dye division to E. I. du Pont, joining the new owner as manager of the branch. In 1924, he left du Pont and returned to work for Givaudan, first as general manager and later as president of their U.S. subsidiary. While with Givaudan he participated in developing and promoting a number of major new products, including hexachlorophene (Givaudan's Compound G-11) which Armour & Co. used in formulating the first truly deodorant toilet soap, *Dial.* Dr. Kunz was also the co-inventor of Compound G-4, an antifungicidal agent widely used during World War II to protect cotton goods from mildew and rot in the Tropics. Among the other products Dr. Kunz helped develop for the cosmetic trade were iso-propyl myristate and iso-propyl palmitate. These products led to creation of an entire industry which manufactures the esters and sells them for cosmetic purposes, and for use in body oils and bath oils.

Dr. Kunz made a major contribution to the Essential Oil Association by giving it a serious purpose, in addition to its social and fraternal function, by organizing the EOA Scientific Committee. The first task of this committee was to draw up specifications for materials in general use by essential oil and aroma chemical companies. Dr. Kunz served as chairman of this important committee until his retirement. He was the recipient of many honors during his long and brilliant career. One which he most appreciated was the award given him in 1950 by the Society of Cosmetic Chemists for his "Outstanding Contributions to the Art and Science of Cosmetics."

MODERN INDUSTRY VETERANS

We could go on and on with these recollections of individuals who have made important contributions to their industry, or who in other cases have simply been striking personalities. If we followed this personal inclination, however, we would unduly lengthen what is already the longest chapter in the book. But, we don't want to leave the subject without at least a brief mention of some of the persons who have figured prominently in industry affairs in more recent years than the "veterans," whose careers were touched on in earlier pages. We will start with the late chief perfumer of IFF, and the present IFF board chairman.

Ernest Shiftan

Mr. Shiftan, who was born in Munich, Germany, in 1903, was one of the world's famous perfumers. His early career was in Europe. In 1924 he became chief perfumer for Agfa Aromatic Chemicals in Berlin, a company which is today Haarmann & Reimer. Then after being a consultant to perfume and cosmetic houses in Paris for several years, he joined the French firm of Ets. Leopold Laserson as a partner in 1932.

In 1939 Ernie came to the United States as chief perfumer for van Ameringen-Haebler, Inc., now IFF. In the course of his U.S. career, he developed some of the world's most famous perfumes. He published many articles and gave many speeches on the subject of perfumery and cosmetics and also trained many new entrants to the industry to become outstanding perfumers. He was a past president of the American Society of Perfumers, and in 1971 received the Prix Internationale de Esthetique et de Cosmetologie from the Academie d' Osmologie et de Dermologie. He died in 1976.

Henry G. Walter, Jr.

Hank Walter, born in 1910, Columbia University, BA 1931, LLB 1934, practiced corporate law in New York City and served as Van Ameringen-Haebler's outside Counsel from 1953 on. The merger of van Ameringen-Haebler and Polak & Schwarz to form International Flavors and Fragrances, Inc. in 1958 involved him more heavily in IFF's business and in 1963 A.L. van Ameringen named Mr. Walter IFF's chief executive, a post he still holds.

Mr. Walter is credited with setting the goals and the styles which have resulted in IFF's growth in sales and profits. His IFF annual reports lifted the traditional veil of secrecy from the industry and increased public awareness of the broad roles of smell and taste — "sex and hunger" — in the world around us.

He linked the social and physical sciences in IFF's expanding R&D programs and was instrumental in setting up the non-profit Monell Chemical Senses Center in Philadelphia for fundamental research on smell and taste. He pursues similar objectives as trustee of the American Museum of Natural History in New York and of the Neurosciences Research Foundation at MIT.

Mr. Walter's active public interests include nutrition, health and public television.

Then, there was **Eddie Kuttroff,** president of Verona Chemical Co. for a number of years. Under his direction, Verona Chemical became one of the major producers of cyclamen aldehyde.

Joe Balsam was chief perfumer for Givaudan Corp. from the middle 30's until his retirement in the 50's. He made a major contribution toward solving the industry's supply problems during World War II through his ability to simulate the aroma effects of imported fragrance oils which were unobtainable at the time.

Pierre Coutin, who was president of Roure Dupont for a number of years, left to organize his own company, returned to Roure to again assume the presidency from 1965 until 1967, and now again heads his own company, Coutin Associates, Inc., Princeton, N.J. Mr. Coutin was president of EOA in 1956-1957.

Al Fiore, a creative chemist at Givaudan Corp., headed EOA's Scientific Committee for several years starting in 1952. Fiore, now retired and living in Teaneck, N.J., was surely one of the best informed figures in the industry so far as technical information was concerned during his active period.

Dr. William Gump, another Givaudan veteran, was co-inventor with Dr. Kunz of G-4 and G-11. He has been retired for several years.

Hans Wesemann was a vice-president and purchasing director of Fritzsche Brothers. He played a major role during WW II in keeping essential raw materials flowing to the fragrance and flavor industry.

Ralph Stevenson succeeded Mike Lemmermeyer as sales manager of Givaudan Corp. when Mike left to organize Aromatic Products, Inc. in 1934. Stevenson had previously been Givaudan's sales representative in Detroit. He continued as sales manager until his retirement in 1954. During the WW II years, Ralph served for a time on the War Production Board in Washington.

Ernst Theimer—Dr. Ernst Theimer joined IFF in 1931 as a research chemist and spent his entire career with them. He was elected a vice-president in 1963 and was in charge of world-wide chemical and fragrance research for IFF. He retired several years ago. He developed numerous processes for the synthetic production of fragrance materials, holds many patents, and is the author of many publications in this field. He and his wife, Jo-Anne, live in Rumson, N.J.

John Cassulo was president of Fritzsche Brothers from 1955 to 1962, and did a great deal to modernize Fritzsche's operations. His promising career in the industry was terminated tragically in 1962 when he was lost at sea in a marine accident off the coast of Peru. A well known sports fisherman, it was believed his fishing boat collided in the dark with a larger craft.

Frederick Haakon Leonhardt, son of Frederick Henry Leonhardt, is now chairman of the board of Fritzsche Dodge & Olcott, having headed Fritzsche Brothers and later

F D & O since the death of John Cassullo, his brother-in-law. Under Leonhardt's leadership, F D & O has continued to prosper, and has recently built a major new aroma chemical manufacturing complex in East Hanover, N.J.

Robert Engel — Bob joined Givaudan-Delawanna soon after graduating from MIT. He was in charge of industrial reodorants, pioneering the development of this type product. He later moved to S. B. Penick & Co., where he was in charge of their perfuming materials operations. In the late 40's Engel joined Trubek Laboratories as part of their management team. He left Trubek in 1959 to go with Fritzsche Brothers. At Fritzsche he directed aroma chemical sales until his retirement.

Joseph Fein — Joe joined Florasynth Laboratories more than 50 years ago after training as a pharmacist. Over the years he has filled any number of jobs at Florasynth, including building programs and management of the perfuming materials department. He has been known specially for his responsibilities in purchasing.

Waldo Reis came to van Ameringen-Haebler, in 1934. He had previously been soap perfumer for Lever Brothers Co. When he retired from IFF in 1965, Waldo was a senior vice-president, in charge of research, manufacturing, aroma chemical sales and purchasing. He served a term as president of EOA in 1953. Waldo lives in Westfield, N.J.

George Cadgene — George started his career at Givaudan-Delawanna. He moved up to director of purchasing before leaving in 1952 to join Trubek Laboratories. At Trubek he was active in both sales and purchasing and also served on the management committee. In 1959 Cadgene joined van Ameringen-Haebler, Inc. as director of purchasing. He later became v-p for aroma sales, leaving the successor company, IFF, in 1968 to become a consultant. George served as president of the Essential Oil Association in 1967-68, winning election for the latter term after the only contested election in the history of the association. He was also president of the Drug Chemical & Allied Trades Association in 1969-1970.

Dr. Martin Szamaltoski was the first plant manager for Givaudan in the 20's when their manufacturing facility was still Burton T. Bush, Inc. A brilliant chemist, of Polish nationality, Dr. Szamaltoski struggled through all the grief of changing ownership of the manufacturing arm of the company from Antoine Chiris, through B.T. Bush, Inc. and eventually Givaudan Corp.

Dr. Max Luthy succeeded Dr. Szamaltoski as Givaudan production manager after the latter's death in 1940. It was under his direction that the company's facilities for production of aroma chemicals grew from the small plant originally built by B.T. Bush for Chiris to the major complex which exists today.

Walter Langsfelder, co-founder of Fleuroma, Inc., was one of the country's great perfumers. His creative ability was responsible in large measure for the early success which Fleuroma enjoyed.

Case Poons was co-founder of Fleuroma, with Langsfelder, shortly after WW II. Poons, who was president of Fleuroma, became president of UOP after Fleuroma was taken over by UOP, but relinquished this job after a few years. He remained active with UOP until comparatively recently, when they decided to sell the fragrance portion of their business to Naarden.

Alfred H. Moeller — Al began his career in the aroma chemical business at E. I. duPont deNemours & Co. In 1939, with the late Albert Nickstadt, he organized Noville Essential Oil Co. They make perfume compounds and also manufacture for their own use several rare aroma chemicals. Al has been very active in the American Society of Perfumers.

He was a leader in organizing the annual meeting of the Perfumers, which attracts outstanding speakers and attendance from all over the world.

Frederick Shoninger was for some years the director of Chiris operations in England, but came to the U.S. after World War II to supervise operations here. After heading operations for about twenty years, he was instrumental in arranging the sale of the entire Chiris complex, including the French and Brazilian companies, to Universal Oil Products. He remained with UOP for several years until a terminal illness dictated his retirement.

Marion Carpenter, chief research chemist for Givaudan for many years, was responsible for many of Givaudan's new aroma chemicals. He is the father of one of the U.S. astronauts.

We resist our temptation to continue further and regretfully conclude this chapter, aware that we must inevitably have failed to mention many individuals and companies whose major contributions to the industry could well have been recognized. Limiting factors include the fact that there were many prominent individuals and companies in the industry we did not know, as well as the impossibility of extending this chapter any further without overbalancing the book. To those whose names have been unintentionally omitted, and to their friends, we humbly apologize.

Chapter X

The Anatomy of the Industry

Fifty years ago, the Fragrance and Flavor Industry in the United States was much less complicated than it is today. Most of the firms in the industry were essential oil dealers. They imported fragrance materials from foreign producing areas, brought to market the limited volume of essential oils of domestic origin, such as mint, citrus and cedar wood oils, etc., in some cases they did a little compounding, but for the most part the finished fragrances and flavors they sold were compounds developed abroad by the French, German, Italian, and Swiss houses they represented.

Then came the rise of the domestic aroma chemical industry, which was one of the by-products of the outbreak of World War I. Up to that time, Germany had been the center of the synthetic chemical industry. When first the British blockade, and later the entry of our own country into the war, cut U.S. importers off from their German sources, the United States had to develop its own chemical industry. U.S. manufacture of aroma chemicals got a big push. Firms like Dodge & Olcott and Fritzsche Brothers had been manufacturing aroma chemicals for a number of years for their own in-house use. Now a number of new companies started manufacturing synthetic aromatics for sale to the trade. First just a few bold, aspiring firms such as Synfleur Scientific Laboratories, Van Dyk & Co., Fries & Fries, Antoine Chiris Co., Morana, Inc. and Florasynth Laboratories went into production. Later, during the 20's, as demand for aroma chemicals increased, they were joined by other new producers, including Felton Chemical Co. and later by van Ameringen-Haebler, Inc., Trubek Laboratories and others.

In the 30's, many essential oil dealers decided that they had better try to be something more than just dealer/importers if they were to survive the depression days. Many of them had represented foreign compounding houses, had acquired some knowledge of compounding, and knew what perfuming and flavoring specialties were in demand by their customers. They started to do their own compounding, and in some cases surrendered their roles as representatives of the foreign houses.

As these changes came about, things started to get fairly complicated. Where previously most companies had been exclusively producers, dealers, brokers or manufacturers of aroma chemicals, suddenly the lines of separation began to grow indistinct, or even disappear. An aroma chemical manufacturer was likely to be, not only a dealer/importer, but a fragrance and flavor compounder as well. He might process essential oils as well as synthesize aroma chemicals. Both dealer/importers and manufacturers of aroma chemicals began to compound. Certain firms confined their activities exclusively to compounding, although they might import from a parent company in Europe most of the raw materials used in their compounds. Still others were straight U.S. offices of European-based companies.

At the risk of giving some of our readers information with which many of them may already be thoroughly familiar, we will outline here the functions of the various types of firms that make up the essential oil and aromatic chemical industry as now constituted. First we will talk of the *Dealer/Importer,* the backbone of the industry fifty years ago, and still a very important factor in getting fragrances and flavors from the places where they originate into the finished products of which they are such important components.

The Dealer/Importer

Dealers are essentially "traders," matching the demands of major buyers of essential oils with the supplies of raw materials offered by basic producers. They operate in both the domestic and foreign markets, and export as well as import. One of their principal functions is to smooth out the peaks and valleys in the markets for the materials in which they deal, minimizing drastic price swings, and trying to maintain stability by keeping supply and demand in balance.

We talked at length about the role of the essential oil dealer with Stephen Manheimer, president of J. Manheimer, Inc., Long Island City, N.Y., one of the most experienced and respected names in the essential oil market. The dealer, says Mr. Manheimer, serves as an adjunct to the buyer's purchasing department, protecting him against spot shortages and sharp and unforeseen price advances. At the same time, the dealer must protect his supplier (the producer) who has to be able to count on a reasonably stable demand and price for a natural crop which may take many months to bring to market.

Some dealers contract with their sources of supply through brokers, while others make regular visits to producing areas, negotiate their contracts directly with producers, and often act as exclusive sales agents for them. The dealer often spends more time developing and solidifying his sources of supply than he does in seeking new accounts, for he must be able to trust the shipper to honor commitments, supply materials that meet specifications of the samples submitted, and keep prices reasonably stable.

Normally, a dealer/importer receives samples of the essential oils he is buying from the exporter, and agrees to a transaction on the basis of that sample. He then executes a letter of credit, assuring the exporter of payment before the shipment leaves the foreign port under this system. If the oil received does not measure up to the sample, he has little recourse. In 1976 a group of dealers in the United States, Europe and Japan sustained very heavy losses when a number of shipments of essential oils from Indonesia proved on arrival to consist mostly of water rather than the oils (patchouli, cananga, vetivert, nutmeg, and citronella) that had already been paid for.

To operate successfully, a dealer should be in position to buy for at least one or two major users of a given material. When negotiating a substantial contract for one of them, at a quantity price, he adds to the order another 25 or 50 drums to go into his own stock. He hopes he will be able to sell job lots of this stock at an expected higher profit per pound than he realizes on the bulk order for the big customer.

A sizeable dealer needs a quality control laboratory to check incoming shipments to make sure they meet specification. Some dealers even do a limited amount of processing, such as fractionating, isolation, decanting, filtering, re-packing, etc. Very few are equipped to handle any chemical synthesis, and when they do, it is usually performed for them by someone else on contract.

There are fewer dealers now than there were years ago. Many firms that used to function primarily as dealer/importers, and whose stock in trade was the major foreign producers they represented, have now found it unprofitable to operate on the relatively small commissions dealers earn, and as indicated earlier, a fair number of them have now moved into the compounding field, where profit margins are substantially higher. They can, however, contract for the entire output of a supplier; or even of a particular oil, thereby assuring a reasonable profit at a stabilized price.

The Broker/Agent

Related to the function of the Dealer/Importer is the Broker/Agent. Some importers deal with their sources of supply through brokers, which is indicative of their kinship. Although there are a relatively limited number of brokers, their interaction in the raw materials marketplace is an important one. The role of the essential oil broker was well described by Bernard P. Champon of L.A. Champon & Co., Hoboken, N.J., in a talk delivered at the International Congress of Essential Oils in San Francisco, in September, 1974. We paraphrase some of Mr. Champon's observations in the following:

The broker, he emphasizes, is an intermediator between buyer and seller. He is usually not given authority to complete a contract, but only to carry on negotiations. If he produces a buyer willing to meet the terms offered by the seller, he has earned his commission even though no final sale may take place.

An offer made through a broker is generally limited in time for consummation (usually 24 hours). If more time passes, the seller is no longer obligated to close the transaction. In a firm or rising market, the seller may well choose not to extend his offer. In a falling market, he will be inclined to extend the time limit.

Agents (brokers) may have several (non-competing) principals, serving them on an exclusive or a non-exclusive basis, and the principals may have different agents for different territories or different materials. So long as brokers or agents act within the scope of their authority, or the apparent authority invested in them, they may bind their principals to third parties. And, as long as the third party is aware that the agent or broker is acting for a principal, whether or not the identity of the principal is disclosed, the agent or broker is not personally liable in the transaction.

All agents, including brokers, must disclose to their principals all material facts related to the performance of their duties, and may not disclose any confidential information they have been given by their principals. It is the broker's responsibility to investigate a new seller or a new buyer, and to obtain bank or commercial references if buyer or seller are not known to each other. It is the offer from the seller, and its acceptance by the buyer through the broker, that creates the contract, not the written confirmation of the verbal deal. The paper work merely confirms and outlines what has already been agreed to, but it is the verbal agreement which is the actual contract.

We conclude this discussion with a quotation from Mr. Champon's talk which spotlights the broker's function in the essential oil distribution pattern: "The broker/agent link between producer or exporter at one end, and the dealer or importer at the other, is an important role in the chain that links raw material producer with finished goods producer. The broker or agent must earn the confidence of both buyer and seller, and must build his reputation in the industry by proper performance of his function."

The Essential Oil Producer

Before we leave natural materials and pass on to synthetic aromatics, we should not fail to mention that there are in the United States and Canada quite a substantial number of *producers* of essential oils. They are for the most part firms producing mint oils, citrus oils, and a few other products of domestic origin such as cedar wood and cedar leaf oils, wormwood and wormseed (oil of chenepodium) oils, etc. There are probably as many as one hundred such firms, some relatively small. A number of essential oil dealers grow at least some portion of the oils they sell. Others distill oils to rigid specifications from herbs imported from all over the world.

Manufacturers of Aroma Chemicals

Next, we go to a very important category in today's fragrance and flavor industry — the manufacturer of aroma chemicals. (See also pp. 135-170). During the first half of this century most manufacture of aromatics was involved with isolation or chemical conversion from natural products. Relatively inexpensive natural raw materials such as citronella oil, cedarwood oil or lemongrass oil were processed to more concentrated, more useful and more efficient products. The aroma chemicals so produced were much more valuable than the original oil. Little if anything was lost. A good compounding facility could use all of the by-products in a wide variety of finished perfuming materials.

For example, citronella oil was processed to concentrate and isolate its principal aromatic ingredients, geraniol and citronellal. Citronellal was in turn processed to hydroxy-citronellal and l-menthol. Geraniol was converted to its esters or to citronellol. The major constituent of lemongrass oil is citral, normally present to the extent of about 75%. The aroma chemical manufacturer either makes an isolate which approaches 100% citral, or manufactures ionones or vitamin A directly from the oil. The refined citral may be used in artificial lemon flavor or for fragrance purposes.

As methods for the manufacture of synthetics improved, ways were found to produce many of these aromatic isolates synthetically. However certain aroma chemicals remain which it has not as yet been found possible to produce synthetically. They still must be processed from natural oils. For instance, eugenol and its derivatives must still be isolated from clove oil, vetiverol is still processed from vetiver oil, acetyl cedrene from oil of cedarwood, and vanillin from lignin (wood pulp waste). With the developments that have come in the chemistry of synthetic aroma chemicals, however, our dependence on natural raw materials has greatly diminished.

That is extremely fortunate, for the demand for natural products has increased sharply, particularly for use in the manufacture of flavors. The entire output of "naturals" is being utilized, and the demand is so great that only synthetics can make up the difference. The fragrance and flavor industry now gets many of its raw materials from the petroleum, sulfite turpentine and acetylene industries. A long list of alcohols, esters, aldehydes, ketones, lactones, etc. are made from synthetic raw materials. As a result, the processing of citronella and lemongrass oils has decreased considerably in the past twenty years. Many of the fragrance materials formerly made from those oils are now made from synthetic sources.

The number of raw materials the aroma chemical manufacturer employs is staggering. He may have to stock materials running into the hundreds of items in order to be in position to manufacture the different aroma chemicals he is called upon to supply. In some cases he may employ ten or fifteen raw materials, and go through six or eight intermediate steps, to produce a single end product.

It is no wonder then that for the most part the big chemical companies have left this business alone. There are not too many carload shipments involved. Dow and DuPont took a tentative look at the aromatic chemical business about twenty-five years ago, and apparently decided that they didn't belong in it. More recently several other large chemical companies such as Monsanto and Mallinckrodt have shown a revived interest in the field, but largely in the compounding area, where profit margins are longer.

Today the larger firms making synthetic aroma chemicals tend to concentrate their interest in a relatively small number of larger volume products, leaving the short run products to the smaller firms, the specialists who are ready to turn out a few hundred pounds, or even a few kilos of cis-jasmone, the dienals, santalyl phenyl acetate or the like for a customer who must have a trace of that particular product for a special formulation in which it is an essential ingredient. *(The reader is referred to Chapter VIII on Synthetic Raw Materials for a more detailed discussion of Aroma Chemicals.)*

Fragrance and Flavor Compound Houses

There are today more companies compounding fragrances and flavors than there are any other category of firms in the essential oil and aroma chemical business. As the name implies, these companies compound finished fragrance or flavor products by mixing a number of materials of either natural or synthetic origin. Years ago, the materials they used were primarily natural. Now, the compounder has access to hundreds of synthetics, which run more uniform in quality, are available in dependable supply, and of such varied character that they give a wider range to the perfumer's or the flavorist's palette. He may employ as many as sixty or seventy distinct materials in compounding a single fragrance, and almost that many in working up a flavor compound. To add to the complexity of the resulting formula, some of the (essential oil) ingredients may contain over a hundred components themselves.

These fragrance and flavor compound houses are sometimes, perhaps, incorrectly, called "compounders." Actually the word is a misnomer as applied to a company. The actual compounder is a production man, a technician, who puts together the ingredients in the complex formula for a fragrance or a flavor that has been created by the perfumer or flavorist. The production compounder is a highly skilled and experienced person. He must have a great deal of practical know-how, must be familiar with the names and types of ingredients, their quality, color and compatibility. He must be familiar with mathematical conversions, must understand the correct order of addition of ingredients in a formula, including the treatment of solids. He must be thoroughly responsible, and perform precisely. He stands in relation to the perfumer, we might say, as a practical engineer does to an architect. In the final analysis compounding is an art, and a necessary adjunct to the acceptance of the finished flavor or fragrance. It is the job of the compounder to solve all the practical headaches, and make sure they don't get through to cause trouble with the finished consumer product.

Some compound houses may also be aroma chemical manufacturers. They thus have at hand from their own production many of the raw materials they use in their formulations. But no aroma chemical manufacturer can begin to manufacture *all* of the thousands of items in use today. Thus there is a great deal of buying back and forth between producers. Some firms are producers of essential oils, but again may themselves produce only a few of the oils they use, and must buy the additional materials used in their formulations from producers, dealers or brokers. If they do not control their own production, and have to buy on spot, they may run into problems of consistency, uniformity or quality.

A compound house may stock as many as 4,000 to 5,000 raw materials, but some of these will be used only in minor quantities. Any overage must be kept in stock, perhaps over a considerable period. Obviously such stocks must be kept under closely controlled storage conditions to guard against deterioration. Thus inventories are usually kept to a minimum on all readily available raw materials. Stocks of materials used in larger volume are bought in larger quantities, and stock is turned over more often.

When a perfumer or flavorist is working with a client or a potential client, whether a fragrance or cosmetic house, a manufacturer of a household or industrial product, a food or a beverage company, et. al., a preferred first step is to discuss with the buyer his profile of the fragrance or flavor desired. (See p. 39-41). If a fragrance, the buyer may want something clean, fresh, natural, invigorating and outdoor-sy. Or it could be just the opposite, or something in between. He may want something evanescent, or something long lasting. At least, with a starting point the perfumer will have a concept on which to focus his creative work. While there are many procedures for developing a successful fragrance or flavor formulation, a typical method of handling will be described.

Responsibility for formulating a new fragrance may be passed along to as many as four or five competitors, each of whom may submit several samples to meet a deadline. Then the competitive submissions are panelled, probably within the company's own personnel, and the choice narrowed down. Sometimes the final choice is made by the buyer's personnel; sometimes the samples are submitted to an outside panel, possibly a panel of magazine readers in the age and sex group that is the target audience, or perhaps to some other impersonal panel set up for the purpose. Often, before the second panel goes into action, the fragrance suppliers may be asked to make certain modifications to any of their submissions that may still be "in the running." Then the "final" versions of the sample fragrances are returned to the evaluation panels for their decision.

One limitation of market testing, says George Cadgene, is that in fragrance preference tests, the winner always seems to be a familiar aroma. "The mass of people," he believes, "don't want to be different, don't want to have it appear that they aren't like others, that they don't have a preference for a fragrance that has been proven to be good. The result of market testing, all too often, is that the new fragrance will beat its competition only when the fragrance is reminiscent of something with which the test panel members are already familiar. This doesn't mean that the winner will necessarily smell like the product with which it is competing. It does mean, however, that no mass marketer is likely to come out with a product to which the public has not already been exposed."

When a fragrance has finally been selected, additional work is involved if the marketer wants to use it for a full product line. The fragrance must be made compatible with each product in which it is to be used, stable, and true to type. Each type cosmetic, whether hand cream, lip stick, bubble bath, cologne, eau de toilette will obviously require an adjustment of the formula. These modifications may or may not precede the introduction of the initial product on the market; it is the marketer's choice.

Despite what has obviously been a major amount of work, skill and art invested in the product up to this point, much additional well planned labor remains before the finished product will be distributed and selling on the retailers' shelves in sufficient quantity to be able to label it as a "hit" or at least a "profitable" product. The single word which best describes this entire effort is "campaign." It takes a well conceived and carefully thought out campaign to come up with a winner in this highly competitive business.

Formulas thus developed are rarely patented, since publication of the patent automatically involves revelation of the formula in the patent application. So formulas are generally treated as trade secrets, although the legal guidelines in this area are poorly defined. A trade secret, of course, does not accord the type protection which accompanies a patent or trademark. A trade secret does not guarantee exclusivity to the originator, and there is no legal support to the premise that an idea, a product, or even a formula must remain for any fixed period the restricted property of the originator. There are many "gray areas" with trade secrets.

The originator or supplier may or may not transfer the exclusivity of a specific formula to a client, depending on a contract to this effect between supplier and client, either written or verbal. Obviously, the marketer's position is strengthened if his position in the market and his purchasing power are sufficiently strong so that the supplier will hesitate to incur his ill will by letting the client's special fragrance be sampled to any other marketer for any purpose. Only if a supplier enjoys a substantial volume of business from the client on an expensive formulation, however, can he be counted upon to protect the client's insistence on exclusivity.

Where patents and trademarks are involved, the law is much clearer, and it is easier for a marketer to protect his rights. Enough cases involving patents and trademarks have been tested in the courts so the rights of the contesting parties have been clarified in considerable measure. Twenty or thirty years ago, patents were rarely sought for any technology in the F/F industry, and most firms depended upon secrecy and "in-house know-how" to protect their positions. Today, however, the industry is becoming much more involved in patents. Developing technology and expert analytical methods have forced the issue, and all F/F companies now employ product, use and process patents.

But there is still a general reluctance to attempt to patent formulas. Too much depends on the subjective reaction of the particular examiner before whom the application will come for decision. If he does not accept the uniqueness, or invention, in combining a mixture of items as grounds for issuance of a patent, the applicant will find that he has revealed his formula to no purpose. The time for patenting formulas may have passed.

In this connection it should be pointed out that there is general acceptance in the industry of the position that a material found naturally as a component of a natural material already in general use in compounding cannot legally be patented. Several applicants have succeeded in having such patents issued, getting them by the examiners by diverse legalistic methods, but as no such case has yet been carried to a contest there is no final determination of whether such patents will be held valid.

One more point should be noted in connection with patents. There is a wide difference between a U.S. patent and a foreign patent. In the United States, a careful and thorough search of the literature and patent records is usually made before a patent is granted. This is not always the case with some foreign countries. Each country has its own requirements for granting patents, and in some foreign countries, notably Germany and France, the requirements preventing infringement are minimal.

Although the legal protection accorded a trade secret is indefinite, the law does prohibit the use by another party of the proprietary information in a Trade Secret if it has been obtained by any illegal method. Breach of a secrecy agreement, unlawful procurement of documents, stealing files or formulas, or obtaining same by bribes or unlawful entry, impersonation, etc. are serious criminal offenses. On the other hand, if information about a trade secret can be gained by examination of a commercial product, by analysis of a sample, by open discussion, or through information disclosed in advertising, there is no basis for legal action against a copier. If the owner of a trade secret expects to maintain his right to secrecy, he will be well advised to have a signed contract, letter of intent, or understanding with any employee, client or other person who can possibly gain knowledge of the secret. He must show by his actions that he intends that the item be kept a secret. He must keep locked files, and labeling must call attention to the secret nature of the idea, product or formulation.

The involvement of the perfumer or flavor chemist with a trade secret owned by the company for which he works results from time to time in serious ethical, and occasionally legal, problems. A situation of the type that occasionally arises might involve a compound house that has developed a special perfume formulation for an account, and becomes the exclusive supplier of the ingredients for the finished compound. Then, as a result of a disagreement of some kind, or perhaps a change in personnel, the "secret" formula becomes known to a competitive supplier. Shortly the original supplier finds that a competitor has taken over the account.

If the original developer of the formula decides to protect what he considers to be his rights, he must then decide against whom he will take action; whether against the disloyal employee who gave the formula to the client, or to another compound house; against his former customer if he feels that the secret formula may have been obtained by illegal means; or perhaps an action against the competitor charging unfair competition for having obtained the "secret" formula in a clandestine fashion. Whom he will sue, if anyone, depends on what he thinks his chances are of winning, and what good will with a possible future client may be lost in the process.

A number of cases of this kind have been tested in the courts over the years, with the decision sometimes going one way, and sometimes another. We know of one major case several years back in which a supplier won a very substantial judgment against a soft drink firm, and then instead of insisting on collecting the financial award, compromised instead on again supplying a major portion of the marketer's flavor requirements rather than accepting the judgment.

The question sometimes arises as to whether it is the creator of a formula, or his employer, who owns a formula.

This point is usually clarified in the contract under which the perfumer or flavorist is employed. Where there is no contract, however, the unwritten rule in the F/F industry is that the employer owns any formula created by an employee. The creator gets the prestige, and normally the remuneration which accompanies development of a successful formula, but ownership lies with the employer.

Certainly the perfumer or flavorist will have the formula indelibly printed on his brain, but the formula is a trade secret. While it would not be considered ethical, or even legal, for him to offer exactly the same formula to another marketer, or to sell the formula as would be acceptable if he held a patent, he can readily modify the formula and offer a slightly different, perhaps even an improved version, to a new employer or to another potential customer without damaging his reputation. However he must always keep his reputation well in mind. To avoid being typed as a "one formula" perfumer, he must be able to demonstrate his skill by an ability to continue developing acceptable new formulas for saleable products.

Another type situation which involves a matter of ethics arises occasionally when a marketer asks for competitive fragrance or flavor submissions, although those asked to submit their products recognize that they have little or no chance of getting the order. Take as a typical case one in which a cosmetic house is looking for a fragrance for a new underarm deodorant. The house buys most of its fragrances from the A-1 Aroma Sales Co., a fact which is rather well known in the trade. Nevertheless, two or three other companies are in this instance asked to offer candidate fragrances for the new product. These compounders do a considerable amount of creative work, followed by laboratory testing of the resulting formulas, and present the results of their labors.

No one is greatly surprised when the A-1 Aroma Sales Co. wins the order. But the "also rans," who have invested thousands of dollars in time, may wonder if they have been taken advantage of. Sometimes they may even get the feeling that the buyer knew all the time who was going to get the order. Perhaps a group of potential fragrance suppliers were invited to submit their suggestions primarily to preserve an appearance of the account's being competitive. Some compound houses occasionally express the opinion that it would be fair for them to be compensated for such development work on a fee basis.

However, they are decidedly in the minority. The industry generally raises no objections to working up formulas on a speculative basis. Quite the reverse. Most compound houses are happy to have the opportunity to put in an offering, even when they realize there is little prospect of them getting the immediate business. But, submission of a candidate formulation involves contact with an account which they would like to sell, perhaps to impress them with what they can do, and pave the way for later business. And, they may also reason, if the product they work up is not accepted, they can always offer it to some one else.

Most companies look on speculative formulation as part and parcel of their business, and are glad to be called upon to submit candidate fragrances. They would not want the prospective customer to pay any part of the development cost, as the customer might then figure that he owned the compound.

A rather common practice, generally considered to be neither illegal nor unethical in most cases, is complying with the request by a marketer to supply him with a duplicate of a compound he is presently buying from another supplier. The marketer often does not know the exact formula of the fragrance he has been using, but for one reason or another, whether personal, logistical or price, he may want to change his source of supply, or at least threaten to do so.

Some purchasers, incidentally, consider this practice unethical and will never resort to it. Rarely, however, will compound suppliers turn down the opportunity to promote their products in such situations unless their previous experience with the prospective account has been unrewarding. A team of perfumer or flavor chemist with a good instrumentalist can often match a formula to 95% or better, although never exactly.

This duplication of sample fragrance or flavor compounds offers some opportunities to both supplier and marketer. However many of the most reputable fragrance and flavor houses view the practice with scant enthusiasm. Experience has taught them that, if business is consummated, it is often at minimum profit, and too often the new submission is merely used to drive down the price the present supplier is getting, while the duplicator gets a "thank you". Experience should guide the compound house on the advisability of accepting such invitations.

There is still another type situation which has been much in the news in recent years — the marketing of so-called "copy cat" products. Many companies advertise fragrances "just like" the best selling brands, but at much lower prices.

The question of whether it is ethical to offer copies of well-known products is one that has long been discussed in the fragrance industry, and one on which there is some difference of opinion. The matter was commented on by Steffen Arctander, fragrance raw materials consultant, in a symposium on "The Future of Perfumery" held at Stratford-on-Avon, England, April 5, 1977. Mr. Arctander is obviously among those who view with alarm the practice of "copying." He recommended that perfumers adopt a Code of Ethics condemning copying, and said that in his opinion the practice contributes to much that is illegal, dishonest and unethical in the fragrance business.

He insists that copying is a "step backward," contributes nothing to progress, and amounts to simply doing over again what someone else has already done. Copying fragrances, he charges, amounts in its worst form to "the commercial stealing of someone else's product, name, reputation and consumer acceptance. Nevertheless he recognizes that any attempt to ban copying "may seem rather hopeless — if the market really demands copies," which he questions.

There are many in the fragrance industry who disagree completely with Mr. Arctander, and who feel that it is quite legitimate to make copies. In completing our coverage of the subject we talked to a cross section of industry leaders to determine how they feel about selling to firms which market copies of well-known products. Their opinions are summarized below.

First, there are the out-and-out fraud products, often sold on street corners by shifty characters who attempt to imitate the bottle, the package, the seal of a famous perfume. The people offering such counterfeit products are thieves, pure and simple, and no one from the legitimate fragrance industry is ever involved in such operations.

But, there are other types of "me too" products which are completely legitimate. Whenever a successful new fragrance or flavor product is introduced and starts to ride up on a marketing skyrocket, every compound house attempts to pull it apart. Competent perfumers or flavorists can readily smell or taste out most of the major components and many of the trace components in the new product. Running up a few model compounds would help to establish quantitative information, and, if a good instrumentalist is assigned to the project, accurate quantitative data can be developed, trace materials identified, and even different sources of generic materials determined. It doesn't take long for a skillful team to develop a reasonably close duplication.

Mind, we do not say an *exact* duplication. Even with the formula in hand and a full knowledge of the source and quality of each component, the finished product will vary slightly from batch to batch. There is an acceptable "wobble" in a series of batches, and of course an unacceptable divergence from the reference sample which will cause rejection of the batch.

Duplication of an existing product by a compound house is completely legal, and as long as the "copy" is not represented to be the original product, no fraud has been committed. The trade secret aspect of the original has not been violated. Essential oil firms openly supply raw materials to the marketers of such products. For example, a marketer may want to offer a straight copy of a well-known product, giving it a different name but claiming it smells the same as the fragrance it seeks to duplicate. Many firms offer such copies. It may be compared to dolling up a Volkswagen automobile to make it look like a Rolls Royce by changing the hood and the trunk. No one is fooled; the customer knows she is buying a copy. She may or may not be satisfied with the result, but no deception has been practiced.

There is a second way in which copies are frequently sold. It is much more common and accounts for a surprising volume of fragrance business. For example, a marketer decides it wants to add a new men's cologne to its line, goes to its fragrance supplier and asks them for "something like Aramis," or some other well-known, fast-selling cologne. The fragrance supplier will give them a fragrance of the *type* requested, which is often difficult to distinguish from the original if the products are studied side by side. It is frankly a copy which recalls the original fragrance, but usually one that has a lower raw material cost so that it can be marketed in a lower price bracket.

What can we say of the strictures by which essential oil houses are, or should be bound? First, one may be certain that the fragrance supplier which created the original formula will not offer duplications for sale. That would not only be a breach of faith with its original customer, but would be stupid into the bargain. But, we must bear in mind that, whenever a good, new fragrance hits the market, perfumers at all the fragrance compounding houses are going to attempt to reproduce it or match it closely. For they know that, very shortly, some marketer will ask them to make a product of the same general character which the new fragrance, now a hit, typifies.

An interesting example of similar "types" appearing in the market in the wake of an initially highly successful product is the case of several fragrance products which closely resembled *Canoe,* all of which were introduced about fifteen years ago following the great success of *Canoe* as a man's cologne. *Canoe* was originally launched in the mid 30's as a woman's perfume, then was picked up in the early 60's as a man's scent, sold in highly functional aluminum containers. The fragrance was well received and several marketers soon decided to launch men's lines of that fragrance type. The result was such highly successful products as *Jade East, Brut* and *Hai Karate,* all big sellers. None were "copies," but all were of the same type, the same family, as *Canoe.*

This marketing of "me too" products is part of the process of satisfying and meeting the overall market demand for fragrance. A woman will smell one of the new fragrances, perhaps in an elevator, and will want to have that fragrance, or something like it. If her budget is limited, she is in the market for a cologne or toilet water, or perhaps a perfume which will be substantially less costly than the original perfume which attracted her attention. She wants *that* new fragrance, and some marketer is going to satisfy her need, at a price within her budget.

In reality, most of the fragrances on the market are not really new. There are so few individual types of perfumes available (See chart on p. 78-79) that it is almost impossible for a perfume compound house and its marketer customers not to work on variants of great original fragrances. That is the bread and butter of the fragrance business, and all perfectly proper.

Difficulties can arise for the duplicator, however, when he is requested to supply a "like" product to a marketer whose packaging, labeling or advertising may become the subject of a trademark or product infringement charge, which could involve the supplier of the fragrance in legal prosecution. The major F/F houses are very proud of their reputations which took years to build, and ethical considerations normally cause them to shy away from soliciting customers who are obviously walking a shaky legal tightrope. When the compounding houses face the threat of legal action as well, it is the general opinion in the trade that they should have the right to preserve their reputations and avoid possible legal responsibility by being allowed to refuse to deal with less than reputable customers without facing charges of restraint of trade.

Chapter XI

Production Processes and Equipment

The methods used to produce fragrance and flavor materials have gone through a long evolution from ancient times. Production methods have necessarily changed along with changes in raw materials to be processed and the finished products sought. Where once only natural raw materials were the sources of F/F products, today synthetic aroma chemicals play a major role in the operation of a F/F factory. And, where once, heating fragrant materials in oil or water, or such relatively inefficient processes as enfleurage or expression by hand were widely employed, today most fragrance and flavor raw materials are produced by efficient, modern, chemical methods.

Early Production Methods

Probably the first technique used for the production of fragrance and flavor materials was the collection of exudates such as aromatic gums and balsams, or spices such as clove buds or cinnamon bark. A number of methods for processing such materials developed, but recorded history is so spotty, particularly in the technical area, that there are few references in the literature to guide us. Based on the meager factual information available, supported by some few references in ancient records, it seems probable that the processes of infusion and extraction preceded by some years the development of the process of distillation.

Fragrant oils and ointments are mentioned in early literature, but these were not necessarily essential oils. There were undoubtedly some pomades, but for the most part the fragrant oils mentioned in the Bible were probably solutions of spices, flowers or herbs in fixed oils such as almond, palm or coconut. How these oils were produced is not clear. They could have been expressed, or perhaps were subjected to a process of rendering with hot water. Heating fragrant materials in fixed oil would readily disclose that the spices or herbs contained volatile components, which could be driven off, condensed and collected. This was an important first step in the development of the process of distillation, but several other indispensable additional steps would have to be discovered before distillation would become a useable production method.

Distillation waters (such as orange flower water, rose water, etc.) were products of importance in ancient Persia, India and Egypt, and later in Greece and Rome. Most students of scientific history believe that the oils and solids (essential oils and plant camphors) which were produced along with the waters in distillation were originally discarded as valueless.

Recent discoveries have shown that distillation was known and practiced in some areas as early as 3,000 years B.C. In 1975, a scientific expedition led by Dr. Paulo Rovesti* researched the Indus Valley civilization and discovered a terra cotta distillation apparatus at least 5,000 years old. Containers for the fragrances and cosmetics of the day were found, along with the apparatus, in the Taxila Museum in Pakistan. These had been taken from the ancient city of Harappa. This discovery predates by ten centuries the Greek and Egyptian technology of about 2,000

*Rovesti, Dragoco Report 3/77

B.C. According to Urdang, oil of turpentine and solid camphor were two substances produced by distillation which were mentioned in Herodotus' history of early Greek civilization.

Distillation was refined by the Arabs, who introduced the process into Western Europe. By the 11th Century this method of separating volatile from non-volatile materials came to be widely used in Italy and France. It is interesting to note however that fragrant fixed oils and ointments were still being used throughout Europe in the 12th Century A.D.

In the latter part of the 13th Century, Arnold de Villanova described the distillation of essential oils for the first time. Subsequently the art was introduced into Europe. However it was not until the 16th Century, in Switzerland, that von Hohenhein recognized the "quintessence," that is the fragrant and therapeutic oils from which the name, "essential oil," was derived.

Since then the practice of hydro distillation and vacuum distillation has been refined and further developed, and today is a major process for producing fragrance and flavor materials from natural raw materials as well as from synthetic sources.

The Enfleurage Process

Earlier we mentioned extractions or solutions of spices and herbs in fixed and fatty oils. We are not sure just when a more refined extraction process, called enfleurage, was developed, but it was based on the simple principle that fats absorb odors when they are exposed to contact with fragrant materials.

In the enfleurage process, flowers are arranged on trays of glass plates coated with neutral fat and allowed to remain in contact with the fat long enough for most of the fragrance to be absorbed. When the flowers are exhausted, the trays are de-fleuraged, by hand, and fresh flowers substituted. In time, the fat becomes thoroughly saturated with fat-soluble components, including the essential oil. At this point the fragrant fat is called a pomade. Treatment

View in "Naturals" plant of Charabot & Cie. in Grasse, France

with alcohol dissolves the essential oil out of the fat. It is subsequently filtered and purified. The alcohol is distilled off, the oil concentrated, and an "absolute" is obtained.

Enfleurage is a woefully inefficient process, requiring a great deal of hand labor and requiring far too much time to be practical in today's economy. The process is now "ancient history," no longer being used commercially.

The process of expression must also have originated before recorded history. Certainly the pressing of grapes and other fruits, to recover the juice was an easy method to adapt to certain F/F products. Since there are few natural F/F materials which contain high concentrations of liquids, yields must have been disappointing, except for the fixed oils which have only limited F/F use. There is little record in the literature on the process of expression until the latter part of the 18th Century, when sponge pressing of citrus oils in Sicily was first described.

Turning to modern day production methods, the basic processes in use today for the production of F/F materials can be generally described as distillation, expression and solvent extraction. They are carried out with varying degrees of refinement, complexity and efficiency, with the process selected and the relative sophistication with which it is applied being influenced by the raw material to be treated, the qualities desired in the finished product, market conditions, and the preference of the buyer.

Distillation

Distillation is the industry's primary production method. As mentioned in the chapter on Definitions (see p. 64) there are several different methods of distillation. These are employed for such various purposes as: to produce essential oils and isolates from essential oils; to concentrate extracts; to refine synthetics; to prepare raw materials for further reaction; to separate out impurities, off-notes and unwanted components, and so on.

If we consider the application of distillation to natural products only, the diverse problems which production people in this industry must solve become evident.

Considering hydrodistillation of natural spices, herbs, roots and the like, it is obvious that we are interested only in the volatile portion of the raw material. Thus we do not distill capsicum, civet, paprika or turmeric, since they contain no volatile portion of interest. If the components of the raw material will tolerate the heat involved in distillation (jasmine and tuberose flowers will not) a type of distillation is selected based on the particular factors involved. If the essential oil crop is subject to rapid deterioration, for instance by fermentation, or if it contains a highly volatile oil (with many low boilers and terpenes), it must be distilled as soon as possible after harvesting. Roses, geraniums and mints ferment; eucalyptus leaves, cedarleaf and pine needle oils are volatile and must be processed rapidly. In such cases, field stills, which in primitive areas could be little more than a modified steel drum, may be used. Occasionally wooden vessels, up to 500 gallons in capacity, or tin-lined stills are used to process raw materials right in the field. However, since transportation has improved markedly even in primitive areas, field stills are much less frequently used today than was the case 25 years ago.

Occasionally we find fairly large stills, of two to five thousand gallons, installed in fixed, but readily accessible locations. Some stills can be dismantled and moved with the harvest. Mobile stills are also in use. A still of this type which finds considerable use in processing mint oil crops is the Cobb still, mounted on a truck body. Widely used in the peppermint growing areas of the Northwest U.S., all that is required to convert it into operation is to attach a steam line to the truck body.

In some areas the essential oil bearing crop is brought to a regional still owned by an oil distributor who collects the crop for processing, or by a cooperative organized by a group of growers. A factory containing a line of stills can distill many tons of herbs in a day. Many crops such as geranium, palmarosa, lavandin, citronella, vetiver, etc., the cultivation of which was once a cottage industry, are now grown on large plantations, harvested automatically, and distilled in large, modern distilleries. The different growing, handling and distillation methods are as varied as the oils themselves. The various production methods employed are well described in Guenther's books on "The Essential Oils."

As mentioned earlier, the method of distillation is determined by the nature of the raw material, the characteristics desired in the finished oil, the standards of quality involved, and the preference of the purchaser. Years ago as many as one hundred essential oils were distilled in the United States from imported plant materials. Items which did not lose their oil content on drying, and which could tolerate shipment without affecting quality, were often brought to the United States to be processed in the modern, efficient equipment available here. Those companies having the facilities and capital to process their own oils had no need to worry about adulteration, or blending, and could be sure of producing oils of consistent high quality. As technology improved in the growing areas, as purchasers upgraded their quality control techniques, and as shipping charges became a more important factor affecting price, however, the economics of the situation finally directed that more oils be processed in the areas where they had been produced. Thus only about thirty or so essential oils are now still processed in the United States. These include clove, cinnamon bark, black pepper, celery and cardamom.

In the production of essential oils it is important to prepare the still charge — the bark, roots, seeds, leaves, nuts, etc. — for the type process to be run. Particle size of the charge is most important. The finer the grind, the greater will be the access of the steam to the oil glands. However, in practice, fine grinding can cause the steam to form channels. If this happens, the full charge will not be exhausted, and the yields will be lower than they should be. The type of plant material in the still charge determines particle size; seeds or berries should be cracked (celery), or ground (black pepper); roots are chopped (ginger); bark also is chopped (cinnamon).

Fragrance manufacturing plant of International Flavors & Fragrances, Inc. at Hazlet, N.J.

Some materials should best be water distilled. Although superheated or dry steam will drive off the oil faster, and must be used for oils which are composed predominantly of high boiling components, the higher temperature and dry conditions can dehydrate certain sensitive components such as tertiary alcohols. Water immersion permits better contact of the oil cells. Water distillation is a milder and slower process, which yields superior products in the case of some materials such as nutmegs or cloves.

Wet steam is normally used where sensitive components are not a factor, where high boiling components are minimal or in low concentration in the oil, or where the oil comes over at a reasonably rapid rate. (Mace and cardamom are such materials). Superheated or dry steam should be used where oils with high boiling points are involved, which would require a large volume of steam and a longer distillation period if another method were used.

When the components of the oil are particularly water-soluble and may not separate completely, the distillation waters may be extracted after separation, using such solvents as benzene, hexane and the like. In other cases the process of cohobation is used. This is a redistillation of the separated distillation waters. The change in concentration and distillation of the oil because of the smaller portion of water allows the additional solubilized oil to be separated. The oil produced by cohobation is usually of high specific gravity, and contains highly oxygenated components which are necessary for a true replica of the natural plant material.

Solvent Extraction

There are of course some F/F natural raw materials that cannot tolerate even the relatively low temperature of boiling water without degrading the sensitive bouquet of the end product. Thus even water distillation cannot be considered for them. These materials must therefore be processed by extraction. A particular solvent is selected to remove the desired components of the plant material. Jasmine and tuberose flowers are extracted with purified hexane or benzene, while animal products such as civet, castoreum, ambergris or musk grains are extracted with a more polar solvent such as alcohol. These items are never hydrodistilled. Some natural plant materials can be processed by either steam distillation or solvent extraction. Examples are rose, lavender or orange flowers, many gums such as galbanum, opoponax, olibanum, myrrh or elemi, and some roots such as ginger or costus.

In some cases it is the heat sensitivity of the material that makes it necessary to select extraction as the production process, but the principal reason for processing by extraction is to make certain that the non-volatile vital components will still be present in the finished product. Typical examples are black pepper and celery, the distilled oils of which are used in fragrance compounding. The flavor value of such oils is limited, since important flavor principles of the distilled oils are non-volatile. Thus these materials must be extracted rather than steam distilled. Extraction produces an oleoresin which contains the oil as well as the desired non-volatile constituents.

Solvent extraction yields an interesting series of products called oleoresins, which possess unique aroma and stability, and are widely used in condiments. Resinoids, balsams and resins, and floral absolutes and concretes are other highly useful extracted products.

Modifications of the solvent extraction process have resulted in the development of numerous special techniques that are not readily classified. Only thirty years ago, the normal yield of oleoresin ginger might run as low as a few pounds a week. This product was made by a cascade process which involved use of a series of ceramic crocks, set in tandem, one higher than its neighbor. The plant material was ground, placed in the topmost crock, and the solvent added. After standing for a day or so, the contents were drained into the next lower crock, and fresh solvent added to the upper crock. As the solution was drained off the lower crock, it was concentrated by solvent distillation and the recovered solvent started on its way again.

Now much higher yields are obtained on modern, stainless steel or glass-lined equipment which can hold tons of plant material. Such equipment can be jacketed to regulate temperature to any desired level, while circulating pumps keep the filtered extract or fresh solvent percolating through the charge, exhausting it fully and rapidly. A predetermined period of steeping may be included in the process. The charge is usually treated several times with fresh solvent, and portions of the extract may be pumped at selected times to a concentrator. The latter equipment may be a flash evaporator or a common still in which the solvent is removed either completely or almost completely, depending on whether a finishing operation is necessary to meet FDA or purchaser's specifications. Usually vacuum distillation is employed to remove the solvent at a low temperature, since heat will almost always cause changes such as resinification or charring of the concentrate. The solvent may be transferred to a rectification still where it can be dried, cleaned, and remade to specifications.

Such semi-continuous operations have made products such as oleoresins, resinoids and concretes readily available. Generally it is not the processing cost, but the cost of the raw material and the relatively low yield which makes these useful materials fairly high cost.

Expression

The use of the expression process in the F/F industry is limited generally to two main areas: the production of citrus oils, and fruit juices for concentrates. Steam distilled citrus oils are usually by-products, but oils of orange, lemon, lime, bergamot, grapefruit and mandarin, etc. are expressed by a variety of methods on a variety of machines. The method varies, depending on the type of fruit and the locality in which it is processed, but, so far as we know, there is no longer any hand pressing of citrus oils.

In the United States, the citrus oils are usually by-products of the frozen concentrate or canned juice industry. Thus production of citrus oils depends on the market for and the availability of fresh fruit. Since the fresh fruit market is the most lucrative one for producers, the amount of citrus oils produced may be limited in the absence of an exceptionally large fruit crop.

The fruit destined for concentrate or juice is graded, washed and sorted, then subjected to high speed extraction. Special equipment presses the fruit, sometimes in a stream of water, separating the juice and oil as an emulsion and setting aside the pulp and peel for other uses. The raw orange juice (water/oil/sugar etc.) emulsion is passed through medium, then high-speed centrifuges. The medium speed centrifuges separate a concentrated oil/water emulsion from the remainder of the juice. The juice eventually goes to a concentrator (film evaporators are one type) where water is taken off. Some essence oil evaporates with the water, is condensed, separated, and may be returned to the concentrate later. The four-fold juice is then frozen to produce the concentrate. The oil/water emulsion from the first centrifugation is finished in a high speed centrifuge, producing a clear, bright orange oil.

Lemon oil is produced in essentially the same manner. Instead of frozen concentrate, the single-fold juice can be stabilized or pasteurized, then sold as canned juice. Single fold juice must be kept cool to prevent fermentation; then it can be shipped by sea-going tanker from producing areas to packaging plants in the distribution areas. Production of the essential oil, however, is generally as described above.

In other areas of the world, where fresh fruit does not bring the price premium it commands in the U.S., production of citrus oils proceeds along different lines, for the citrus oil may be the principal product rather than a by-product. In the Mediterranean area for example, Italy, Spain, Greece and Israel produce citrus oils by a combination of Italian and American methods. They use machines which rasp the peel of the whole fruit, or penetrate the oil sacs in the peel with short, sharp needles. The released oil is washed off with water, and then finished by centrifuging in an operation similar to the American method. Where juice is not a factor, as with bergamot or bitter orange, only the peel of the whole fruit is treated.

In the United States other fruit juices than citrus are also produced by expression. Starting with the earliest harvest of such fruits as strawberries, peaches, grapes, plums, apples, cherries, blue berries and raspberries, the processing of fruit juices for use in the manufacture of flavor compounds is almost a year-round process. There is a sort of seasonal concentration on each fruit as high yield areas come into their peak harvest periods, but the processing of fruit for juice goes on almost year-round except for the coldest winter months. Even then stored juice may be concentrated to keep the factories operating year round.

Production of fruit juice is almost completely automatic. The fruit is graded, washed and may be depitted. It is then pressed by screw presses or roller pressed, the skins, seeds or pits and other hard or pulpy material screened out, and the raw juice is then enzyme treated (except for juice for jam) to degrade the pectins and starches. The juice may be pasteurized and cold stored, or may be concentrated immediately. The concentration equipment in a modern plant is highly engineered to remove only water, and to concentrate the fruit sugar acids and flavor components to the maximum extent. Film evaporators have a large throughput, and essence recovery units are attached to them to condense the volatile flavor components. The essence recovery units must be modified for each individual fruit essence, since the flavor components vary with the fruit. The concentrates are not fermentable if the sugar content is above 70° Brix (low water, high acids), or if a small amount of alcohol is added as a stabilizer. These fruit concentrates may be fortified with the essence, or may be used separately. They are the normal basis for flavor compounds whenever fruit flavors requiring at least some naturals are desired.

Users of fragrance and flavor products should be aware that products made by different processes will have different chemical composition. Even a process modification may change the character of the resulting end product somewhat. We have already mentioned the differences in character between water-distilled or dry steam-distilled oils. In the case of citrus oils such as lemon, a difference will be noted in the end product depending upon whether or not juice has been in contact with the oil in the course of the production process. When the whole fruit is pulverized, and a mixture of juice and oil results, the end product will have one character. On the other hand when the peel is rasped, and there is no contact between juice and oil, an oil of a different character is produced.

Also, a steam distilled lime oil will contain but little citral, but much *a*-terpineol, while an expressed lime oil will be high in citral, but practically no *a*-terpineol. It will also be high in coumarins, psoralens and waxes. Ylang ylang concrete is low in sesquiterpenes, while the steam distilled oil is high — around 65% sesquiterpenes. Orange flower absolute and its precursor concrete may have essentially the same components as the distilled oil (oil of neroli), but in much different proportions.

Essential oils and natural essences or other natural products are conglomerates, — that is they are composed of a variety of chemical entities of different structures and degrees of complexity. Some of these are desired, *per se*, others are unwanted because they cause off-notes or result in unstable products, etc. To cope with either or both of these situations, some essential oils are rectified. The rectification process chosen is usually vacuum fractional distillation. Chemical treatment might also be used. For example, clove leaf oil might be treated with dilute potassium hydroxide to remove the phenol (eugenol as the potassium salt), which is water-soluble. The unreacted oil remaining is separated, dilute acid is added to regenerate eugenol from its potassium salt, and the purified eugenol can be rectified if necessary. Citral is separated as the water-soluble, sodium sulfite addition product, the oil is removed, and the citral isolate then regenerated to produce a pure citral. Both eugenol and citral could be separated from the original oils by distillation, but the quality of the finished product would not be as fine.

In many cases a particular component of an essential oil may be separated out by any practical physical or chemical method, then used as a raw material for further chemical treatment. A favorite example is the fractionation of citronellal and geraniol out of citronella oil. Citronellal is then further processed to hydroxy-citronellal or to menthol through pulegol. The geraniol can be further processed to citronellol, dimethyl octanol, or the esters of the three alcohols. Citral produced from lemongrass oil was at one time the major source of *b*-ionone, from which Vitamin A was then synthesized. *d*-limonene, a by-product of orange oil production, is converted to l-carvone, the major component found in spearmint oil. The terpenes from cedarwood oil, *b*-cedrene and thujopsene, are the source of acetyl cedrene, a valuable woody fragrance note made by treatment with acetyl chloride, or other similar processes.

This discussion should clarify the point of how and why synthetics came into increased use as natural oils became scarce or expensive. The industry has long been accustomed to the concept of processing a relatively inexpensive material through a series of steps to arrive at more valuable end products. Thus essential oils or their components have been saponified (crude geraniol/geranyl acetate from citronella oil/to geraniol), esterified (linalool/from bois de rose oil/to linalyl butyrate), hydrogenated (citral to nerol), oxidized (3-methyl butanol/from fusel oil/to isovaleraldehyde), hydrated (hydroxycitronellal from citronellal), hydrated (guaiacwood oil to guaiene), or subjected to many other processes.

As a result of the tremendous progress made in synthetic organic chemistry over the past fifty years, it has now become possible to duplicate synthetically many products which could be made previously only from natural sources. The entire line of products derived from geraniol, linalool, citral and citronellal may now be made by three different synthetic routes. Vanillin, once available only by the oxidation of isoeugenol, can now be produced by chemical methods from lignin, a by-product of the wood pulp industry. Work is under way to make synthetic eugenol from catechol, a coal tar phenol. Similarly new methods are being worked on for the production of heliotropine, available up to now only from safrol, mainly from *Ocatea cymbarum.*

The number of aroma chemicals produced by the fragrance and flavor raw materials industry is so large that it is actually difficult to keep track of them. Some experts in the industry have placed the number as high as four thousand or more. In addition many so called "in-house" F/F compounds are manufactured which never enter into commerce, as well as specialty items of unrecorded composition. There is, without question, no other industry that produces a wider variety of end products or intermediates.

New Federal legislation, to be enforced by the Environmental Protection Administration, will require that all products manufactured, processed, used, stored, or imported by any chemical processor be registered with the E.P.A. Perhaps in the enforcement of these regulations some government agent will make a list of the products of the F/F raw materials industry and count them. This would then give those in it, as well as those who will have the responsibility for regulating it, some idea of the tremendous complexity of this industry. And, over and above this first list of actual products already in production are the hosts of new products and their scale-ups for pilot plant production.

The chemicals encountered in this industry run the gamut. They include hydrocarbons from one carbon to as many as fifteen, or occasionally up to twenty-five, exhibiting many degrees of unsaturation, aliphatic and olefinic, polycyclic, aromatic, alicyclic alkyl-aryl rings, and heterocyclics, each with many possible group and positional substituents and isomers, aldehydes, ketones, alcohols, esters and lactones, ethers, amines, acids, and nitrogen and sulfur containing molecules. Bifunctional molecules are not uncommon.

It goes without saying then that a wide range of processes and reactions is employed in producing and processing these endless items. A complete knowledge of modern organic chemistry is necessary for any company intent on remaining a leader in any area of the production of synthetic aroma chemicals. To these other requirements must be added the special demands of this industry, which involve precise control of odor and flavor, concern over toxicity problems, and the maintenance of food grade purity standards where required.

We list some of the more important reactions employed in manufacturing in the F/F industry, including the names of typical products of these reactions:

Saponification — used to produce rhodinol (the alcohols) rectified out of geranium oil after the esters have been saponified (using dilute potassium hydroxide).

Hydrolysis — production of benzyl alcohol by the reaction of benzyl chloride with dilute potassium carbonate.

Hydrogenation — using nickel, platinum or palladium catalysts, low pressures (11-300 psi) and high pressures (1200 psi, even as high as 2,000 psi on occasion). Phenyl propyl alcohol is made from cinnamic aldehyde. Methyl cinnamate is reduced to methyl cyclohexyl propionate, then trans esterified with allyl alcohol to allyl cyclohexanepropionate.

Reduction — non catalytic hydrogenations using sodium borohydride, lithium aluminum hydride, aluminum isopropylate, or sodium in alcohol. Cinnamic aldehyde is converted to cinnamic alcohol, or carvone is converted to dihydrocarvone by this method.

Esterification — is used to produce various esters from alcohols. The method may vary, and several variations are available depending on the alcohol. If a primary or secondary alcohol, it can be esterified directly. If tertiary, it would require anhydride, mixed anhydride or a hydrogen acceptor to prevent dehydration. Isoamyl butyrate can be made by direct esterification. For production of linalyl formate from linalool, a special method is required.

Trans Esterification or Alcoholysis — is employed to replace a portion of an ester (the alcohol moiety) with a high molecular weight alcohol. Benzyl cinnamate is made from methyl cinnamate or phenyl ethyl salicylate can be made from methyl salicylate.

Aldol Condensation — normally joins two carbonyl aldehydes and/or ketones to produce a hydroxyl intermediate which subsequently dehydrates to produce

Another view in IFF's Hazlet, N.J. manufacturing plant

an unsaturated carbonyl. Benzaldehyde adds to acetaldehyde to form cinnamic aldehyde. Citral adds to methyl ethyl ketone to form pseudo methyl ionone, then finally methyl ionone on cyclization.

Oxidation — may be accomplished either by catalysis or by chemical methods. Silver gauze in a hot tube (higher than 350° C.) will catalytically convert all the aliphatic alcohols to the corresponding aldehydes. The use of chromium trioxide or potassium dichromate will convert isosafrol to heliotropine. More recently ozone is being used as the chemical additive for this reaction.

The use of various other reactions could extend this list on and on. It will suffice, perhaps, to mention just a few of these special reactions, dispensing with descriptions and examples. They include:

- Darzens glycidic ester synthesis
- Meerwein-Pondoff reaction
- Oppenauer oxidation
- Freidel-Crafts synthesis
- Grignard reaction
- Cannizzaro and Cross-Cannizzaro reactions
- Nitrations
- Chloromethylations
- Hydrochlorinations
- Cyclizations
- Hydrations
- Dehydrations
- Epoxidations
- Alkylations
- Acylations

Equipment

The equipment used to process the myriad compounds this industry manufactures, by the extensive variety of reactions it employs, is usually fabricated from stainless steel or glass. Some monel, nionel, and in years past lead-lined and wooden reactors were used. Glass-lined equipment is the most inert, but cannot be used when processing strongly alkaline reagents because the alkali would attack the glass. It is also expensive and fragile. Since the aroma chemical plant must be prepared to process such a wide variety of products, the equipment used is mostly multi-purpose, readily converted from one operation, or one product, to another.

A reactor-still is the most popular piece of equipment. In such equipment a reaction can be run and the solvent or product distilled off, or the product washed, extracted, drained, etc. Equipment ranges in size from as small as 30 gallons or so up to perhaps 4,000 gallons capacity. In some cases production may be finished on a laboratory bench in a 22 or 50 liter flask. Large volume synthetics (production of above 500,000 lbs.) may have their own equipment line for semi-continuous operation. Essential oil stills often have a capacity of as much as one or two thousand gallons, and hold up to 12,000 lbs. of plant material. Extractors have an equally large capacity. Bulking tanks may be large enough to contain as much as 100 drums of oil. (Consider the tremendous investment a F/F manufacturer has in material in production, plus finished product inventory. Even with moderately priced oil, at $6 a pound, 5,000 gallons, or 35,000 lbs. in a bulking tank represents an investment of $210,000).

Vacuum distillation equipment is the most important in the plant. A variety of columns and column packings generally produces an efficiency of 10-30 theoretical plates. Certain terpene separations are run through columns having a 60-80 theoretical plate efficiency.

Reactors and stills are heated by steam, oil or Dowtherm, or directly by natural gas in completely contained apparatus. Vacuum pumps are important because it is essential to have a low vacuum to reduce the heat required for distillation, and to prevent charring or degradation of sensitive components. The vacuum is usually produced by 3- or 4-stage steam ejectors or high efficiency vacuum pumps.

Pressure vessels or autoclaves are used most commonly for catalytic hydrogenation. They range in size from fifty to 4,000 gallons in capacity. Autoclaves are also used for flavor reaction products, hydrolysates and the like, as well as for alkylations and oxidations where one of the reactants is a gas.

It is sometimes preferable to be able to offer in solid form F/F products which are initially produced as liquids. In such cases they may be made into dry solubles, or mixed or absorbed on salt, starch, cerelose, etc. They may also be spray dried. In this operation the liquid F/F product is mixed with a solution of gum acacia, gelatine, dextrin or starches, then homogenized, pasteurized, and forced through a small orifice or through an air-turbine atomizer to produce a fine spray. The spray is injected into a conical shaped drier where a counter current of hot, dry air, heated electrically or by a gas heat exchanger, swirls the particles, drives off the moisture, and encapsulates the F/F liquid.

The size of the capsule can be regulated by the size of the sprayed droplet. The heat removes the water, with but little temperature increase in the capsule. The fine capsules are collected at the bottom of the drier and screened for particle size. The amount of F/F product in the capsule is usually 20-30%. It is a free flowing powder.

Solids are readily handled in the aroma chemical plant. They are purified by controlled temperature recrystallization, then separated on centrifuges, washed, spin dried and discharged for further drying. The melting point of the solid determines the method of drying selected. Many solids are vacuum dried and flaked if crystal size or shape is not important.

It should be apparent that production in the F/F industry is highly technical and well engineered. All equipment is explosion-proofed, even to the telephones. F/F factories must abide by the most rigid safety regulations, meet FDA and OSHA standards, and comply with the environmental clean air and clean water requirements of the Federal E.P.A.

Almost all companies in this industry are involved with production of some sort, only the pure compounding houses and brokers being exempt. Since many companies market finished products which require a large number of components, some of which they must purchase from competitors, sooner or later they develop their own specialties. No aroma chemical company produces all its own raw materials. There is a high degree of interdependence among members of this industry.

In view of this dependence of aroma chemical producers on one another, reputation is a most important asset for any company. Word is quickly passed along as to the dependability and product quality level of every supplier. Because the industry is so interdependent, there are few production secrets, and each firm has a stake of its own in seeing that its competitors solve any production problems they may encounter.

Chapter XII

The Scientific Side

by Ernst Theimer PhD.

Although the use of natural materials as flavorings and fragrances dates from antiquity, organized attempts to produce, study, improve, manufacture and market products for these purposes did not emerge until the middle of the 19th century. It was then that the flavors and fragrances industry was born.

In the early days, the main thrust was towards the improved utilization of known natural raw materials, but the burgeoning science of organic chemistry, which had already achieved successes in the dyestuffs industry, was quickly extended into the field of odorous compounds. Soon flavor and fragrance companies underwent a transition into chemical processing concerns, first slowly and tentatively, then with an increasing momentum until now natural products constitute only a minor source of fragrance materials, at least on a volume basis. This is not to say that natural products are unimportant, for without them many of the fine perfumes made today would not be possible, but at the same time the ever-increasing stream of quality-controlled fragrance materials which flows from the industry would be beyond its capacity without the predominant use of synthetics.

Incidentally, our emphasis on fragrance in the following discussion should not be construed as a conscious neglect of flavors, but rather as an attempt to put both types of products into the picture together. Although flavor materials have their own special characteristics, requirements and problems, it should not be overlooked that, except for the four primary elements of taste, namely sweet, sour, salty and bitter, and some ancillary effects such as "mouth feel" and trigeminal stimulation, flavor owes its character mainly to its effect upon the olfactory system, that is the odor sensation evoked. In this context, flavors can be assumed as included in the subsequent discussion wherever fragrance is mentioned.

Another unifying concept for flavors and fragrances is that both taste and odor are stimulations of the chemical senses, since the appropriate nerves receive primary signals from the receptor cells in the nose and mouth when either of these senses is activated by the incursion of *molecules.* By contrast, the other senses, touch, hearing and sight, are activated by *energy* stimuli.

Unfortunately for the early flourishing of the flavors and fragrances industry, these concepts were at the time entirely unknown. The mechanism by which the ofactory system functions was shrouded in mystery, as also was the taste process. Even now, only a beginning has been made in developing knowledge in this area, while, surprisingly, the mechanisms of the physical senses have been unravelled far more completely. The relationship between molecular structure and odor is an area of study which has occupied the attention of scientists only in recent years. It was an entirely unknown concept in the early years of the industry.

It is scarcely surprising, therefore, that early progress was entirely empirical in nature. Research, the exploration of the unknown, extended in four major directions.

1. Process improvement in the production, isolation and separation of natural essences.

2. Investigation of the components of natural essences, and determination of their chemical structure.

3. Synthetic production of naturally occurring fragrance compounds.*

4. Discovery of new fragrance products, not found in nature, and development of commercially feasible processes for their manufacture.

The first area is discussed elsewhere (Chapter XI pp 269-90) and need not concern us here. The other three areas suggest the groundwork for a master plan in the organization of a research department, which could well serve as a cornerstone in the structure of an establishment manufacturing aroma chemicals. It is a generally accepted working formula that the rate of growth, and in some cases even the survival, of an industry, and the component companies in that industry, depends upon the fraction of the sales dollar that is plowed back into research effort. Some industries by their very nature do not require and, indeed, may not be able to afford a large research program. Others, for example the pharmaceutical industry, lie at the other extreme. The flavors and fragrances industry has taken a similar stance, and its investment in research (and development) has been surprisingly large historically. This may account for the overall rapid growth of the industry.

The precise definition of research expense is a matter of some interest. If research is taken to include the cost of creating finished fragrances, involving the work of perfumers who blend the necessary ingredients to produce the final products, our industry's research cost may be higher than that incurred in most other industries on a percentage basis. But even if the research cost calculation is restricted to the scientific phase, involving nothing more than the development of basic ingredients, the total still works out to a very substantial sum.

As in all programs involving skilled, highly-trained personnel, about two-thirds of the F/F industry's research

*the word, "compound,"is used in the chemical sense of a discrete molecular species, in other words a single chemical. Unless this is clearly understood, confusion may arise by confounding this word with the much used term, "compound," meaning a mixture of materials intended for fragrance purposes.

cost goes for salaries. In addition, unusually large capital expenditures are required for special, highly sophisticated instruments, which, when used correctly, enable scientists to probe effectively into the composition of mixtures and to determine the structure of pure compounds. These analytical tools give a well-staffed and outfitted research organization a substantial advantage over more modestly programmed groups, even when the latter are intellectually on an equal level.

This situation did not always obtain. In the early days of the industry, research was on a more primitive, free-wheeling and even intuitive basis, and the individual creative chemist had more chance to star. His tasks might include all three areas (2, 3 and 4 mentioned previously), which today are each generally assigned to a separate specialty group. Let us examine each of these areas in turn, as they were handled before and after the advent of modern instruments.

Composition of Natural Essences

Separation was usually performed by fractional distillation, which was more or less complete depending upon the efficiency of the fractionation column. Thirty theoretical plates was a goal seldom achieved, since most essential oils are unstable at the high reflux rates and the consequent long residence times required in the still at elevated temperatures. Vacuum fractionation, used to bring operating temperatures down to tolerable levels, lowers column efficiency drastically. Large samples (several hundred grams or more) were needed as still charges, but often the amounts available were inadequate.

The fractions from the distillation could be analyzed by physical means: refractive index, specific gravity, optical rotation, melting and boiling point, and for chemical composition.

carbinol (hydroxyl) group by acetylation

carbonyl (aldehyde or ketone) group by oximation or combination with sodium sulfite or bisulfite to form water-soluble addition complexes
ester group by saponification,
and so on.

These were functional group, not specific compound analyses. Since the original separation was usually incomplete, all these data were only approximations.

Molecular weight could be calculated, hopefully to ± 2 per cent on pure compounds, by determing the effect on the melting point of camphor of an accurately weighed amount of the unknown.

Elemental analysis for C, H, N, Cl and S by combustion gave the empirical structure of a compound with reasonable accuracy. This, together with the MW (molecular weight) functional group information and physical constants was often sufficient to identify a compound *provided it was already in the literature.* This elemental analysis, which in the early days required several hundred milligrams, was refined by Pregl's microanalytical techniques to give equally accurate results on samples one thousand times as small. Weighed on analytical microbalances to hundredths of milligrams, the samples could be analyzed in a matter of minutes. To analytical chemists of the 1930's it must have seemed that the millenium had arrived.

All this background information, which today is to a large extent of merely historical interest, although many of the techniques are still in use for special situations, serves to emphasize the revolutionary impact of the new techniques developed during the past quarter century which have completely changed the science of chemical analysis. To understand precisely how these new techniques work, the student requires at least an advanced degree in physics and in physical and organic chemistry. To appreciate what they have done in changing the industry requires only a mature consideration of the results that can be obtained by using them intelligently.

The best fractional distillations are great improvements over those of past decades. Modern techniques, employing molecular distillation, spinning band columns and others with more efficient packings, better vacuum pumps and leak-proof systems, all combine to provide better separations on smaller amounts of material. Although distillation can now provide tenfold better efficiency (up to several hundred theoretical plates) on as little as several grams of still charge, the chemist will rarely, if ever, use this method of separation for other than preparative or production purposes, for he now has at his command the technique of gas liquid chromatography.

GC, also abbreviated as GLC or GLPC, separates mixtures with efficiencies reaching a million theoretical plates. Sample size can be as little as a fraction of a microgram (10^{-6} gram) and the time required is usually a matter of minutes. It is truly an analyst's dream, but as with all good things tremendous pains and effort have had to be expended to achieve best results. Correct use of the technique remains an advanced science and, some say, practically an art. Here, a rough picture must suffice.

The sample to be separated, often in a solvent, is vaporized rapidly into a hot chamber of flowing gas — usually helium or nitrogen for technical reasons and this mixture is carried through a tube (column) containing a liquid phase absorbed on the surface of particles of a solid support. The components of the mixture comprising the original sample are taken up by the liquid phase and revaporized constantly as the mixture passes through the tube, which is held thermostatically at a desired temperature. The instrument — the gas chromatograph — can be, and often is, programmed automatically to a continuous or intermittent temperature rise, for efficiency reasons.

Since each re-evaporation at points on the liquid phase constitutes a separation based on the difference in boiling point and/or polarity characteristics of the components of the sample, a sufficiently long tube can usually give complete separations. If necessary, the partially separated portions of especially difficult mixtures can be trapped and re-chromatographed. Thermal hazard — the degradation of structure due to heat — is usually not significant, since time of exposure is much shorter than in distillation, and the temperature can be controlled within wider limits. Materials of construction also play a part. The original columns were of metal, but these have now largely been replaced by tubes of inert glass.

Obviously, considering the minute amounts of material involved, nothing will be seen to emerge from the end of the column. This is, however, no problem. At the exit of the column is a detector which responds to the presence of vaporized material in the carrier gas. Generalized detectors measure the amount by the change in thermal conductivity compared to pure carrier gas. Specialized detectors respond to particular atoms or groups only — halogen, sulfur, nitrogen, amino groups, etc. — and are therefore analytical instruments in themselves.

The detectors are coupled with pens traveling on recording charts. These produce gas chromatograms to tell the story of the GC analysis for subsequent study. The abscissae record time of elution, while the ordinates (peak heights) indicate the amount of response for each peak, or separated component, as part of the total. Sharpness of a peak depends upon the efficiency of the column. Incomplete separations will result in overlapping peaks. Formerly the quantitative composition of mixtures was calculated by measuring the area of the peaks, or counting the squares under the curve on the quadrilled chart paper. This is no longer necessary. GC instruments can now be plugged into computers which print out clearly the time of elution and per cent by weight or molecular fraction of each component, even if the separation is incomplete or not even recorded by the pen of the recording chart, as long as the separation actually exists. These charts are usually valuable as visible guides, particularly in cases involving new materials. To prevent pens from going off-scale, instruments are fitted with automatic attenuators which eliminate the need for constant attendance during a run. In fact, maximum use of these costly instruments is made possible by round-the-clock automatic operation with automatic injection of samples by power-actuated hypodermic syringes, and instrument programming.

For purposes of quality control, it is obvious that a GC chromatogram approaches a fingerprint in its ability to determine the structure and monitor the purity of simple fragrance materials and the integrity of complex mixtures. But such a chromatogram by no means represents the extent of GC potential as an analytical tool. By varying the nature of the liquid phase, its concentration, column packings or linings, column length, oven temperatures and flow rates, the relative elution times of the components can be altered. A chromatogram from a non-polar column will usually be different from one made from a polar column for the same sample.

Thus GC alone makes control of fragrance materials, which are essentially volatile products, a much more precise technique, and disposes with ridiculous ease of what were previously considered refined attempts at "sophistication." It supplements the still-used wet analytical determinations mentioned before, it monitors the integrity of natural products, and permits precise control of synthetic fragrances, no matter how complicated. Used by manufacturers and buyers of fragrances alike, it has tightened specifications on products for the entire industry. When used in the study of materials with unindentified components, it has become the first essential step in separating these into pure isolates for identification.

Investigation of unknowns

This is an area of research which requires not only the separation of mixtures into their components, but also identification of the components, which may or may not be known compounds. Whereas GC is usually sufficient by itself in quality control, (by comparing the quantitative relationship of peaks), other disciplines are required if the product being studied is an "unknown." Here also modern instruments have in great measure replaced the wet analysis methods of classical chemistry. These instruments delve into molecular structure by various methods.

Spectrometers — Depending upon their composition, organic molecules absorb light rays of various wavelengths along the electromagnetic spectrum. If they absorb rays which lie in the visible range (about 400-800 nanometers), they are said to contain chromophores, and the compounds are colored.

Some organic molecules contain groups which absorb rays of shorter wavelength (about 200-400 nanometers). These can be detected by the use of UV spectrometers, which measure the amount of ultraviolet light absorbed by the correctly positioned sample as the wavelength of the incident light is varied over the desired range. The first UV instruments employed for analytical purposes were manually operated, wavelength by wavelength, and the data were gathered in tables which could be graphed. Soon these early models were replaced by recording instruments which delivered the graphs directly.

Although UV spectra are relatively simple to obtain, (given the expensive spectrometer) and require only a minute amount of recoverable sample, they are useful for only a small percentage of molecular species, those containing UV "chromophoric" groups. These include conjugated systems such as unsaturated ketones or esters, aromatic (benzenoid) structures, and unsaturated lactones such as coumarin. It is because of their property of absorbing UV rays, the skin-burning rays, that many of these materials are used as the active ingredients in sunscreening preparations. These can be tested rapidly by UV in the laboratory in preliminary testing.

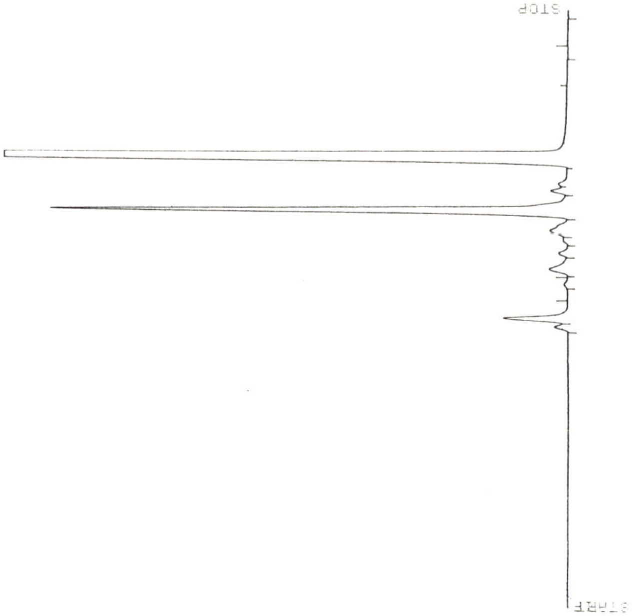

Gas chromatogram of commercial grade Geraniol 96/98

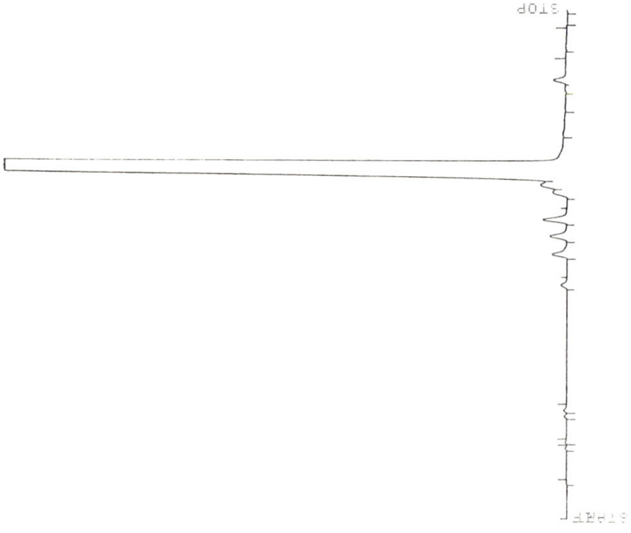

Gas chromatogram of Alpha Terpineol FCC

It must be emphasized, however, that the artificial UV light in the laboratory has differing spectral characteristics from sunlight. In addition, compounds which appear to be very effective when examined for short periods in UV-transparent cells in the spectrometer may fail under actual use conditions. One reason is that the energy absorbed by the molecule as UV radiation can destroy the molecule by rupturing bonds, rendering it worse than useless. Other molecules get rid of the same energy by re-emitting it as higher wavelength radiation (fluorescence), harmless to the skin. Such a material would prove far more effective than laboratory tests would indicate.

In contrast to the specialized utility of UV spectrometry as a diagnostic tool, IR (infra red) spectrometry is of great value for structure elucidation of organic molecules. All absorb some IR rays, and the absorption areas are narrower, often characteristic of specific groupings within molecules, and, similar to UV, proportional in intensity to the amount of chromophore present. When recorded on chart paper, the IR spectrum is often sufficiently characteristic to identify a pure compound by comparison with a standard spectrum. As libraries of standard spectra become ever more complete, this method increases in value.

If the compound is an unknown, this identification by comparison will fail, but functional groups and other important features of the structures are revealed. For example, hydroxyl group absorptions can scarcely be mistaken, aromatic groups can often be classified as mono, di, or tri substituted, and even whether ortho, meta or para in position, a detail once quite beyond the ability of other than the most complicated chemical degradation reactions to reveal.

As a quality control technique, IR spectrometry can be used in some cases to determine quantitatively the amount of a given compound in a mixture, provided no interfering substances are present. The per cent of citral in lemon oil, for example, can be determined from the conjugated

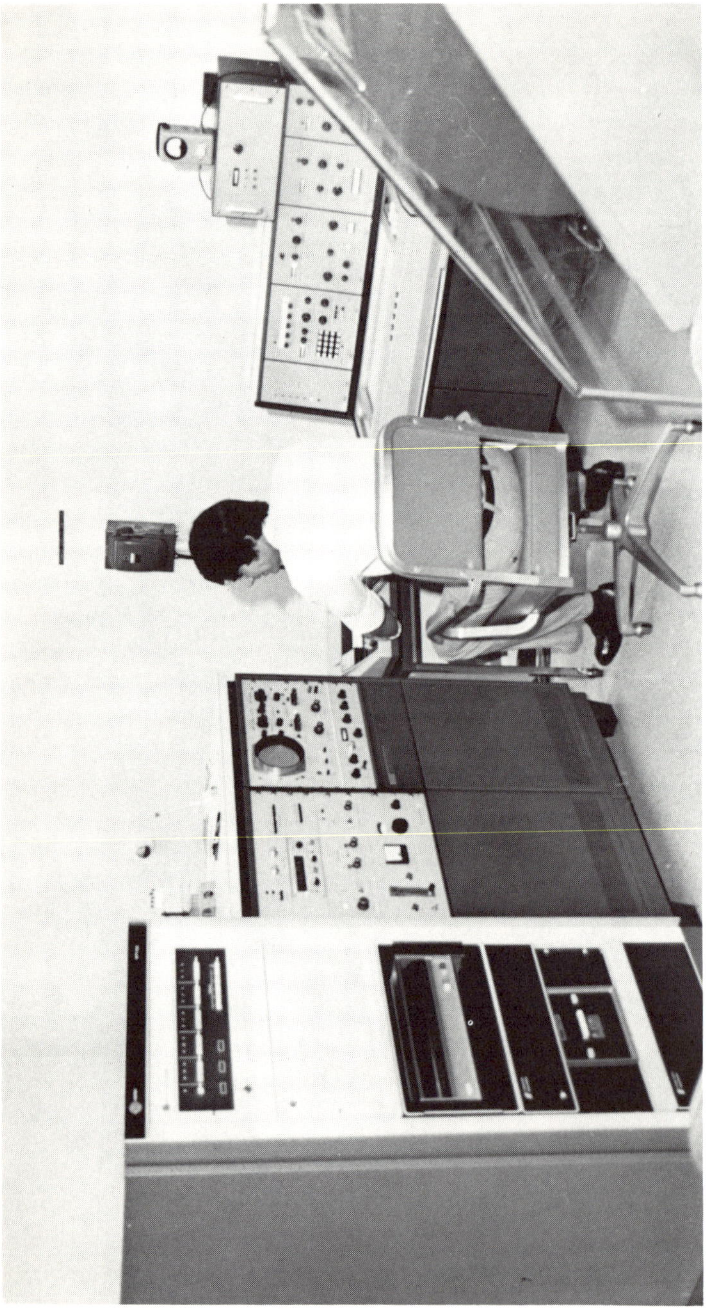

A modern NMR Laboratory.

carbonyl absorption, specific for citral in the mixture, and the analysis is more accurate and rapid than chemical methods such as oximation. In addition, much less sample is required, and even that can be reclaimed since the technique is non-destructive.

A specialized type of IR spectrometer, sometimes useful in unusual cases to supplement standard IR, is the Raman spectrometer. This highly sophisticated device utilizes the laser beam as energy source, and can reveal certain absorptions not seen in ordinary IR spectra. The instrument is part of the armamentarium of the best equipped laboratories.

The spectrophotometers discussed up to this point measure the response of molecules to the impingement of rays of the electromagnetic spectrum, but there are other spectrometers which are based on another principle and which reveal chemical structure with far greater precision and in greater detail. They are based on the magnetic properties of the atoms themselves. Since an atom consists of a nucleus and electrons, each can be examined separately. Electrons may be studied by scanning their spinning with ESR (electron spin resonance) spectrometers, but this is of limited use for structure elucidation, and finds its main use in physico-chemical work. It need not concern us here. By contrast, the study of the nucleus by NMR (nuclear magnetic resonance) spectrometry is by far the most powerful tool for structure elucidation ever devised. It is not unusual for an NMR spectrum to reveal the complete unambiguous detailed structure of an unknown compound in a matter of minutes on a fraction of a milligram of sample (which can then be recovered).

The technique is, in principle, not complicated. Many atomic nuclei have their spinning characteristics altered by subjecting them to a strong magnetic field. The sample in the field can be scanned by high-frequency radio waves and each nucleus of a species pinpointed by its response to the scanning field as shown on a recording chart. Most important to the value of this for structure analysis is the fact that the frequency at which the nucleus resonates, and therefore shows up on the chart, *depends upon its environment within the molecule.*

What this means in practical terms can be understood only generally in terms of a lay-oriented discussion, although we will attempt to explain. Let us take as an example a typical molecule — that of geraniol, with empirical formula C_{10} H_{18} O. Oxygen does not show up on NMR, nor does carbon of atomic weight 12. However the natural abundance ratio of carbon isotopes in nature is 99-1 for carbons 12 and 13, and the carbon 13 isotope, which *can* be seen on NMR, can be seen in its various chemical orientations. C 13 resonance spectrometry is being used currently with great success in spite of the fact that this isotope comprises only 1 per cent of any material.

Here, however, we shall disregard carbon and discuss only NMR spectra due to the hydrogen nuclei. Since these are protons, the spectra are called proton magnetic resonance (PMR) spectra, and they alone are sufficient for the most detailed insights into molecular structure. Incidentally, hydrogen and carbon 13 show up at entirely different places in such spectra, and there is no overlap, thus no interference.

Returning to our consideration of geraniol, we note that the geraniol molecule has three methyl (-CH_3) groups, a primary carbinol (-CH_2 OH) group, and various other hydrogen on carbon groups, including two on double bonds. *Each group* will show as a peak at a different position in the spectrum. Each will be of an intensity equal to the number of hydrogens it represents. Each will have characteristics depending on its chemical environment, so will be a singlet, doublet, or other multiplet depending upon the number of hydrogens on *adjacent* carbon atoms. The hydrogen of the (-OH) group is in a class by itself, and the other two hydrogens of the carbinol group will be in a special position because their carbon is also attached to oxygen. Even the olefinic (on double bond) hydrogens are different, since one is closer to an oxygen atom than the other. This hydrogen can be differentiated from the corresponding one in nerol, since one is trans and the other cis to methyl across the double bond. Thus the close isomers, nerol and geraniol, can be easily differentiated.

This is, of course, only an inkling of what NMR can do, and has done, to revolutionize organic chemical analysis. Perhaps it will serve to whet the appetite for further knowledge. An anecdote will readily illustrate what a powerful tool modern instrumentation is. In the early days of NMR, two perfumers from competing fragrance houses were having a few friendly beers. In rather a mellow mood, one showed the other a smelling blotter which had been touched with a new fragrance material from their research department. The other was much taken with the product and asked timidly if he could have a small sample. This of course got him nowhere. "Show your group that paper, if you like," he was told, "but that's the only sample you'll get." The next day a chemist from company No. 2 extracted the paper with deuterochloroform and ran an NMR. Although a sample in those days had to be considerably larger than now, the small amount obtained was sufficient, and within a few hours the exact structure of the compound was known. NMR is one reason why new discoveries are now generally protected by patents, rather than relying merely on secrecy as before.

Supplementing the technique just described is mass spectrometry (MS), which fragments molecules by ionization into groupings, sorts them according to weight, and records them. This gives a spectrum which tells much about molecular structure and, like IR, can be used as a "fingerprint" identification, provided again that a standard spectrum of the compound in question is available for reference. In this case the spectra are strictly comparable only if they were run under very similar conditions.

Since the injected molecules are broken down, this is a destructive analysis, but the amount of sample needed is so small (a fraction of a microgram is ample), that this becomes academic. A portion of the molecule gets to the recording site whole, and this gives the molecular weight accurately. How accurately depends upon the degree of sophistication of the MS machine, but even the simplest instrument gives the result to the nearest whole number, which is much better than the old molecular weight by freezing point method, mentioned previously. The more sophisticated spectrometers, called high-resolution, give the MW to three decimal places: This is not just gilding the lily. Atomic weights are not whole numbers, and although we may be accustomed to thinking of CH_4 as 12 + 4 = 16, the same as oxygen, actually H = 1.008, and the greater precision of the instrument allows differentiation between these moieties. MS even separates isotopes, so that chlorine does not show as 35.5 (roughly), but as three parts chlorine 35 and one of chlorine 37. This usually is enough to establish the presence of chlorine in the molecule and to distinguish it easily from any other halogen. The old Beilstein copper wire — flame test is still used at times to establish whether a product is FFC (free from chlorine), but it requires a full drop of liquid and is not always reliable. However one would scarcely use a $100,000 instrument and valuable time for routine testing unless absolutely necessary. The new analytical techniques must be used intelligently.

This is perhaps best illustrated by another anecdote. A freshly-minted Ph.D., working on his first assignment in industry, had separated and identified all but one of the components of a rather complex reaction product. One minor component still needed to be completely characterized. His university training had been thorough in GC, MS, IR, UV, NMR, and even Raman, and by these means he had resolved the entire problem (in about a month) except whether there was a double bond in the molecule. (Tetra substituted double bonds are exceptionally difficult to spot, and often even Raman is unavailing). Finally, in despair, the young Ph.D. took the problem to his supervisor, who had not forgotten that there was such a thing as chemical analysis. He added a drop of bromine in carbon tetrachloride to the unknown and the bromine was immediately decolorized. There was a double bond.

The major components of most of the important essential oils and other natural products had already been identified in the early days of the industry, using the cruder methods then available, but the new instrumentation has made possible fruitful study on more complex and less stable fragrances, and especially flavors, which have many hundreds of trace components, some quite unstable when isolated in pure form. It was by then no longer necessary even to isolate the individual compounds by trapping from GC. Combination devices such as GC-MS can separate and identify a complex mixture while the analyst stands by and observes.

A small sample, usually a fraction of a milligram, of mixture is injected into the port of such a machine, which usually has a capillary glass column for the GC separation. Such columns, which are routinely drawn into coils in the laboratory are several hundred meters in length and achieve the highest possible degree of resolution. As each component elutes and shows its presence as a peak on the chart, a small amount is by-passed into the MS part of the machine, and its spectrum recorded. The spectra can be taken so rapidly that a number can be completed during the time that the peak is eluting, even when this is but a few seconds. If the spectra at the beginning and end of the peak elution time are identical, it is likely that the peak represents a pure compound.

This technique will process a mixture of hundreds of components within a few hours. The data from the run are fed into a computer which reports the analysis by separated peaks on a quantitative basis. At the same time, the MS data are compared by the computer with the spectra of reference compounds stored in its memory. If this library is complete and the compounds are all known, the entire analysis has been accomplished and is reported in plain language as a series of identifications. Usually, some unknowns remain.

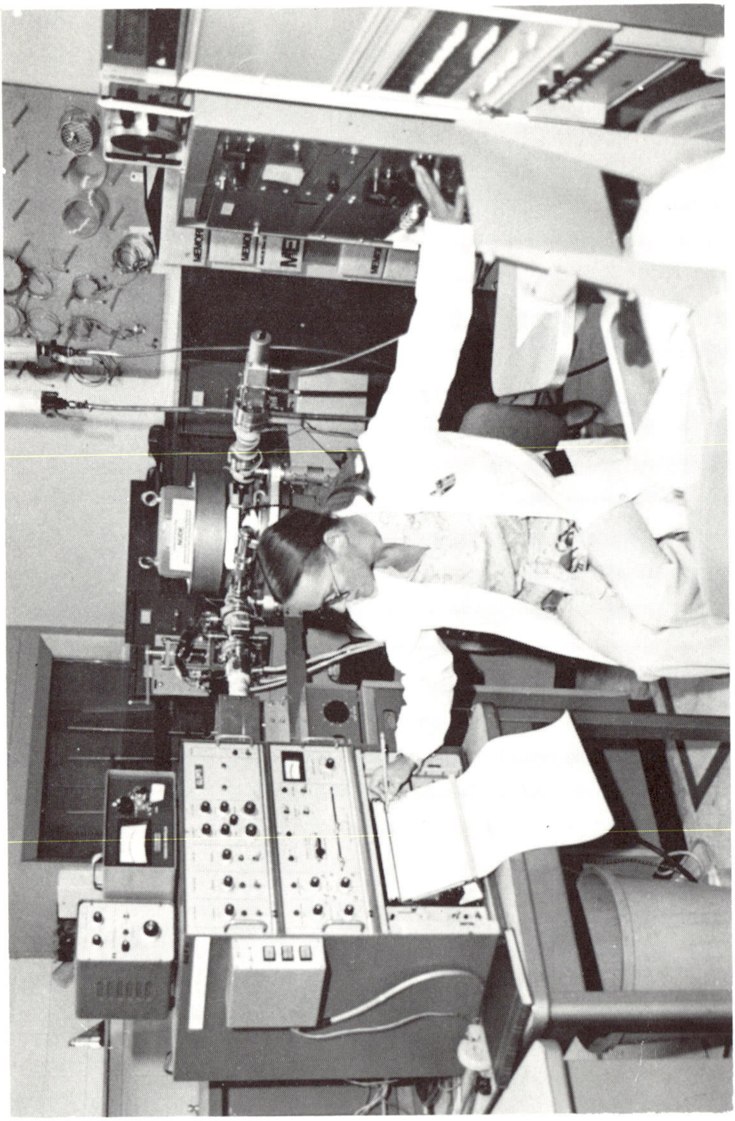

Computer interfaced GC-MS

These unknowns may be studied further by the other spectral techniques, some of which are already being coupled with their own GC columns and connected with reference spectra-supplied computers. Herein lies the developing part of the analytic function at the present time.

Polarography is an analytical method suitable for use in analyzing one or possibly two components of a complex mixture, the composition of which is not known. The technique of polarography, has been known for many years, and has frequently been used for analysis of inorganic systems. Measurement of oxidation-reduction potentials permits determination of metallic ions in very low concentrations (parts per billion). However, most organic systems contain interfering substances which preclude use of this technique.

Nevertheless, polarography may still be used in special cases involving oxidizable or reducible organic compounds, provided a suitable procedure is first worked out for the particular mixture. For example, content of an antioxidant in a mixture can be assayed quickly with an accuracy of parts per million. This has been made possible by modern, improved instrumentation, including a three-electrode cell and accurate metering pipettes.

For mixtures which do not lend themselves to alternate analytical procedures such as GC or TLC (thin layer chromatography), polarography sometimes becomes the method of choice.

Synthetic Production of Naturally Occurring Fragrance Compounds

The role of advanced analytical techniques in clarifying chemical structure is perhaps best exemplified by the ionones. These aroma chemicals made their impact in the early days of the industry, as synthetic fragrances, discovered accidentally, whose structures were but imperfectly understood.

In their search for synthetic routes to the valuable but expensive violet aroma contributed by irone to orris root,

Tieman and co-workers condensed citral with acetone and cyclized the resulting compound to "ionone." The product, although not irone, proved to have valuable fragrance quality. Realizing that ionone had one methyl group less than irone, the researchers substituted methyl ethyl ketone for acetone in the synthesis. This gave "methyl ionone," isomeric with, but different from irone, but also a highly useful new aroma chemical.

In the early days of these products it was known that the intermediate products from the citral-ketone condensations were acylic ketones of high refractive index and low specific gravity. Cyclization to the final products raised the specific gravity and lowered the refractive index more or less, depending upon the strength of the cyclizing acid. Ionone could be purified by combining with sodium sulfite and extracting the non-ketonic impurities before regenerating the ionone. Methyl ionone, however, reacted only partially with sulfite or bisulfite, and although the regenerated methyl ionone was now clean and pure, it had changed in character from the original reaction mixture, not necessarily for the better. These sulfite combination reactions were of no use for analytical purposes, since they were very slow and incomplete at best. Such was the state of the art in 1930.

By 1940 it had been found that the reaction with hydroxylamine was quantitative and could be used to assay the purity of ionones as far as the total ketone content was concerned, but with no regard to the ratio of isomers. Physical constants and boiling points were of adjunct assistance in this regard.

Today there is no longer any mystery about these compounds, and it is interesting to look back at what caused the original confusion. Cyclization of citrylidene acetone (pseudo ionone) gives a mixture of three isomers, alpha, beta and gamma ionone, in proportions dependingupon the conditions. Alpha or beta can be obtained fairly pure, but gamma is at best a minor product, although this is no serious problem since its fragrance value is not much different from that of the alpha isomer.

The situation with the methyl ionones is complicated by the fact that methyl ethyl ketone can react in two ways with citral to give eventually two series of methyl ionones — normal and iso — each possible in the three forms, alpha, beta and gamma, according to the position of the ring double bond. This is by no means the limit of the isomeric possibilities, since some of these isomers exist as racemic mixtures of their chiral *d* and *l* forms, which may be separated as discussed elsewhere. Also there exists the probability of cis and trans double bond isomers. All these are mentioned merely to emphasize the complexity of fragrance chemistry, for the nose is capable of appreciating all such differences. This point need not concern us here. It is quite sufficient to deal with the problems of commercial significance.

In the case of the methyl ionones, these involve mainly four of the possible six isomers enumerated, since the gamma double bond isomers can be disregarded. Alpha normal methyl ionone (1), beta normal methyl ionone (2), alpha isomethyl ionone (3) and beta iso methyl ionone (4) all show up as components of the usual reaction product. To make matters even more complicated, they were originally designated as alpha, beta, gamma and delta methyl ionones respectively. The important alpha isomethyl ionone is still often referred to in the trade as gamma methyl ionone.

Isomer (1) combines with sodium bisulfite; (3) does not. This accounts for the loss of odor character when mixtures are chemically "purified" in this way, for (3) is now recognized as the most valuable isomer. In fact it is so desirable that in the days before distillation techniques were sufficiently effective to separate (3) as completely as desired, it was instead isolated by forming the solid semicarbazone with semicarbazide reagent, washing and regenerating. This was possible because (3) reacts preferentially to the other isomers with this reagent. However, the cost of this purification was prohibitive for other than the "fanciest" intended uses.

Today, routine GC permits total separation and quantitative determination of all the isomers of methyl ionone in mixtures within minutes. The analysis also determines any ionone or non-ketonic impurities present from the process. The various isomers have long since been subjected to spectral analysis and their exact structures confirmed. GC has been of value also in research on process modification to improve yields of the desired isomers.

As a final note, it is interesting to consider that irone, the pursuit of which led originally to the discovery of the ionones, is now available commercially from a process which is also used for manufacturing ionones. Although prized for special effects, irone has not, as might have been imagined, put the ionones out of business. In fact, its use is at rather a modest level, considering that it can be obtained at only a slight premium over the price of what are after all "ersatz" products.

The synthesis of naturally occurring fragrances in the laboratory, followed by their commercial production, has followed naturally from the recognition of their chemical identities. It is obvious that improved analytical abilities are a great boon to the organic chemist, but it may not be as apparent that their appropriate use has changed the entire thrust of the research technique. In years past, the standard procedure for determining optimum conditions for a given process was to run numerous experimental batches at various conditions of concentration, time, temperature and so on, and to "work up" the end product by distillation or otherwise for yield and purity. Each run had to be on a multigram scale to give a workable amount of product. The cost of such a study in terms of materials used, and more important in man hours, can be imagined.

The enlightened research chemist knows, however, that the function of reseach is to produce data, not product. He will therefore make a run, which can be on a very small scale to save time, and at periodic intervals withdraw a small sample of the reaction mass. After suitable preparation, perhaps by neutralizing the acid or base catalyst, this sample is analyzed by GC together with any required spectral techniques. In this way a series of samples from one run gives the entire history of the run in terms of yield of desired product (which may have reached a maximum and then dropped off), as well as the rate of formation of by-products. The cost of gathering any particular set of data has thus been cut by a factor of at least ten.

Discovery of New Fragrance Products

In the search for new fragrance products to supplement those found in nature on the perfumer's palette of working materials, the same factors come into play. There are also new criteria which have developed with the enhanced ability to examine chemical purity and composition. These criteria vary with the nature of the product. They may best be perceived by examples.

Tertiary Butyl Phenol — When tertiary butyl phenol became available as an inexpensive chemical intermediate, it was quickly seized upon as a likely fragrance raw material. Hydrogenation readily gave para-t-butyl cyclohexanol, and the acetate was patented as a woody, fruity aroma compound. It was quite some time before it achieved its potential as a large scale ingredient in soap perfumes. At this point *Vertenex* was analyzed chiefly for purity. UV easily showed the presence of unhydrogenated material (aromatic ring) and GC the presence of the ortho isomer. However GC also showed that *Vertenex* was not a single material, but in fact a 70-30 mixture of trans and cis isomers. This assumed importance when the isomers were separated. It was established that the cis (30% component) had more "lift," and, more important, had several times the odor strength of the trans.

This led to attempts to "improve" the isomer ratio. The hydrogenation process, which gives at first all cis isomer, uses nickel catalyst, but the cis is rapidly converted to the 70-30 equilibrium mixture by the alkali necessary to promote the hydrogenation. Several methods have been patented to produce high cis content product, but, as is so often the case, when this product became commercially available it did not replace the regular product, but instead assumed a modest role as a specialty item for certain applications. Nevertheless it was GC that drew attention to the whole isomer question in the first place.

Rose Alcohols — Fifty years ago "geraniol" was the product of distillation of Java citronella oil after the lower-boiling terpenes and citronellal were removed and the remainder stripped of residue. It was also the main component of palmarosa oil, from which it could be isolated by distillation. But these two "geraniols" were quite different, and by no stretch of the imagination would a perfumer equate them for use. The reason is of course that, whereas geraniol from palmarosa is fairly pure (by old standards), "geraniol" from citronella was about 30% citronellol, and could contain up to 10% sesquiterpenes (not even alcohols), and still be acceptable. Acceptance was based on odor, physical constants such as specific gravity, acetylation value (which showed about 90 per cent as terpene alcohol), and various empirical tests such as atmospheric (!) distillation from a flask over a free flame without a column (!), during which a given percentage had to distill off in the 232-233° range. What this meant in terms of fragrance value remains a mystery, but those were the customer's specifications. In fact, the introduction of better columns in the stills used for making this "geraniol," which in fact produced a much purer product, led initially to rejection.

A large part of the "geraniol" manufactured today comes from turpentine via beta pinene and myrcene. It is essentially pure $C_{10}H_{18}O$ but consists actually of almost equal amounts (60/40) of geraniol and its cis isomer, nerol. (When desired, these can be separated and sold as distinct products, but for many purposes this is unnecessary). Nerol was for many years a particularly prized fragrance compound, commanding a high premium. It was isolated from expensive neroli or petitgrain oil.

It was also separated from mixtures with geraniol (produced by the reduction of citral) by the sloppy, inefficient treatment with anhydrous calcium chloride, which combines reluctantly and incompletely with geraniol, but does not combine at all with nerol. An enriched nerol can thus be obtained.

Now that nerol is as available as geraniol, it is generally recognized that the two isomers have about equal fragrance value, although differing somewhat in character, so that each is preferred for special effects. Whether the former great desirability of nerol was due to trace components from its origins remains moot.

The third important rose-odored terpene alcohol is citronellol, which, having an asymmetric carbon atom, exists in both dextro and levo forms. It is this fact that has raised the question, long debated, as to whether chirality influences odor.

Since the dextro and levo (d and l) forms of any compound are identical in all respects except shape (as, for example, a pair of gloves), they would of necessity produce the same odor perception unless the olfactory sensors were also chiral. Since nothing was known about these sensors, the question could not be answered except by observation, and in this there was much confusion. Many authorities claimed that reported differences between *d* and *l* forms were so small that they had to be the result of "impurities." Others stoutly maintained that there were indeed fundamental differences in odor between the pure isomers themselves. Lending weight to the latter position was the fact that perfumers were willing to pay large premiums for "rhodinol," a terpene alcohol long thought to be a chemical entity. When its composition was established as being a mixture of geraniol and levo citronellol, this served to enhance the prestige of the latter, for it was known that mixtures of geraniol with dextro citronellol did not give the same fragrance effect as did rhodinol. There must be a real difference, it was argued, for otherwise why would cost-conscious perfumers pay five times as much for rhodinol as for the other alcohols?

There were, however, counter indications. Dextro citronellol can be made very pure by chemical reduction of *d*-citronellal from Java citronella oil. Racemic citronellol (optically inactive because it is a 50-50 mixture of *d* and *l*) is readily made by the hydrogenation of geraniol. But racemic citronellol is by no means intermediate in value between *d* and *l* citronellol (the latter *ex* rhodinol), as would be expected from its composition. Then came the conclusive refutation. It became possible to make pure *l*-citronellol from optically active pinene *ex* turpentine by the steps of hydrogenation to pinane, cracking to dihydromyrcene, and hydration *via* metal-organic reagents. The resulting *l*-citronellol is of high purity both chemically and optically, and it does possess a somewhat different character — lighter and perhaps fresher — than the pure dextro. For this reason it enjoys a modest volume of use at a reasonable premium, but as a replacement for rhodinol it has been a disappointment. Clearly the answer must lie elsewhere than in chirality alone.

The question of the value of chirality having been muddled for so many years, it is not surprising that only minor importance has been assigned to this property in finished products. Instead, its major role has been in the quality control of raw materials and some intermediates derived therefrom. For the most part essential oils are optically active because of their biochemical origin, and the simple test of measuring their optical purity has long been a standard analytical measurement, where applicable. A few grams of the oil, or the solid dissolved in a suitable solvent, are placed in a glass-ended tube of known length, and the ability of this sample to rotate the plane of polarized light is then measured quantitatively in a polarimeter. The light source is generally a sodium lamp which emits intense yellow radiation from which the monochromatic d line is polarized by the Nicol prisms of the instrument.

By this procedure, a number of important materials have been and still are assayed routinely. In the past, when it was the only criterion, it would fail to detect adulterations of valuable oils such as Bourbon geranium (from which rhodinol is produced), provided the rotation was adjusted appropriately. Thus the addition of cheap racemic citronellol might be compensated by adding the correct amount of some levo rotatory adulterant, preferably odorless, to return the mixture to its original rotation. Nowadays GC would reveal such fakery easily. The purity of citronellal distilled from citronella oil was most easily measured by rotation, since the co-boiling components of each fraction from the still have smaller or zero rotation. Even today, beta pinene from turpentine, the raw material for terpene alcohols, citral, and the ionones, is checked for rotation as part of the regular analytical procedure. This is in spite of the fact that practically all the beta pinene is first converted to myrcene and thus loses it asymmetry and therefore its optical activity.

However, the majority of fragrance materials derive from synthetic sources or from racemizing processes and are optically inactive. There are a few exceptions, mostly in the flavor sector. They include menthol (peppermint), nootkatone (grapefruit), and *d* and *l* carvone (caraway and spearmint respectively), but in the strictly fragrance area chirality is generally disregarded.

At first it might appear that the proponents of the "differences due to impurities" school of thought have prevailed, but recent findings prove otherwise. It is now definitely established that chirality can be a major factor in odor perception. The proofs cannot be presented within the scope of this discussion, but it can be said that some of the anomalies seem to have been explained. One may theorize that the usually small odor differences between chiral isomers is due to the types of structures which have been used as test substances. These molecules have in general been flexible, with many degrees of freedom, which allows them to twist into conformations compatible with the receptor sites for odor perception. More rigid molecules, those containing one or more rings, show greater odor differences between their chiral isomers, and in extreme cases exhibit odorous vs. odorless properties. At present, this is of theoretical interest only, for with very few exceptions rigid odorous compounds derived from natural sources can be obtained in one chiral form only. Those made purely by synthesis are necessarily racemic. In addition, the techniques of resolution (separation of *d* and *l* isomers from racemates) are expensive, laborious and time-consuming, and to date not enough is known of the relative value of optical isomers to warrant extensive research being undertaken in this direction.

Meanwhile, a beginning has been made in the laboratory. The semirigid molecule, alpha ionone, has been resolved and reported by two separate investigators who used the same laborious procedure. That they reported diametrically opposite odor evaluations served only to muddle the matter further. Recently, however, (unpublished data) a far simpler method has been used to separate the chiral isomers of not only alpha ionone but also of alpha methyl ionone, and the differences in odor, particularly in the latter compound, are remarkable. Nevertheless, until commercial separations become feasible, these isomers must remain laboratory curiosities.

For this reason, also, the improvements in the instruments for measuring optical rotation find application mostly in areas outside the fragrance industry. It is now possible to obtain a complete ORD (optical rotary dispersion) graph from a recording instrument on a fraction of a milligram of sample in minutes only. Such a graph gives the rotation value over a whole range of wavelengths instead of at just one, and since the values change greatly and not always continuously, much can be learned about chemical structure from such an analysis. It is not the lack of instrumentation, but the lack of appropriate compounds to be examined, that limits the usefulness of the technique for fragrance materials. On the other hand, the rising interest in "naturals" in the flavor area is increasing the significance of chirality as a criterion of materials used for such purposes. Measuring this function is actually the only rapid, facile way to determine the natural origin of a material.

The Nose As Analytical Instrument

In spite of the great influence that laboratory instruments have had upon research and quality control in the fragrance field, it must never be thought that they have replaced the human nose and its olfactory system as the final arbiter of fragrance value. True, instruments have made things easier, but only as adjuncts. They are, when used correctly, rapid, sensitive, and precise, but they cannot substitute for human judgment. In addition, the nose is much more sensitive than the finest GLC yet devised. However, human perception of odor is often colored by subjectivity, and there are certain physiological defects in even the most highly trained olfactory system. If disregarded, this can lead to serious errors. It is by the careful combination of the two methods of testing that best results are obtained.

Since the value of a fragrance depends upon its effect on the nose, the ultimate test must obviously be by that member. It is not enough for a product to meet all analytical tests, if the nose of an experienced specialist says "no." To lighten the load on the creative perfumer, whose chief role is the creation of finished fragrances, special panels of odor evaluators function in quality control of raw materials, intermediates. and fragrance products, to make sure that all of them pass required odor standards, usually by comparison with reference samples. These evaluators need not have the same abilities in the creative sense, but they generally have the most sensitive and perceptive noses in the business. They of course use GC systems in their own laboratories as adjuncts, but their odor judgments are based mostly on the classical "smelling paper" studies in which the odor of the material being evaluated is observed over a period of hours or even days. This reveals much of importance which might not be readily apparent from a cursory sniff.

Nor must the judgment factor be minimized. A valuable raw material may on first inspection appear unacceptable, but if this is due to a trace of "top note" which quickly disappears and does not affect the over-all fragrance value, the material can be passed. On the other hand an oil too low by 5 per cent in an important odorous component may appear satisfactory on a blotter, but would underperform in a fragrance. Here GC can be more helpful than the nose. Different criteria apply to raw materials to be used in synthetic processes rather than directly as components of finished fragrances. The former usually require only the correct content of the desired compound necessary for the synthesis, and the absence of interfering foreign materials.

Often the analytical laboratory in the production department can make all the tests required on such material. This production analytical laboratory, which is a major department in any well run fragrance manufacturing plant, makes use of all the techniques for quality control. It performs an exceptionally important function in any fragrance establishment. Relying heavily on GC to supplement its testing of physical properties and its wet chemical analyses, the production analytical laboratory generally leaves the NMR and MS analyses to the research department, since these pertain more to unknowns than to regular products. However both UV and IR are regular techniques for quality control in the plant, and such spectrometers are part of the usual analytical laboratory equipment.

Returning to the odor evaluation panel, it might seem that its stamp of approval is the final hurdle which must be surmounted by a fragrance product, but such is far from the case. The approved materials may of course be sold as such, but in most cases they are next mixed (compounded) according to rigidly controlled formulas to produce fragrance "compounds." These "compounds" must themselves pass all the quality control standards, which we need not repeat here. It should be emphasized, however, that certain mistakes, the result of human error, are occasionally picked up at this point by the odor control group. Often GC will detect such mistakes, since an incorrect weighing, or even an incorrect ingredient, will be obvious by comparison with a standard GC chart for the "compound" in the reference file. Sometimes judgment is required. What may appear a gross aberration on GC may actually be of no importance whatever. For example, a particularly powerful ingredient may be used normally as a 10 per cent solution in an inert solvent, such as diethyl phthalate. This time, for some reason, the inert solvent used happened to be benzyl benzoate. This makes no difference in the performance, but it will show up glaringly on GC, cause much confusion, and possibly result in rejection by the customer. Having started to employ GC and other control devices, buyers have become much more demanding in their specifications.

Some of the fragrance products passed by odor control also go to the perfumer, who may create a new fragrance of which he is very proud. But even though the perfumer may be a supreme master of his art, his judgment is necessarily subjective. Before the finished product is shown to a customer, it must pass another quality control point, approval by a panel of experts for odor evaluation. This involves a judgment as to whether the fragrance can be expected to meet the requirements of the customer in his intended application, and also whether the product, if submitted, will pass the ultimate test of favorable comparison with submissions from other would-be suppliers.

Such an evaluation is of course more subjective than all previous tests, but in one aspect it may reveal a subtle flaw, the result of a difference in odor perception between the expert creator and the average consumer, who will be the ultimate arbiter in success of the product when it reaches the market place. Such a difference in odor perception may be due to greater sensitivity of the expert, but just possibly to the reverse. It has been found in recent years that there are deficiencies in the olfactory perception mechanisms of certain individuals which make them unresponsive to specific odors. These lacks are called specific anosmias; they can occur in anyone, no matter how sensitive he or she may be to odors in general. Fewer than fifty such specific anosmias have been catalogued to date, a very small percentage of all odors. Furthermore, each anosmia generally affects only a small percent of the population, and no individual will normally exhibit more than two or three anosmias at most.

Anosmias would not even be worth considering except for the significant fact that recently an anosmia has been found which affects 50 percent of the total population, and from which perfume experts are not exempt. Since this particular anosmia is to the "urinous" odor, it would not seem to pertain at all to fragrances, except that it now seems that this odor is closely related to the important "amber" type odor. Similar anosmias, incompletely documented to date, are probable. In any case, this is a fascinating subject for future investigation. Meanwhile, a perfumer should neither deny the existence of an anosmia, nor be ashamed of having one, since it will never detract from his abilities so long as he recognizes that it is present.

The ideal fragrance exerts its desired effect upon the odor sense and is inert in all other respects. Unfortunately, things are not always so simple. Molecules are all more or less reactive and can change by interaction with other molecules either of the same or other species. Incorrectly formulated fragrances can change and deteriorate, a process speeded by elevated temperatures. For this reason some of the more reactive materials are best stored cold when pure, or else kept as solutions. The fragrances may also react with the functional products they are intended to perfume, such as soap bases, talcum powders, etc. They may diffuse through the wall of the package and be lost or at least changed by selective permeation. They may discolor, cloud by solid formation, or give other undesirable side effects. All these possibilities must be considered and guarded against before the fragrance can be considered acceptable for the intended use. This requires an applications laboratory. This unit not only considers all these factors, but also advises on suitable formulations as part of its customer service function. Such a laboratory, geared largely to the cosmetic products in which fragrances are such an important ingredient, is an important part of every complete fragrance establishment.

Even after a fragrance compound has passed all the above tests, it is still not "home free." The compound can also react unfavorably with elements of the human body; it may not be physiologically inert. One manifestation of this interaction is the variation in behavior of perfumes on individuals because of their different "body chemistry." This effect can be guarded against if the individual makes sure to select an appropriate fragrance, but the problem is not so easily solved when the fragrance is one being used in small amounts in a functional product. Fortunately, in such cases, the idiosyncratic reaction is normally of secondary importance, but each individual component of a fragrance must still have passed muster as being non-irritant, non-sensitizing, and non-toxic in a broad spectrum of animal testing. As governmental requirements for safety constantly tighten, and consumer protection agencies exert increased pressures, the importance of the toxicological and pharmacological testing laboratories of fragrance houses has grown steadily. These laboratories have responsibility not only for the behavior of individual compounds but also their effect in mixtures in the finished product.

What applies in fragrance use *per se* is even more important when the intended use is in flavors, for in this case ingestion must also be taken into consideration. For a product to be acceptable it must meet the regulations of the U.S. Food and Drug Administration, and the requirements of each country where marketing is planned must also be taken into account.

Truly the flavors and fragrances industry has grown mightily since its beginnings as a mixer of plant extracts, not only in size but also in responsibility and reliability. This growth has been due in large part to its ability to satisfy greatly expanded world-wide demand for its products by the application of science and technology.

Chapter XIII

The Practical Aspect of Quality Control

In the practical sense, quality control involves all the departments in a fragrance and flavor company. In the previous chapter we dealt with the subject of quality control from the scientific side, considering the role in maintaining product quality of the numerous scientific instruments and procedures that have been developed. Here we approach the subject from the practical side.

Usually we think of the responsibility for quality control falling on two groups of technicians, — the subjective and the objective group. In the subjective group are the flavor and fragrance quality control laboratories, which compare flavor and fragrance against established standards, and perhaps examine the gross physical characteristics of a material such as color, fluidity, clarity, etc. In the objective control groups are the analysts who measure the physical and chemical characteristics of the materials, and the instrumental groups that compare the various physico-chemical properties against reference samples, check molecular structure, and obtain quantitative and qualitative information. The work of the objective group is not related in any way to the measurement of flavor or fragrance of the material being evaluated.

Perhaps the best way to observe the interplay of the various departments of a flavor and fragrance company in keeping close tabs on quality is to follow the course of a raw material through the company to its final destination as an ingredient in the customer's product. Much of what we will discuss applies to synthetic as well as natural materials, and to flavors as well as to fragrances.

First and most important is the selection of a reference standard for each raw material. It is fortunate for our industry that years ago at least one of the leading buyers of fragrance and flavor raw materials took the time and spent the money to go to all the major areas of the world where aroma raw materials are produced to take authentic samples of actual field distillations. These were carefully preserved, and analyzed to establish the odor, as well as physical, chemical and instrumental specifications to guide buyers in quality selection. Dr. Ernest Guenther of Fritzsche Brothers was one of the pioneers in this work. His interest and experience helped establish essential oil production in many areas — for instance, production of lemon and orange oils in California, the distilled lime oil industry in Mexico, and production of lemon oil in Greece and Israel.

In the typical essential oil house, almost all departments are involved in selecting standards, and all have an equal share in assuring that raw materials meet those quality standards. The Purchasing Department must find a source of supply for all raw materials the company needs, and its task is to bring the highest quality materials into the plant at the most reasonable cost. Purchasing obtains samples of materials for approval, then routes the samples to those who will measure quality, subjectively and objectively.

If quality is equal, the task of selecting the most favorable price is a simple one; but there are other factors Purchasing must consider. Is there enough of the lot available? Where is it located? Is the supplier reputable? What are the details of delivery? customs duties? packaging? shipping costs? method of payment? The purchaser may have to deal through a particular broker for a particular oil, because this broker has an exclusive source which is able to offer the best rose or galbanum for a particular use. The standard grade raw material may be adequate for all other compounds, but this specialty may require the best. Sometimes the reputation of the supplier is worth a modest premium. Satisfactory past experience with the producer, the broker, or the exporter is one way to ensure raw materials of consistent quality.

When new sources develop for raw materials, the contact between producer and purchasing agent is extremely important. The purchasing agent establishes the type material he requires for his perfumer or flavorist and passes the word along to the supplier. It may be some such comment as "Your lime oil is burned, — you distilled at too high a temperature," or "It is too terpeney — you forgot to extract the water, or distilled at too low a temperature, or too quickly."

The evaluation the purchasing agent gives the supplier may have originated with the chemists or odor evaluators on the Q C staff, or there is a good chance that research chemists or instrumentalists may have been asked to evaluate the material if the purchase is one of any size.

The Accounting Department may also get into the act. Their comment might be, "Yes, that's a good quality, we guess, that is up to you. But the finished product in which this fragrance is to be used can't afford that high a raw material cost and still stay competitive. Somebody will have to make a substitution. Can't you look for a cheaper source? But remember, if you have to compromise on quality, the customer must be advised."

Or, after the perfumer has selected a beautiful jasmine, Accounting may decide that the price is too high to justify even the obvious high quality. Accounting tells Quality Control to convince the perfumer to choose another jasmine, pointing out that the slight difference in fragrance doesn't justify the major difference in cost.

In another interplay between departments, the flavor chemists may tell Accounting that the quality material they have elected to use has twice the strength, at only one-third more in cost, thus is a better buy. Perfumers and flavor chemists get involved in raw material selection in strange ways. The quality control laboratory may have rejected a sample of lavandin because it was thin, and contained too much cineol and camphor. The perfumer has prevailed upon Purchasing to continue buying this grade in spite of a poor compromise between quality and cost because in the particular finished product being made the impact of cineol/camphor is very necessary, and the lavender character is not that important. The control chemist enters the discussion, his opinion being that it would be better to add cineol and camphor to the compound if necessary, but he doesn't go along with buying a low grade oil. Often the requirements and recommendations of the perfumer will prevail in such a discussion against all the other departments.

From time to time new reference standards are established for particular raw materials, but restricted to their use in specific compounds. Compromises are sometimes made, but it is a constant challenge in an F/F company to be able to count on a consistent supply of best quality raw materials at a stable price. Stocking too many grades of the same type material invites problems and opens the door to errors. Hence it is highly important to have a good quality standard for every item purchased, and to try to stay close to it in all stock.

There are of course many reasons why natural products vary from year to year, from supplier to supplier, and from one geographical area to another. There are natural variations that occur. A difference in rainfall, temperature during the growing season, varying sunlight, harvesting conditions, a change in processing equipment — all these and other factors may change the nature of the essential oil produced. The work of Dr. Guenther and other students of the subject referred to earlier demonstrated clearly that natural raw materials are subject to certain variations which must be accepted as normal. This is not to say that reference standards may be disregarded. They should still be maintained as the quality of product for which the buyer is seeking, although for some reason it may not be available that year, from that seller, or perhaps from that area.

Sometimes the variations may be minimized by blending according to odor or chemical specifications. For instance a 30 drum lot of oil patchouly may be available. Past experience has shown that often every drum shows a variation in quality. That isn't too difficult a problem to solve, but when 300 drums of citronella oil are available, every one of which must be checked, that is a much larger problem.

But how do we handle the drums of patchouly? The first step is to check the aroma of each drum. If too weedy and cade-like, the lot will probably be refused without further consideration. But, suppose half of the drums are satisfactory from an odor standpoint, good enough to be used as is. Those fifteen and three other borderline drums may be treated to reduce and standardize the color, and be accepted odorwise; or they may be treated and rectified to remove most of the color, and accepted odorwise. If they pass the rest of the Q C procedure, meeting physical, chemical and instrumental specifications, these drums will be taken in, bulked, and the remainder of the lot rejected. The bulked lot will of course be re-evaluated.

The stock computer may show that the plant needs ten drums of patchouly to be rectified for low color. Since stock has gone below minimum, the new drums are shipped to the plant, manufacture is scheduled and completed, and quality checked in process by the plant control laboratory for odor and other specs. If accepted, the finished batch is bulked, checked again before shipment to the compounding laboratory, and samples sent to corporate quality control for odor and specifications evaluation. That final approval is the most important one. Only after it has been issued is the new material assigned to stock, to be used as needed.

An order may be in the house for a fragrance compound requiring several "keys," one of which calls for natural (untreated) patchouly oil. The senior compounder checks the products that have been set up for him to make sure they match his reference standards. He notes any changes or possible errors, then compounds the keys (partial compounds), evaluates them, and forwards them to the Fragrance Q C laboratory. Each key is checked against a standard and also against a few of the most recent batches. After the key has been approved, the compound is then finished according to the formula. Some of the patchouly oil which has been decolorized by rectification is part of the compound. Lot numbers, names and codes are all checked to be sure each item has been approved, and the components weighed in. The total must agree in weight with the formula.

The compound is blended by thorough mixing, then checked by the compounder. If he considers it satisfactory, he again sends a sample to Fragrance Q C for approval against a new standard and recent productions. The evaluation may indicate it requires further mixing, since it smells too fresh. After further mixing, the sample is approved, and the material labeled and drummed for shipment.

We have described a typical fragrance compound production operation. The same steps would apply to flavor production. At any step in the process it may be necessary to have GC or IR curves run for any one of a variety of reasons. Strict control at every step obviously requires the experience, the skill and the time of a varied team of experts all of whom may be required to support a single finished product.

It is evident that in the F/F industry team work and communication are the routes to successful operation. In many projects, the entire company can get involved, and in certain projects even a single item may require direct contact with a host of special skills. And, it must be kept in mind that an F/F house may manufacture anywhere from 2,000 to 4,000 raw materials and at least the same number of finished products, all requiring adherence to the highly complex system described above. It is certainly an understatement to say that this is a highly complex, involved and specialized industry.

Chapter XIV

Product Safety and Government Regulations

One of the new factors in the pattern of American life over the past twenty-five years has been an increasing concern for public safety, accompanied by a parallel growth in the imposition of governmental controls. With the growth of product safety consciousness, numerous laws have been passed requiring manufacturers of everything from talcum powder to toasters to see that the products they sell comply with safety standards. A corollary development has been the multiplication of government agencies charged with supervising those manufacturers to see that they comply with the standards.

Recognizing that aromatic chemicals and related substances used in foods and cosmetics would inevitably become a subject for governmental concern, some of the leading figures in the essential oil and aroma chemical industry started to consider, as long as twenty-five years ago, what they could do on their own initiative to organize a voluntary, industry-supported program of research and safety testing of fragrance raw materials.

In 1954, Dr. Ernst Theimer of International Flavors & Fragrances, Inc., proposed to the Essential Oil Association that a cooperative program of research on essential oils be started. While no immediate action was taken on his proposal, this sort of initiative led, in 1966, to organization of the Research Institute for Fragrance Materials, which has since then coordinated a voluntary, industry-supported program of research and safety testing of essential oils and fragrance materials. (See p. 348).

The rising concern of legislators over product safety has closely followed the growth of the Consumerist Movement. Starting back in the 20's, with publication of books like "Your Money's Worth" by Stuart Chase and F. J. Schlink, followed by "100,000,000 Guinea Pigs," and the widely distributed magazine, "Consumer Report," we have had with us since then dozens of self-appointed champions of the American consumer. They have vied with one another to protect said consumer against real or imagined threats to his safety, or his pocketbook. The trend toward consumer protection has grown at a tremendous rate over recent years, with one of its least acceptable features being commercialization of the movement by its leaders. Many of them have enriched themselves, or gained positions of considerable national prominence, by their role as aggressive consumer advocates.

The legislative arm of our government, both on a national and a state level, has responded as might have been anticipated, reacting to the pressure from consumer organizations, special interest groups, ecologists and the press. As a result, legislation aimed at protecting the consumer has been enacted at an unprecedented rate. Literally thousands of bills have been introduced into Congress and State legislatures, designed to protect the consumer, at times against hazards which exist largely in the minds of the framers of the legislation.

A current case in point is the recent edict by the U.S. Food & Drug Administration which, if allowed to stand, will prohibit the use of essentially all aerosol sprays employing chlorofluorocarbons as propellants within a year or two. This allegation that escape of such propellants to the upper atmosphere threatens to damage the "ozone layer" has never been proved, and is at this point based on nothing more substantial than an unproved theory. Yet, with the wide publication of the charge in the press, and the constant hammering at public officials by ecologists, it has been enough to cause several state legislatures, and now a Federal agency, to ban the use of an important and highly useful type of aerosol product. This latest restriction by the FDA will drive many products off the market, deny consumers the use of preferred product types in the anti-perspirant, hair spray and cologne categories, cost workers thousands of jobs, and make it necessary for marketers to spend hundreds of millions of dollars in formulating and introducing a whole series of new products to the market.

In another victory of government regulation over industry, the Food and Drug Administration began in April, 1977, to require the listing of all ingredients in cosmetic and cleansing products (to be listed in decreasing order of volume present in the finished product). FDA had first proposed this regulation in May, 1975, but its enforcement was held up for almost two years as manufacturers of soaps and cosmetics fought the proposed regulation through the courts, charging that such listing of ingredients would prove cumbersome, and could result in disclosure of trade secrets. The Court of Appeals finally ruled that the regulation was to be enforced, starting April 15, 1977. In a concession to marketers of the affected products, they are permitted to list "fragrances" and "flavors" as such, without revealing the composition of the formula.

In addition to this new labeling requirement, the F.D.A. is also reported to be preparing to enforce another regulation which would require manufacturers who are not conducting "adequate" safety testing to label their products with a warning that the safety of the product had not been tested. There is some question as to whether or not such a regulation would be enforceable, as there is no law requiring manufacturers to release safety testing data. The authority the F.D.A. possesses allows it only to remove from the market cosmetic and other products which have been proved to be "adulterated or misbranded."

Senator Thomas Eagleton (Dem. Missouri) has for several years been trying to get Congress to pass a bill which would bring the cosmetic industry under strict Federal control. His proposed legislation would require product registration and formula disclosure, and would give F.D.A. staff the right to inspect laboratory and production facilities. Manufacturers of cosmetics would be required to establish the safety of any new cosmetic product before putting it on the market, and safety test data would have to be available for inspection by F.D.A.

The Eagleton bill seems to be no closer to reaching the floor of the Senate now than it has been in previous years, but it is interesting to note that the measure satisfies neither critics nor defenders of cosmetics. Industry spokesmen point out that for an industry that sells over six billion units a year, a record showing a minimum number of adverse reactions to its products on the part of users (a little over 15,000 annually, limited largely to rashes and allergic reactions), is nothing to be ashamed of. The attitude of many Congressmen and Senators on this issue, incidentally, seems to favor the industry position. It is well reflected in a statement by Rep. Paul Rogers (Dem. Florida), head of the Health Subcommittee of the House Commerce Committee, who said, "We've looked at the Senate's hearings for 1974 and 1975, and they contain nothing that suggests there are health consequences that require immediate action. Until we get that, we wouldn't want to legislate."

By contrast, the Nader Health Research Group consider the Eagleton bill to be so weak that it would be useless for Congress to pass it. A Nader spokeswoman, Anita Johnson, sharply criticized the Congressional record on regulation of cosmetics, saying, "Until someone else on the Hill shows leadership, cosmetics are going to go unregulated as usual. Putting a new cosmetic product on the market is as easy as selling a paper clip. All you have to do is find someone who will buy it."

One material sold by a major manufacturer of aromatic chemicals, hexachlorophene, has brought the industry highly unfavorable publicity while at the same time exposing users of products containing it to a potential safety hazard. Developed by Givaudan Corp. in the early 30's, and widely used since then as an ingredient in soaps, including use as an antibacterial agent for washing babies, hexachlorophene was reported (BIBRA Bull. 1971, 10, 217) as having caused rats to develop brain lesions when treated with high dosage levels. Following this report, F.D.A. issued an order requiring precautionary labeling, and then placed drastic restrictions on use of hexachlorophene.

The most recent product of the fragrance and flavor industry to have aroused the concern of the Food & Drug Administration is saccharin. In April, 1977, F.D.A. banned its use as a food additive, while suggesting it might be permitted as an over-the-counter drug. Public outcry from millions of buyers of diet drinks was so strong that effective date of the order has been postponed for 18 months.

This whole subject of product safety is one that is necessarily in a flux. What is true today may not be true tomorrow. Thus the reader is cautioned to treat all statements merely as hopefully true at the time they were made, but by no means immutable.

The first national legislation in the United States covering foods and drugs was the Food and Drug Law of 1906. At the time, many states had laws regulating foods and drugs, but there was no Federal law. With many food and drug products being distributed nationally, it became obvious that a uniform national law would be desirable, so that residents of every state would be protected against unsafe products, and national marketers would be assured that an article acceptable for sale in one state would not be banned in another. The 1906 act was a relatively simple law by modern standards. The right of Congress to enact the measure traced to its role in controlling commerce between the states. Thus, the 1906 act prohibited introduction into interstate commerce of foods and drugs which were "misbranded or adulterated." It also prohibited use of a number of harmful chemicals, and established sanitary requirements for food processors.

Almost thirty years later, under the administration of Franklin D. Roosevelt, protection of the consumer came to be one of the presidential priorities, and a new and stronger law was proposed to replace the 1906 act. In addition to foods and drugs, The Federal Drug and Cosmetic Act of 1938 also covered cosmetics and therapeutic devices. It set standards of identity for the composition of foods, and prohibited the inclusion of any poisons in food. It also established labeling requirements for foods, drugs and cosmetics. This measure remained the law of the land until it was amplified and corrected by the Food Additives Amendment of 1958.

One incident which led to the rewriting was a problem that arose with coumarin in 1954. Tests showed that coumarin had anticoagulant properties, and as the Federal Drug and Cosmetic Act banned the presence of poisonous materials in foods, the F.D.A. banned the use of coumarin as a flavoring ingredient. At the same time they prohibited the use of the tonka bean extract.

This principle of a material being "poisonous per se" had been debated at considerable length for some years prior to rewriting the 1958 Act. It is obvious that there are many materials such as salt, bromine, arsenic, potassium, chlorine, etc., that are poisonous in concentrated form, yet completely safe to ingest in reduced concentration — even in some cases essential to human growth and well being.

Another factor which complicated matters at this time was the development of new analytical procedures (including perfection of the gas chromatograph) which made it possible to identify even minor traces of contaminants in a sample (down to parts per million). Tests on most any potential ingredient in a food or cosmetic product might thus show that the material contained recognizable traces of poisons. Did this call for automatic exclusion of such a raw material from use? If so, it might soon be found equally logical to ban water as a drink unless distilled, for traces of many different poisons are of course identifiable in all water supplies.

It became obvious that the Federal Drug & Cosmetic Act needed to be rewritten. The Food Additives Amendment of 1958 was the result. This amendment contained the celebrated Delaney Clause* which provided that no substance could be present in any food product, regardless of concentration or condition of use, which had been shown to cause cancer in man or animal. It was under this clause that, in 1960, the F.D.A. banned safrol for use in foods. Tested on laboratory animals at extremely high concentrations, safrol had proved to be carcinogenic.The ban applied not only to safrol but also to the closely related chemicals, dihydro-safrol and iso-safrol, and to any essential oil such as oil of sassafras or oil camphor sassafrassy containing safrol.

The 1958 Food Additives Amendment introduced a new phrase into the vocabulary of toxicologists — "generally recognized as safe," or "GRAS." The amendment permitted use in foods, drugs and cosmetics of an extensive list of materials "generally recognized as safe." This was defined to mean "general recognition of safety by experts qualified by scientific training and experience to evaluate the safety of such a substance." But who were these experts to be? Were they to come from industry, from government, or from the academic world?

The Flavor and Extract Manufacturers Association, an organization which was seriously concerned with the

*Section 409(c)(3)(A) of the Act as added by the Food Additives Amendment of 1958 and amended by the Drug Amendment of 1958 and amended by the Drug of 1962 states:

> *Provided,* That no additive shall be deemed to be safe if it is found to induce cancer when ingested by man or animal, or it is found, after tests which are appropriate for the evaluation of the safety of food additives, to induce cancer in man or animal, except that this proviso shall not apply with respect to the use of a substance as an ingredient of feed for animals which are raised for food production, if the secretary finds (i) that, under the conditions of use and feeding specified in proposed labeling and reasonably certain to be followed in practice, such additive will not adversely affect the animals for which such feed is intended, and (ii) that no residue of the additive will be found (by methods of examination prescribed or approved by the Secretary by regulations, which regulations shall not be subject to subsections (f) and (g)) in any edible portion of such animal after slaughter or in any food yielded by or derived from the living animal.

manner in which this GRAS designation would be interpreted, injected itself into the problem at this time by setting up the FEMA Expert Panel. This panel consisted of a group of experts whose training clearly established the members as equipped to make the judgments which were called for. As members of this panel FEMA named recognized scientists in pharmacology, toxicology, food chemistry, etc., and gave them the responsibility for reviewing data on the vast number of flavoring materials. The Panel put together a list of the materials in general use, together with figures on use levels, and applications in different types of foods. Then, based on the data, the Panel classified many of the materials as GRAS.

The Food Additives Amendment contained a clause (the "Grandfather Clause") which was supposed to permit use without further testing of any ingredient then in use. However, if a material had not been in use prior to 1958, or if it was considered to be a food additive, it had to undergo the test procedures prescribed in the act.

There was much disagreement at the time on whether it was the responsibility of F.D.A., or of industry, to make these classifications and the issue is still unresolved, even today. Finally, in 1966, F.D.A. agreed to accept as "safe under specified conditions of use" many of the flavoring materials previously classified as GRAS by the FEMA Expert Panel. Up to that time only a limited number of materials had been accepted by F.D.A. as GRAS. But, in 1966, the F.D.A. accepted* a number of other materials as free from specified tolerances, considering them to be "safe when used under good manufacturing practices." The phrase "under good manufacturing practices," was understood to mean the normal levels used by industry as reported in the survey made by FEMA.

*Originally this recognition of the safety of certain materials was covered in two sections of the regulations — 121.1163 for natural materials and 121.1164 for synthetic materials. However the regulations were later rewritten, and this acceptance is now covered in renumbered sections — 172.510 for natural materials and 172.515 for synthetics. This is just another example of the confusion created by government regulations, which themselves are subject to change.

Even after the Food and Drug Administration had agreed to this expansion of the list of GRAS materials, the status of a number of important flavoring materials still remained unsettled. For example, an unrealistic tolerance was proposed by F.D.A. for methyl salicylate, which did not reflect levels of industry use. A top level of 3,300 ppm was proposed for the chewing gum industry, well below the normal use level, which incidentally had never resulted in any reports of deleterious effects. But, while the proposed tolerance was discussed at length, it was never officially put into effect.

In addition to the sections of the Federal Drug & Cosmetic Act concerned with flavors, other raw materials are the subject of other portions of the act. The Color Additive Amendment of 1960 regulates the use of color additives, either natural or artificial. Then there is the Drug Act Amendment of 1962 which provides that drug manufacturers may deal in the flavor and fragrance industry only with companies which are registered with the F.D.A.

There are other regulatory agencies than the F.D.A. which exercise, or seek to exercise, control over fragrances and flavors. The Hazardous Substances Labeling Act requires that a warning statement appear on the label of every consumer product if there is any hazard in connection with its use. Since many consumer products contain at least some small percentage of flavor or fragrance, essential oil suppliers are thus often queried as to the safety of the aroma ingredients they supply. However the Act specifically exempts fragrances and flavors (Section 2(f)(2).

The fragrance and flavor industry, like all others, is also concerned with the Occupational Safety and Health Act (OSHA) which went into effect in 1971. This act prescribes standards of safety to be observed by firms in any manufacturing industry. A provision of the act that has been particularly repugnant to industry generally is one which permits Federal inspectors to demand admission at will to any establishment the inspector elects to visit. Recently, however, (April, 1977) a court decision in a suit brought against OSHA by an Ohio company gives hope that this odious "right to inspect at will" privilege may be vacated.

Another governmental agency whose regulations affect the fragrance and flavor industry is the Department of Transportation. DOT has a logical concern with the flash points of materials shipped in commerce. The fragrance and flavor manufacturer uses hundreds of different ingredients in his formulations, which must be reported to DOT. The task of collecting flash point data for all these materials was one that called for group action. It was turned over to the Scientific Committee of the Essential Oil Association, whose members assembled the data for each essential oil and aroma chemical in general use in the industry. This still left to the individual companies the task of determining additional flash points for their own fragrance and flavor mixtures.

Many of the areas in which government agencies seek to regulate the fragrance and flavor industry are concerned with matters other than product safety; some eager bureaucrat, for example, may be trying to make sure that the consumer gets what he thinks he is paying for, and is not cheated by having one material substituted for another. An obvious example is the Fair Packaging and Labeling Act which became effective in July, 1974 (some parts will not become completely effective until January 1, 1978). This Act requires that, in labeling flavoring materials in finished food products, the packager must indicate whether the ingredients are of "natural" or "synthetic" origin. The government's concern here is obviously a matter of economics rather than safety. It seems equally clear that supplying such information on the label will confuse rather than help the average consumer, and that such a regulation is clearly unnecessary.

We believe that it is the desire of every legitimate manufacturer, just as it is of "consumerists" and legislators, to see that the consumer is enabled to purchase a safe and effective product at a fair price. Wherever and whenever any safety factor has been involved, the fragrance and flavor industry has, on its clear record, taken action on its own initiative to serve that purpose. But in the face of mounting over-regulation by so many government agencies, whose edicts sometimes conflict directly with one another, the typical fragrance and flavor supplier can perhaps be pardoned if he questions whether super-regulation really serves the interests of the consumer.

It may be many years before we will know for sure whether the consumer has really benefitted from the excessive concern for his interests volunteered by consumer activists during recent years. Over-regulation inevitably means reduced profits for marketers, and can well result in the disappearance of products from the market, bringing unemployment to the workers who produced them. Those who pass new laws and write stricter and stricter regulations should always keep in mind that, in the final analysis, it is the consumer who pays any increased cost in producing the product she buys. The salary of every government inspector, and every government file clerk, tens of thousands of whom have been added to government payrolls in recent years, must eventually be paid by the buying public.

Chapter XV

In Union

In addition to the Essential Oil Association of U.S.A., of which almost all firms in the essential oil and aroma chemical business are members, there are a number of other organizations in this field that are of active interest to both producers and users of fragrance and flavor raw materials. Some, like the Research Institute for Fragrance Materials (RIFM), the International Fragrance Association (IFRA), or the Fragrance Foundation, have as their members *producers* of fragrance and flavor materials, while others like the Cosmetic Toiletry and Fragrance Association (CTFA), the Soap and Detergent Association (SDA), and the Flavor and Extract Manufacturers Association (FEMA), have *users* of fragrance and flavoring materials as their members. Let us call the roll on the most important of these organizations.

The Essential Oil Association of U.S.A., Inc.

60 East 42nd St.
New York, N. Y. 10017

This association was organized in 1927. Originally called the Essential Oil Dealers Association, Francis T.

Dodge of Dodge & Olcott was the first president. Dues were initially $5 a year. When A. L. van Ameringen was elected president in 1930, he proposed an expanded program of activities, accompanied by a dues increase to $15 a month. He was warned that such high dues might "drive half the members out of the association."

In spite of objections, the reorganization went through, a managing director was hired, at the princely stipend of $125 a month, and a more ambitious program of activities adopted.

The association had started out with about twenty-five charter members. Membership built up slowly, but the association leaders were very particular as to the type firms they would accept. Several membership applications were turned down because the company applying was "not strictly an essential oil house," or because some member questioned whether the applicants were "actually importers of essential oils as a major part of their business." There appeared to be a desire to confine the membership to essential oil importers and dealers, and for a time there was some question as to whether firms that were primarily producers of aroma chemicals would be admitted.

The EOA and most of its members took a strong stand against the practice of "extending price protection against decline" in writing contracts for the delivery of essential oils, recognizing that such a practice protects only the buyer, leaving the seller defenseless, and nullifying the significance of his "contract."

Another firmly held position, at least by the great majority of the association's members, was that it made no sense to stay open for business on Saturdays. EOA adopted a policy on this point, but there were occasional complaints that even some of the big wheels in the association were known to violate the policy.

The association's chief function, as stated in their application in 1933 for an Essential Oil Industry Code under the National Industrial Recovery Act was stated to be "to promote fraternal relationship and goodwill among producers, dealers and distributors of essential oils and allied products, to discountenance all customs not in accord with sound business principles, and to take such action as may be proper for the conduct of our industry on a sound basis."

During the great increase in association activity which accompanied the industry code program under NIRA, the Essential Oil Association lost its first managing director, Raymond E. Bell. He was succeeded by Ray C. Schlotterer who continued in the post for many years.

In 1933, EOA started the custom of scheduling its annual meeting during the holiday season. The first such social event, with dinner and entertainment, was held at the Hotel Astor in January of that year. But, social events were only one side of the EOA program. The Technical Committee, in particular, performed a very useful function. Its members worked with the publishers of the "U.S. Pharmacopoeia" and the "National Formulary" to keep specifications on essential oils and aroma chemicals up to date, recommending changes in specs and test methods, as required.

Dr. Eric C. Kunz, president of Givaudan-Delawanna, Inc., (predecessor to Givaudan Corp.), served with distinction for a number of years as chairman of EOA's Scientific Committee. The Scientific Committee did a great volume of work in setting up specifications for essential oils and aromatic chemicals, leading in 1946 to the publication of an EOA Index of Standards. Standard test methods were also developed.

As the activities of the Scientific Section became more extensive, it was decided, in 1946, to have the Section hold its own meeting. The practice has continued since then, and the Scientific Section of EOA normally holds its annual meeting each year in January in New York City.

When World War II broke out, there were serious problems in getting supplies of perfuming materials out of Europe, and some essential oils became almost unobtainable. When the government imposed restrictions on the use of tin, the supply of containers also became a major problem for essential oil dealers. After the U. S. entered the

Annual meeting, Essential Oil Association of U.S.A., January, 1950

Photo courtesy Gerry Danco

war, essential oil dealers and aroma chemical manufacturers had to cope with orders from the Office of Price Administration, the War Production Board, and the Office of Defense Transportation.

During this period EOA set up a special committee to look after matters of specific interest to manufacturers of aromatic chemicals. This shortly became one of the association's major committees, giving producers of aroma chemicals a forum in which to discuss their particular problems, and a specific arm of the association through which to act in attempting to find a solution for them. EOA also established Import and Export Committees to act in these two areas.

The outbreak of the Korean War brought serious problems for the essential oil industry to which the EOA responded. In December, 1952, the United States Government prohibited trade with China and North Korea. This cut off the supply of many raw materials for the fragrance industry. Dealers were forced to seek supplies in other areas of the world or to develop substitute products.

In the case of menthol, which became a very scarce material, the growing of *Mentha Arvensis* in other parts of the world was encouraged. As a result, a new industry was developed in Brazil, and this country is now the largest supplier of natural menthol. In the case of such products as cassia and anise oils, the industry had to turn to synthetics to satisfy the demand.

In 1949, the by-laws of EOA were amended to allow for associate membership by qualified firms in foreign countries. Today, membership in EOA stands at 79 active and 33 associate members. Clyde T. Nissen is executive secretary of the association.

Research Institute for Fragrance Materials

375 Sylvan Avenue
Englewood Cliffs, N. J. 07632

With the increasing concern that has been evident over product safety in recent years, suppliers of fragrance and flavor materials became aware at an early date that it would be in the best interest of all parties concerned — manufacturers and users of these materials, as well as the consuming public — for producers to undertake a voluntary program of high-level research and safety testing of these materials.

Two industry leaders who took an important role in getting the testing program under way, back in 1955, were A. L. van Ameringen of IFF and Charles C. Bryan of Firmenich & Co. Their efforts were successful, and the Research Institute for Fragrance Materials was established in 1966, with the support of a major portion of the industry. Other prime movers in organizing RIFM were Alfred Moeller of Noville Essential Oil Co., Frederick Leonhardt of Fritzsche Dodge & Olcott Co., Raymond Picard of Rhodia Inc., Ernest Durrer of Givaudan Corp., E. Poons of Universal Oil Products, and Kenneth Voorhees of Ungerer & Co. Today RIFM operates as an international non-profit organization, whose members include 53 companies representing the major factors in the industry in the United States, Europe and Japan.

One early decision in setting up RIFM proved highly important. This was that each industry delegate must be the chief operating officer of his company. With none but top level decision makers participating, it is possible to arrive at immediate, firm decisions.

Raw materials are the only items tested by RIFM. They are selected on the basis of the following criteria: (1) they must be representative of the material *in actual use* by the industry; (2) they must conform to the specifications and standards, if any, of the Essential Oil Association of the USA (EOA) or of the International Fragrance Association (IFRA); (3) they must be supplied to the RIFM without indication of the supplier, with name and an identification number only; (4) they must be accompanied by gas-chromatographic, ultra-violet or infra-red curves to 'thumb-print' the materials.

Specifications and standards in the monographs must be those of the EOA, where available. The levels of usage reported are the results of industry-wide surveys.

The chief parameters selected for toxicological screening of all these materials include tests for allergenicity and phototoxicity on human skin; tests for general toxicity in animals by the oral and dermal route, and other procedures where required by an International Panel of Experts drawn from outside the industry.

Several allergens and phototoxic materials have been reported. The ubiquity of fragrances in cosmetics, household goods, insect repellents, etc., makes the responsibility of such an institute a significant one, and it is hoped that this ingredient approach, responded to zealously by the industry, will eliminate potential offenders from fragrances world wide.

RIFM has now completed the evaluation of 776 of the fragrance raw materials in greatest use, representing about 90% of the volume of fragrances. The program is continuing on the smaller items. Of these 348 have been published in *Food and Cosmetics Toxicology* (1973, *11*, 95-115, 477-495, 855-876, 1011-1081; 1974, *12*, 385-405, 517-537, 703-736, 807-1016; 1975, *13*, 91-112, 449-457, 545-554, 681-923).

Included in the monographs are average and maximum levels of usage of the materials in soaps, detergents, creams and lotions and perfumes; regulatory status; measurements of acute toxicity, irritation, allergenicity and phototoxicity; as well as citations of pertinent literature.

These monographs are available without charge to interested parties either from: International Fragrance Association, 8 Rue Charles-Humbert, 1205 Geneva, Switzerland, or from: Research Institute for Fragrance Materials, Inc., Englewood Cliffs, N.J.

The Fragrance Foundation

116 East 19th St.
New York, N. Y. 10003

The Fragrance Foundation is a non-profit, educational organization founded in 1949 by a group of individuals identified with the perfume industry in an effort to develop a better understanding of the role of fragrance in our lives.

Annette Green has been executive director of the Foundation for the past fifteen years.

The major purpose of the organization is: to further the education of the public regarding the manufacture, use and application of fragrance; to assist manufacturers, suppliers, distributors and retailers of fragrance in their efforts to maintain and enhance the public's image of the fragrance industry and its products; to engage in research of fragrance and related products; to compile and disseminate statistical data and trade information relating to fragrance; and to perform all other acts appropriate to the promotion of fragrance and the development of public interest in fragrance products.

The Fragrance Foundation is governed by a board of directors of fifteen members. The officers, president, executive vice-president, vice-president, secretary and treasurer, serve on the board. The services the Foundation offers members include the following:

Information Center - A clearing house for the industry and the consumer, as well as for press, students, libraries, etc. on all aspects of fragrance.

Fragrance Library - A comprehensive library encompassing a collection of books and articles on fragrance. Also a full-color film which may be rented by members.

Special Studies and Reports - The Fragrance Foundation has established a liaison with the Syndicat Francais de la Parfumerie in France, so that trend information may be reported to members of both organizations. The Foundation also has a liaison with RIFM.

Public Relations - The Fragrance Foundation initiates fragrance news and responds to press requests.

Publications - The Foundation publishes the booklet, "The History, The Mystery, The Enjoyment of Fragrance" (75¢ a copy). Another publication is "The Compleat Fragrance Directory" ($30.00 a copy).

Fresh-Up Training Ideas for Salesgirls - The Foundation publishes a bulletin, "Salesgirl Fresh-Up Sales Training Ideas of the Month", offering ideas to spur the sale of fragrance products, often based on seasonal or holiday themes. The letter also helps to educate salesgirls on the background, uses and display of fragrances.

A highlight of recent years of activity by the Foundation has been the annual Recognition Awards Ball . . . an "Oscar" evening for members of the fragrance industry. Some of the awards presented are for: "Most Exciting Women's and Men's Fragrance Introduction"; "Retailer of the Year"; "Outstanding Person of the Year"; and the "Hall of Fame Award". Winners are selected through ballots sent to newspaper and magazine editors, members of the Foundation, and retailers across the country.

Another major project is a series of educational seminars dealing with such subjects as fragrance trends, marketing of fragrances, package design, product safety, etc. Some deal with technical or specific subject matter. Other seminars deal with subject matter of a more general nature, and are open to non-members.

International Essential Oil Congress

Beginning with a session in Calabria, Italy, in 1957, a series of International Essential Oil Congresses has been held at roughly three-year intervals, with the most recent Congress being the successful gathering in San Francisco, in September, 1974. Over nine hundred atended the San Francisco Congress, more than twice as many as had ever attended any of the five previous meetings. The 7th Congress was held in Kyoto, Japan, in October, 1977.

After the first Congress in Italy, later Congresses were held in Grasse, France, in 1959, in Plovdiv, Bulgaria, in 1964, in Tbilisi, U.S.S.R., in 1968, then in Sao Paolo, Brazil, in 1971. There is no specific international body that organizes or conducts the Congresses, the host country being selected on a volunteer basis. The sessions are open to representatives of any essential oil or aroma chemical producer interested in participating. In general, the ten associations which are members of the International Fragrance Research Association (IFRA) all urge their members to attend the Congresses, but it is significant that three of the host countries for previous Congresses, Bulgaria, U.S.S.R. and Brazil, are not members of IFRA.

I	Calabria, Italy, 1957 attendance 70	IV	Tbilisi, Sept., 1968 U.S.S.R., attendance 300
II	Grasse, France July, 1959, attendance 135	V	Sao Paulo, Brazil, Oct. 1971 attendance 400
III	Plovdiv, Bulgaria, May, 1964 attendance 200	VI	San Francisco, California, U.S.A., Sept., 1974 attendance 950

VII Kyoto, Japan, October, 1977

At the San Francisco Congress in 1974, representatives were present from over 40 foreign countries, including groups from France, Germany, Spain, Japan, Italy, the United Kingdom, Egypt, the U.S.S.R., Bulgaria, Brazil, Indonesia, Madagascar, Algeria, Chile and Poland.

The host country normally organizes tours of its country's essential oil producing areas, as well as visits to factories manufacturing aroma chemicals or processing essential oils. Sightseeing tours are scheduled and social events included in the program.

James Rogers of Fritzsche Dodge & Olcott, Inc., and Ira Kapp of Felton International, Inc., were co-chairmen of the VIth Congress, held in San Francisco, in September, 1974. Mr. Rogers also served as Scientific Chairman of the Congress.

At the San Francisco Congress over 200 scientific papers were presented, reviewing the latest scientific advances in the cultivation of fragrance and flavor raw materials, as well as their manufacture, testing and use. Proceedings of the Congress, with abstracts of all technical papers, are made available to registrants in connection with the registration fee.

The chairman of the VIIth International Congress, held in Japan, in 1977 was Mr. Kenji Nakanishi of Takasago Perfumery Co. Mr. Yasamasa Kato of Ogama & Co. was secretary-general of the Congress. The sessions were held from October 7 through October 11, 1977, in the International Congress Hall in Kyoto, Japan

International Fragrance Association

49 Square Marie-Louise
1040 - Brussels, Belgium

IFRA is a world-wide association for the Fragrance Industry which was organized at a meeting in Brussels, Belgium, Oct. 17, 1973. Member associations include the national associations of fragrance manufacturers of:

Belgium	Netherlands
France	Spain
Germany	Switzerland
Italy	United Kingdom
Japan	United States

Dr. G. Waldvogel of Givaudan & Cie., Switzerland, is president of IFRA. The general secretary is R. Verbeurgt, Brussels. Each association is represented on the board of IFRA by one member, and also has two representatives on the Technical Advisory Committee. Representatives for the United States are W.G. Dempsey of IFF, board member, and Ira Kapp of Felton Intl. and Dr. Jan Stofberg, Polak's Frutal Works, technical advisor.

One of the basic purposes behind organization of IFRA was the formulation of a world-wide Code of Practice for the Fragrance Industry. Through this Code it was hoped to arrive at a uniform approach toward ingredients of fragrance formulations and their testing, specifically:

Common attitude with regard to traditional as well as new captive ingredients.

Common attitude with regard to the testing of compositions.

Common attitude with regard to adverse reactions reported, and organization of a national and international monitoring system for such reactions.

Common attitude with regard to the testing results of individual ingredients obtained with RIFM.

Common attitude with regard to the proper use of perfume ingredients by the customer industries.

All member associations of IFRA have also adopted a Code of Practice which establishes standards of good operating practice for firms in the fragrance industry.

Principal areas covered are: standards of good manufacturing practice; adequate training of personnel; maintenance of premises and sanitation; quality control and storage of fragrance ingredients; labeling and packaging; adequate laboratory facilities; maintenance of safety.

A scientific advisor is responsible for the functioning of IFRA, and is responsible for the agenda and working documents for all meetings of the General Assembly, the Board of Directors, and Committees. Dr. F. Grundschober, 8 Rue Charles-Humbert, 1205 Geneva, Switzerland, is serving currently in this capacity.

The responsibilities of the Technical Advisory Committee are to establish, update and maintain guidelines for the use of fragrance ingredients.

Monell Chemical Senses Center

3500 Market St.
Philadelphia, Pa. 19104

This research center, located on the campus of the University of Pennsylvania in Philadelphia, Pa., was established in 1968 as the nation's first institute devoted exclusively to research on the senses of taste and olfaction. The program of the Center is devoted exclusively to research and teaching that will lead to a better understanding of the functions and mechanisms of taste and smell. Research is conducted at a fundamental level. It includes chemical, behavioral, physiological, biochemical, nutritional and neurological approaches.

The Essential Oil Association of U.S.A. Inc. makes a grant of $5,000 a year to the Monell Center, which is used to support postdoctoral fellows. The Center does not perform contract research for industry, but has a group of industrial sponsors who support the Center financially. The sponsors also exchange information with the Monell staff and in some instances provide specialized equipment and materials which are employed by the Center in its research. The sponsors include: Firmenich, Inc., Givaudan Corp., Hercules, Inc., Intl. Flavors & Fragrances, Inc. and Monsanto Flavor Essence, Inc.

The J. Manheimer Memorial Lecture Fund has been established at Monell to bring a scientist of international reputation to the Center each year.

American Society of Perfumers

Box 181
Pearl River, N. Y.

The American Society of Perfumers was established in 1947 in New York City, its first members being 21 individuals who represented 18 companies dealing in essential oils and aroma chemicals. Membership in the society has always been by invitation only. There are currently about 300 members of the organization.

The Society has given perfumers a professional status, and has provided a forum in which perfumers could meet with one another, interchange ideas, and talk with each other about the common problems of their profession. Through the Society, perfumers are enabled to speak freely with one another, in language that is common to those in the profession, about subjects that are peculiar to the perfume industry.

The existence of the American Society of Perfumers has made possible the collaboration of workers in adjacent fields, such as cosmetic chemists, biologists, etc. It is possible to hold meetings with members of other technical associations, broadening the horizons of the perfumer and bringing to the attention of the industries using fragrances the highly technical and intricate nature of the work involved in perfumery.

The Society holds an Annual Symposium, which is one of the highlights of the fragrance industry. Topics discussed range from the creation of fragrances to their marketing, from the physiology and psychology of odor to the use of instrumentation in odor analysis.

The society holds regular meetings (normally every third Wednesday, September thru June) and also schedules several social functions. The organization has also sponsored government-industry conferences, as well as courses on perfumery and symposia on odor and related subjects.

The society makes a major contribution to the spread of knowledge about fragrances, as it helps to keep its members and other perfumers informed of the latest developments pertaining to fragrance. Through its existence there is an agency to speak for the professional perfumer, and present his viewpoint. The organization has also established contact with similar societies in England and France to foster the mutual exchange of professional ideas.

The first president of the Society was William H. Dunney Sr., now deceased, who was employed by Antoine Chiris Co., Budd Aromatic Chemical Co. and Ungerer & Co. The second president was the famous perfumer, Dr. A. T. Frascati, who served with Givaudan, headed the perfumery department of Ungerer & Co. in 1929, was perfumer for Max Factor & Co. in 1933, then from 1936 to 1948 was technical director of Firmenich & Co. He later organized Perfumery Associates, then in 1955 became a consultant to George Lueders & Co. Dr. Frascati, at age 80 or thereabouts, is now living in retirement in Florida.

The Soap and Detergent Association

475 Park Ave. South
New York, N. Y. 10016

This organization, formerly known as The Association of American Soap and Glycerin Producers, Inc., was organized in 1926. Roscoe C. Edlund was the first manager of the association, continuing in this capacity for 21 years. He was succeeded in 1948 by Roy W. Peet. In 1963 E. Scott Pattison became manager. His title was changed to president in 1967. He was followed by Charles G. Bueltman in 1971, and by the current president, Theodore E. Brenner, in 1972.

The association was organized "for the purpose of collecting and circulating information valuable and useful to the public throughout the United States with reference to soap and kindred products and for the purpose of investigating the nature and use of glycerin and kindred substances and disseminating information with reference thereto".

In 1962 the name was changed to its present form, The Soap and Detergent Association. The office has always been located in New York City.

SDA operates through five divisions: Fatty Acid Division; Glycerine Division; Household Division; Industrial and Institutional Division; and the Technical and Materials Division. There are 129 member companies, which SDA says represents over 90% of the soaps and detergents produced in the United States.

Among the accomplishments of SDA has been its major role in helping to popularize the concept of cleanliness. SDA also serves as the communications crossroad for the soap and detergent industry, channeling information to and from consumers, industrial and institutional audiences, the media, and all levels of government.

As part of its public information program, SDA issues brochures, newsletters and other printed material, some of general consumer interest, and other releases on more technical subjects. Many radio scripts have been prepared, and a number of motion pictures have been produced. The latest such film, released in 1976, is titled "Commitment to Safety". It documents the industry's research to ensure that its products will be safe for humans and the environment, as well as effective in use.

Statistical data are compiled periodically. The association also processes thousands of requests from the general public for information about soaps and detergents. SDA also monitors Federal, state and local legislation, keeping track of and bulletining members on any regulations and laws affecting manufacture, marketing or use of the industry's products.

Research is another important area of interest. SDA conducts and guides research related to environmental quality and consumer safety. New test methods on biodegradability were developed under SDA sponsorship. Ongoing research is currently being conducted to determine the effect of product constituents in water and sewage treatment processes, as well as on aquatic life and water resources. Research is also proceeding on the causes and controllability of eutrophication, and on methods of removing all nutrients from wastes.

One of the outstanding research accomplishments of the soap and detergent industry, working in conjunction with SDA, has been the nationwide voluntary conversion to readily biodegradable detergents. As the result of a voluntary changeover by the entire industry, completed in 1965, all household laundry products used throughout the U. S. are now of the readily biodegradable type.

Society of Cosmetic Chemists

50 East 41st St.
New York, N.Y. 10017

The Society of Cosmetic Chemists was organized in May 1945, in New York City by twelve charter members, the objective being "to establish a medium through which scientific knowledge of the cosmetic and toiletries industry can be disseminated, and to improve cosmetic and cosmetic scientists' professional standing." First president of the SCC was Maison G. deNavarre. Dr. Emil G. Klarmann, the first program chairman, succeeded deNavarre as president in 1947.

The Scientific Programs, presented annually in December, represent one of SCC's most important activities. Papers are read on a wide range of topics important to the cosmetic chemist, with many of them subsequently being reprinted in the Journal of the Society. In 1954 the Society added a Fall seminar to its program. These Seminars are now held in May of each year, and normally consist of a symposium on one topic or on several closely related topics. As the Society now has chapters in New

York, Chicago, California, New England, the Mid-Atlantic states, St. Louis, the Ohio Valley, the Twin Cities, the Southwest and the Golden Gate district of California, the role of host chapter for these seminars is rotated.

The Society of Cosmetic Chemists is also affiliated with the International Federation of Societies of Cosmetic Chemists, which since its organization in 1959 has held an International Scientific Congress every two years.

Membership in SCC is open to "those individuals who are interested in the objectives of the Society and meet the requirements for Active, Junior, Fellow, Honorary, Emeritus or Emeritus Fellow members as provided in the Bylaws." One of the requirements is adherence to the SCC's Code of Ethics. The Society of Cosmetic Chemists has over 2,200 members.

The SCC presents a number of awards to outstanding members of the cosmetic chemists' profession. The SCC Medal Award is given annually to an individual actively engaged in the art and science of cosmetics, irrespective of citizenship, residence or membership in the SCC, who is normally recognized for accomplishments in the cosmetic field over a period of years, rather than for any single accomplishment.

The SCC Literature Award honors the author of a scientific report appearing during the previous year considered to be of the greatest potential value in improving cosmetic technology. The SCC Award, sponsored by International Flavors & Fragrances, Inc. for $1,000, is given to the author of the "best scientific paper" presented and published in the Journal. Another SCC Award, sponsored by Perry Bros. Fragrances for $1,000, is presented to the author or authors who have contributed significantly to cosmetic science and technology in the area of microbiology. A similar Award for $1,000, sponsored by Shaw Mudge & Co. is given to the author of the best paper on "Contributions to safety in the use of toiletry and cosmetic products." Miranol Chemical Company sponsors a new SCC Award of $2,000, for the "best published contribution to the understanding of biological and cosmetic effects of topically applied surface active agents."

There are also a series of Education and Scholarship awards to stimulate scientific interest and participation in the field of cosmetic science, primarily through educational facilities, as well as the SCC's Award of Merit, which recognizes a member for distinguished contributions, achievements and service to the Society.

As mentioned earlier the SCC publishes "The Journal of the Society of Cosmetic Chemists," which is described in another chapter. (See page 379). The SCC also maintains a Library, which is housed in the offices of the Society.

Flavor & Extract Manufacturers' Association

900 17th Street, N.W.
Washington, D.C. 20006

F.E.M.A. was organized in 1909 in Baltimore, Md. to look after the interests of the Flavor Industry as they are affected by legislation and governmental regulation. The major goals of the association are:

The creation of better flavors for the benefit of the consumer.

The assessment and evaluation of regulatory changes affecting the flavor industry.

Assisting members with a broad variety of support services.

F.E.M.A. developed the GRAS safety concept for flavor ingredients under the Food Additive Amendment of 1958, and in 1960 established an Expert Panel for the determination of GRAS status for flavors. The association assists its members in dealing with flavor labeling regulations, and to that end publishes a Flavor Labeling Handbook. F.E.M.A. has also worked with other groups such as I.O.F.I. and Codex Alimentarius in developing specifications for flavoring materials.

F.E.M.A. holds two meetings each year; the Annual Convention, in the Spring, and the Fall Symposium.

Annual Dinner, Manufacturing Perfumers' Association, 1916
Photo courtesy Howard Isermann

The Cosmetic, Toiletry & Fragrance Association

1133 15th St., N.W.
Washington, D.C.

The original predecessor organization to CTFA was the Manufacturing Perfumers Association of the United States, formed in New York City in 1894. MPA was begun by a group of industry leaders desirous of expanding the market for their products. The name of the organization suggested the preponderance of perfume and fragrance manufacturers in the cosmetics industry at that time. Bowles Colgate was the first president of MPA, Alfred G. Wright was vice-president, and Henry Dally, secretary-treasurer.

In 1916, importing companies separated from MPA and formed their own organization, The Perfumery Importers Association, Inc. (P.I.A.). Their interests diverged from those of the domestic companies, because of the tariffs imposed on imported cosmetics and raw materials, and it was to look specifically after these interests that they formed their own organization.

In 1918, the Manufacturing Perfumers Association changed its name to the American Manufacturers of Toilet Articles (A.M.T.A.). In 1923, AMTA sponsored the nation's first Perfumery Show, with 75 American companies exhibiting their cosmetic products. The association also gave a grant to the U. S. Dept. of Agriculture for research designed to encourage the cultivation of floral raw materials in the U.S. They also sponsored a course on perfumery and cosmetics at Columbia University. In 1924 the association had 115 active and 105 associate members. Gilbert Colgate of Colgate & Co. was president, and Northam Warren of Northam Warren Corp. was vice-president.

The 1929 Crash and the Depression which followed seriously affected the cosmetic industry along with all others. Sales dropped drastically, and the number of firms in the industry was cut almost in half. The expense and difficulty in obtaining natural raw materials prompted increased research to develop alternative materials, and many new synthetic materials came into use during this period.

By the early 30's the divergent interests of AMTA and PIA had been resolved. Members of the importers group joined AMTA and the name of the group was changed to Associated Manufacturers of Toilet Articles. The PIA continued to function until 1968, but after 1933 most of its members were also members of AMTA and succeeding organizations.

AMTA was succeeded by the Toilet Goods Association in 1935. The industry had set up a Code Authority under the National Industrial Recovery Act in 1934, but when NRA was declared unconstitutional many of those in the cosmetic industry felt that the programs that had been started under the Code should be continued. Thus they structured the staff and the program of the new association, TGA, to emphasize practical service to members. After the Federal Food, Drug & Cosmetics Act was enacted in 1938, TGA set up a Board of Standards to assist manufacturers of cosmetics in bringing their labels into conformity with the new law. The association also developed a series of publications designed to keep members informed of government and other activities affecting the cosmetic industry. An annual scientific conference was also started, providing a forum for presentation of papers on new scientific developments affecting the cosmetic industry.

To describe membership and activities more adequately, the name of the association was changed in 1971 to the Cosmetic, Toiletry & Fragrance Association. Now headquartered in Washington, D.C. at 1133 15th St., N.W. 20005, CTFA is the leading trade association for the cosmetic industry. It represents more than 200 manufacturers who market 90% of the industry's products, and a comparable number of firms which supply ingredients, raw materials, packaging and services used in the manufacturing and marketing of cosmetics.

Activities of the association, which have expanded steadily over the years, are concentrated in several areas:

> *Government Regulations* - CTFA monitors federal, state and local legislative and regulatory activity affecting the cosmetic industry, and has sponsored many industry-initiated programs in self-regulation.
>
> *Science* - The CTFA scientific program is directed to assuring quality manufacturing practices and consumer safety. CTFA has 21 scientific committees, composed of leading scientists in the industry, which work with the association's own scientific staff to initiate, structure and oversee vital research. In August, 1976, CTFA began an ongoing Scientific Ingredient Review (CIR) to examine all ingredients used in cosmetics manufacturing in order to arrive at an objective scientific opinion on the safety of these ingredients under conditions of use.
>
> *Communications* - In addition to communicating with members through the usual channels of bulletins, newsletters, audiovisuals and a quarterly Journal, CTFA issues technical publications, including a manual interpreting Federal packaging and labeling regulations. The association also holds work-shops, seminars and special briefings for members.

The organization had its first professional director in 1935 when Charles S. Welch, formerly vice president and general manager of Houbigant, Inc., and later of Helena Rubenstein, Inc., was named executive secretary of the newly formed TGA. His title was later changed to executive vice-president. Welch's role was a natural extension of his previous role as executive secretary of the Toilet Goods Industry Code Authority established under NRA. Mr. Welch served TGA until his death in 1942, when he was succeeded by Stephen L. Mayham. Mr. Mayham, widely known in the industry as editor of "The American Perfumer", joined TGA in 1939 as director of its Board of Standards, then succeeded to the post of executive vice-president on Mr. Welch's death.

Under Mayham's leadership, TGA was successful in the 1940's in averting potentially crippling wartime restrictions on raw materials used in the manufacture of cosmetics. During his tenure, TGA also took its first major step into public relations activity, producing a history of the cosmetics industry in America and a series of educational projects interpreting the industry to the public.

Mayham was succeeded in 1967 by James H. Merritt, who had previously served as executive vice president of the National Association of Chain Drug Stores, and as secretary of the Proprietary Association. Merritt supervised transfer of CTFA's offices from New York to Washington in 1968, a move made necessary by expanding government activity affecting the cosmetic industry. During the ten years since he took over, TGA has increased its membership by approximately 10% in spite of numerous mergers; has experienced an "explosion" of scientific programming, which now represents a major part of the association's activities; and has expanded communications to assure that information generated by Association programs will be delivered speedily, not only to CTFA members, but also to individuals and groups outside the association which have an interest or an involvement in activity of the cosmetic industry.

Chemical Specialties Manufacturers Association

1001 Connecticut Ave. N.W.
Washington, D.C. 20036

The association out of which CSMA developed was organized in 1914 as the Insecticide Manufacturer's Association. First members were a group of manufacturers of agricultural insecticides who were unhappy over what they considered to be unduly harsh enforcement of the Insecticide Act of 1910, which regulated their products. Within a year several disinfectant manufacturers joined the organization, which dictated a name change to the Insecticide and Disinfectant Manufacturers' Association. First president was Frank Hemingway of Frank Hemingway & Co., who served from 1914 to 1919.

By 1920, manufacturers of agricultural insecticides had dropped out of the association, leaving the disinfectant group in ascendancy. Harry Cole of Baird & McGuire, a disinfectant manufacturer of Holbrook, Mass., was elected president. He was a prominent figure in the association for many years, taking over as secretary in 1923 and holding that post until 1933. During the lean years, when I&DMA was struggling to stay alive, headquarters were in Baird & McGuire's office.

Through the 20's, as pyrethrum insecticides became popular new products, a number of companies manufacturing household insecticides joined the association, and in time took at least as important a role as the disinfectant manufacturers. About 1929, the association spearheaded a successful, industry-wide effort to have the Terry patent on pyrethrum fly sprays declared invalid. This was followed by a big expansion in household insecticide sales.

Much of the attention of the members of I&DMA was concentrated on legislation during this period, and on product standards. The association pressed for uniformity in state laws regulating the industry's products, and favored adoption of a single Federal standard.

In 1932 the name of the association was changed to the National Association of Insecticide & Disinfectant Manufacturers (NAIDM). At this time there were 106 members, and membership had been widened to include manufacturers of potash soaps and sanitary products as well as disinfectants and insecticides. John Wright of Zonite Products Corp. succeeded Harry Cole as secretary in 1934 and served until 1938. Following Dr. Wright, Ira MacNair of MacNair-Dorland Co. was secretary until 1941.

In 1942, the association had its first paid executive, Dr. H. W. Hamilton, who had earlier (1928-'29) served as president of I&DMA while in the employ of White Tar Co. Hamilton directed the association's affairs until 1959, and exerted a strong, positive influence on development of the association during that period.

The name was changed again, to the Chemical Specialties Manufacturers Association, in 1949, and a divisional set-up adopted. There were five divisions, for manufacturers of: Insecticides, Disinfectants, Aerosols, Soaps and Detergents, and Automotive Products. The reason why aerosols chanced to be the concern of CSMA was that the first aerosol product developed was an insecticide. Thus CSMA meetings were the arena in which discussions were conducted, and standards adopted, for pressurized containers, valves, spray patterns, etc. It would have been just as logical, or as illogical, to have had pressure-packages become the concern of a paint association, had the first aerosol been a spray paint, or a cosmetic association if the first aerosol had chanced to be a hair spray or a spray cologne. But, it was an insecticide. Over the years CSMA committees have done so much work on aerosols that aerosol fillers, and marketers of aerosol chemical specialties, have remained loyal to CSMA, and the Aerosol Division is by far the largest of the divisions of CSMA.

Perhaps because CSMA and its predecessor trade organizations operated for so many years without the services of a paid executive, the tradition of a "do-it-yourself" association developed. Members are very active in committee work, and give thousands of hours of their time to association activities. Dr. Hamilton always counseled new members to take an active role in the association, reminding them that an association member gets out of membership in an organization only as much as he puts into it.

It was during Hamilton's tenure as association manager that CSMA instituted the policy of holding two expanded board meetings each year in addition to the annual meeting in December and the mid-year meeting in May. Much of the association's business is conducted in the committee meetings at these four yearly sessions. In recent years the mid-year meeting has been held regularly at the Drake Hotel in Chicago, and the December meeting on a revolving basis, every third year in New York City, Miami Beach, Fla. and Washington, D.C.

When Dr. Hamilton retired as secretary of CSMA in 1962, he was succeeded by Al Mulliken. All through this period the office was in the Chemists Club building on 41st St. in New York City. When Mr. Mulliken retired in 1974, he was succeeded as executive director by Ralph Engel, and the office was moved to its present location at 1001 Connecticut Ave. N.W., Washington, D.C. The move was made to bring the association office closer to the seat of the Federal Government, to assist it in monitoring legislation, and keeping members advised on what Congress and the regulatory agencies are doing.

Cosmetic Career Women, Inc.

614 West 51st St.
New York, N. Y. 10019

This organization, the only one in the cosmetic industry whose membership is limited to women, was started in 1954 by Kay Colton and Kay Fitzpatrick, with the encouragement of Steve Mayham, then executive vice president of the Toilet Goods Association. The purpose of the organization, in the words of the late Mildred Blum who wrote the slogan of the CCW, was and is "To promote better acquaintance among women in the cosmetic industry and to insure participation by women in improving the standing of the industry in the business community."

CCW is unique in that it has no president or other officers, but is operated by a Board of Governors which numbers ten members. Membership in the organization is by individual rather than by company, and memberships are not transferable. Membership is limited to women who have served for at least three years in an executive capacity in a company engaged in the industry, or as executive secretary to an official of an industry company.

CCW sponsors luncheons and meetings designed to inform its members, to publicize the industry, and to serve charitable purposes. An annual feature is the Christmas luncheon, to which members and their guests bring hundreds of gifts which are donated to institutions for under-privileged hospitalized children. CCW also runs grooming classes for blind children, and serves as a clearing house to supply information to schools and colleges on how to get a job in the cosmetic industry.

S.O.C.M.A.

1075 Central Park Ave.
Scarsdale, N.Y. 10583

The Synthetic Organic Chemical Manufacturers Association was founded in 1921. It presently has 81 active members and 10 associate members. Meetings are held at the Hotel Roosevelt in New York City the first Thursday of each month, except for May, July and August. A two-day Spring business meeting is held each year during the month of May. The annual meeting luncheon and the annual dinner are held annually in December.

Among the purposes of the organization are the following: to encourage the manufacture in the United States of all kinds of organic chemicals; to disseminate scientific and general information about the industry and its products; to cooperate with governmental agencies; to promote the highest scientific and business standards; and to take such collective action as may be proper for the perpetuation of a strong, independent organic chemical industry in the United States.

Sales Association of the Chemical Industry

50 East 41st St. (Room 81)
New York, N.Y.

This organization, earlier known as the Salesmen's Association of the American Chemical Industry, was founded in 1921. Its first president was Burton T. Bush, then general manager of Antoine Chiris Co. SACI has approximately 700 members today. Its primary objectives are: to increase selling efficiency (in the chemical industry) and to foster highest sales ethics; to encourage wider acquaintanceship and good fellowship among those in the industry; and to engage in other activities to foster the growth and welfare of the chemical industry.

SACI runs several golf tournaments each summer. Another yearly social feature is the annual Christmas Party. SACI publishes a humorous publication annually, "The Chemical Peddler." On the serious side, the organization has sponsored several chemical marketing conferences and serves as a clearing house for sales openings in the chemical industry.

Paul B. Slawter, Jr. of Rosston, Kremer & Slawter, Inc., 79 Madison Ave., New York, serves as administrative assistant of SACI.

C.I.B.S.

This organization, whose initials stand for Cosmetic Industry Buyers and Suppliers, was organized in 1947 by a group of salesmen and buyers who used to meet informally at Zum's Brau Haus at 54th St. and 3rd Avenue in New York City. Arthur Mulligan of Lehn & Fink was one of the prime movers in getting C.I.B.S. started and was the first president of the organization.

The basic purpose of C.I.B.S. is to allow people in the various branches of the cosmetic industry to get better acquainted with one another — buyers, raw material sales people, technical people, representatives of packaging firms, and individuals connected with finished goods houses. There are now approximately 300 members.

Luncheon meetings are held regularly throughout the year, one of these being a Ladies Day luncheon in May. The annual meeting for the election of officers is held in November, and a dinner dance during the Christmas season.

Each year C.I.B.S. makes an award to the originator or originators of the most beneficial paper submitted in the previous year to the scientific section of the C.T.F.A. The recipient of the award is selected in cooperation with the Editorial Board of the Scientific Section of C.T.F.A.

The principal location of C.I.B.S. is the business address of the Corresponding Secretary.

B.I.M.S.

This organization, whose members consist of Buyers, Importers, Manufacturers and Salesmen in the drug, cosmetic and pharmaceutical industries, was organized in 1932. Charles Dorr of Harriet Hubbard Ayer was one of the founders, and Martin Shultes was the first executive chairman. Shultes held this post until 1948 when Peter L. Forsman of C.H. Forsman Co. took over. Forsman turned the post over to George E. Davidson of Anchor Brush Co. in 1965.

Applicants for membership must be active in the industry and must be sponsored by an active member. They are then voted on by the Executive Committee. There are five members on the Executive Committee, and twelve members on an Advisory Committee. Membership of B.I.M.S. currently consists of 165 active members, and 35 honorary members, retired former active members.

The group holds four golf outings each summer, and an annual dinner each year. Guests are permitted at all events except the September outing, which is for members only.

Headquarters of the organization is at the office of the executive chairman at 2105 Evergreen Lane, Pt. Pleasant, N.J. 08742.

Chapter XVI

We Share Our Knowledge

There are a number of publications through which producers as well as users of flavor and fragrance materials keep themselves informed on what is going on in this fast moving, dynamic field. Included among the long list of trade and technical publications which circulate widely throughout the industry are "The Journal of the Society of Cosmetic Chemists," "Cosmetics & Toiletries," "Soap/Cosmetics/Chemical Specialties," "Aerosol Age," "Drug & Cosmetic Industry" and many others.

Several manufacturers of aroma chemicals issue their own publications, distributing them without charge to selected lists of users of aroma chemicals. These publications include the "Dragoco Report" and Roure Bertrand Dupont's "Recherches."

Finally, there are a number of highly useful technical books, including "The Essential Oils" by Dr. Ernest Guenther of Fritzsche Dodge & Olcott, "The Givaudan Index," published periodically by Givaudan Corp., and several books by Steffen Arctander and Paul Bedoukian on perfume and flavor raw materials, of both natural and synthetic origin.

In this chapter we offer brief descriptions of the most widely read and most useful of these publications.

The American Perfumer

1031 South Boulevard
Oak Park, Illinois 60302

Although this leading trade magazine of the essential oil and aroma chemical industry was split into two magazines in 1976, "Cosmetics & Toiletries" and "Perfumer & Flavorist," we will tell its story under the name by which it was known for so long. The first issue of "The American Perfumer and Essential Oil Review" was published in March, 1906, by Ungerer & Co., New York City. The magazine was initially a house organ for Ungerer, with most of its advertising support from European suppliers of fragrance specialties who were represented in the United States by Ungerer & Co.

Louis Spencer Levy was the man who developed the magazine for Ungerer, and he was its first editor. In August, 1906, Mr. Levy acquired the magazine from Ungerer and became owner and publisher under the name, The Perfumer Publishing Company.

Maison G. deNavarre first appeared as a contributor to the "American Perfumer" in 1932, then started writing a monthly column under the title, "Desiderata," in 1933. On the staff were Bill Lambert, who joined the magazine in 1921, and S.L. Mayham, in 1926.

In June, 1935, Mr. Levy, who was in ill health, sold the magazine to Robbins Publications. Mayham became managing editor and Lambert advertising manager. Mayham stayed with the "Perfumer" until 1937, when he went with the Toilet Goods Association as executive director. Lambert took over as editor and remained with the magazine until his retirement in 1958.

Howard Moore, one of the principals in Robbins Publications, had general supervision of the "American Perfumer" as his responsibility. The new owners named Harland J. Wright vice-president of the publishing company and Wright was active in the management of the

"American Perfumer" for a number of years. During the 40's, James Moore gradually took over active management of the company from his father.

The original name of the magazine was "American Perfumer and Essential Oil Review." In 1956 it was changed to "American Perfumer and Aromatics." By this time the magazine was owned by the Moore Publishing Company which was a direct descendant of Robbins Publishing. Allured Publishing Corporation purchased the magazine from Moore in 1960 and soon shortened the name to "American Perfumer." Stanley E. Allured was publisher and M. G. de Navarre was editorial director.

The name of the publication was changed to "American Perfumer and Cosmetics" in 1962, and to "Cosmetics and Perfumery" in 1973. Then, in 1976, as indicated earlier, the publisher decided to split the magazine into two publications, "Cosmetics & Toiletries" and "Perfumer & Flavorist," to indicate the industries and functions for which the two magazines are published.

Soap/Cosmetics/Chemical Specialties

101 West 31st St.
New York, N.Y. 10001

The first issue of a monthly magazine then called "Soap" was published by MacNair-Dorland Co. in September, 1925. The principals were Grant A. Dorland and Ira P. MacNair, who had previously worked together at "Drug & Chemical Markets," MacNair as editor, and Dorland in advertising sales. Dorland had also published the 1924 edition of the Annual Directory of the American Manufacturers of Toilet Articles.

Both men had a wide acquaintance in the essential oil trade, so it was not surprising that manufacturers of perfuming materials were generously represented among the advertisers in the first issues of "Soap."

In addition to editorial coverage of the soap industry, the new magazine also treated what has come to be known as the chemical specialties field, which includes potash soaps, household insecticides, disinfectants, floor waxes, polishes, etc. When the first issues of "Soap" appeared, there was another magazine called "The Soap Gazette" covering the soap field. It ceased publication early in the 30's.

There have been several changes in the name of this publication over the intervening years. It was changed to "Soap and Sanitary Chemicals" in 1946, then to "Soap and Chemical Specialties" in 1954. The most recent change, to "Soap/Cosmetics/Chemical Specialties" occurred in 1971.

Ira MacNair was the first editor of "Soap," and continued to direct the editorial side of the magazine for many years. He died in 1974. Frank Reilly was editor for about twenty years until his death in October, 1976. He was succeeded by Janet Donohue, formerly associate editor.

MacNair-Dorland Co., of which Grant A. Dorland is president, publishes three other trade magazines in the chemical field in addition to "Soap/Cosmetics/Chemical Specialties." Thomas Morgan is vice-president and general manager of the company, and James D. MacNair is secretary-treasurer.

Aerosol Age

200 Commerce Road
Cedar Grove, N.J. 07009

"Aerosol Age" was started in 1956 by Wayne Dorland, Eleonore Dorland and Ralph Dorland. Ralph Dorland subsequently sold his interest in the magazine to the other two partners. "Aerosol Age" is a monthly magazine whose function is to make available in one publication information about the field of pressure packaging. First introduced during World War II for packaging insecticides for military use, the aerosol dispensing principle was extended in post-war years to a wide range of personal and household products, drugs and pharmaceuticals, industrial and automotive specialties, paints and foods. "Aerosol Age" is read by marketers of these products as well as by contract packagers.

In 1970, Wayne and Eleonore Dorland sold the magazine to a publishing firm headed by Robert Hundley, and the magazine is now published by Industry Publications, Cedar Grove, N.J. Mrs. Margaret Hundley is editor.

The aerosol dispensing principle has meant a great deal to the fragrance industry. Almost all products dispensed as aerosols include some fragrance in their formulas, and the tremendous expansion in the sale of aerosols over the past 25 years has been accompanied by an equally impressive expansion in the sale and use of essential oils and aroma chemicals used to perfume these pressure-packed products. The president of a leading manufacturer of aroma chemicals, in an article in the Feb. 1976 issue of "Aerosol Age" (pg. 27) said, "If I were asked what has contributed most to the popularity and advancement of fragrances in the past 20 to 25 years, I would have to say that the aerosol package did more for the increase in fragrance consumption than any other single factor."

Aerosol fillings reached the impressive total of almost three billion containers in 1973. Between that year and 1975 there was a drop of 15 to 20% in sales, under the impact of unfavorable economic conditions, plus the unfavorable publicity the industry received when it was charged that fluorocarbon propellants, which are used to charge certain aerosols, may in time have an unfavorable effect upon the environment.

Recent restrictions placed on the use of fluorocarbons have had an unfavorable short-term effect on aerosol sales, but the industry's ability to convert to propellants other than chlorofluorocarbons has probably been underestimated. In May, 1977, Precision Valve Corporation, Yonkers, N.Y., released details on a new propellant system based on use of hydrocarbons which they predict will give a new impetus to aerosol packaging. They say that as a result of their development of a new type valve, it will now be possible to dispense as aerosols materials which could not be handled in previous aerosol packages, and that costs will also be reduced. They suggest the name "Aquasols" for these new aerosol product types.

Drug & Cosmetic Industry

757 Third Ave.
New York, N.Y.

"Drug & Cosmetic Industry" is one of two magazines that came from the break-up of "Drug & Chemical Markets." The latter monthly trade paper was founded in 1914 by Williams and Oliphant Haynes, and served an important role in keeping the infant chemical industry informed all through the years of World War I and the post-war years. In the early 20's Ira P. MacNair was editor of "Drug & Chemical Markets," and for a time Grant Dorland was advertising manager. When they left the Haynes organization to start their own publication, they were succeeded by Thomas Farrell as editor and Frazer V. Sinclair as advertising manager.

In 1926, Haynes split his magazine, selling the drug portion to Farrell and Sinclair, who renamed it "Drug Markets." The chemical portion became "Chemical Markets," subsequently "Chemical Industries." After it was later sold to McGraw-Hill, it became a weekly and was renamed "Chemical Week."

In 1932, Farrell and Sinclair changed the name of their magazine to "Drug & Cosmetic Industry," reflecting their intention to give major editorial attention to matters of concern to the cosmetic industry, which was then emerging as a much stronger and distinct industrial group. At the same time the drug industry was changing from one having a strong botanical base to an industry dealing in the new chemical-based drugs being developed by German and Swiss companies.

In 1936, Sinclair and Farrell bought the cosmetic retailing magazine, "Toilet Requisites," and subsequently changed its name to "Beauty Fashion." They sold "Beauty Fashion" to another publisher in 1965.

When Farrell retired in 1960, the Farrell-Sinclair partnership was dissolved. Mr. Sinclair became sole owner of the magazine, and took over the post of editor. Upon his death in 1965, his brother-in-law, Paul Alexander, who had been advertising manager, was named publisher. Donald

Davis succeeded Sinclair as editor. In September, 1976, "Drug & Cosmetic Industry" was sold to Harcourt, Brace Jovanovich, Inc., New York.

Journal of the Society of Cosmetic Chemists

50 East 41st St.
New York, N.Y. 10017

The first issue of the Journal was published in July, 1947, the magazine having been started primarily as a result of the energy and enthusiasm of Maison G. deNavarre, who served as the first editor and continued in this post until 1962. The magazine is the official publication of the Society of Cosmetic Chemists and has first right to publish any paper presented at a meeting of the Society. The Journal publishes papers concerned with cosmetics or one of the sciences underlying cosmetics, as well as other papers of interest to its members.

Articles for publication in the Journal fall into the following general categories:

Review Articles — Intended to present recent advances in a specific area related to cosmetics.

Original Articles — Covering original laboratory investigations in cosmetics and related areas.

General Articles — Articles of a scientific or technical nature concerned with newer analytical techniques, developments in dermatology, toxicology, etc.

Short Preliminary Communications — Reporting new findings of sufficient importance to warrant prompt publication.

Dr. Martin M. Reiger succeeded Maison deNavarre as editor of the Journal in 1962. Dr. Karl Laden became editor in 1967, followed by Dr. Paul A. Sanders in 1972 and the present editor, Dr. John J. Sciarra, in 1973.

Household & Personal Products Industry

26 Lake St.
Ramsey, N.J. 07446

This publication was started in 1964 by Rodman J. Zilenziger and Ralph E. Dorland under the name

"Detergent Age." Dorland retired in 1972. The title of the publication was changed to "Detergents & Specialties" in January, 1969, then to its present name in December, 1971.

HAPPI gives editorial coverage to soaps and detergents, cosmetics and toiletries, cleaners, polishes, waxes, sanitizers and related household and personal products. Regular departments include a Washington Report, an International Report, a Formulary, a New Products Section, a review of Test Markets, as well as feature articles and news.

Hamilton C. Carson is editor.

Cosmetic World

"Cosmetic World" is a newsletter which was founded by the editors of "Beauty Fashion" in 1966. The publication issues 24 times a year, the first and third week of each month. It reports on directions, trends, and personnel changes in the manufacturing and supplier ends of the business. In addition, it reports on buyer changes and significant policy pronouncements with respect to retailing in the cosmetic industry.

Beauty Fashion

48 East 43rd St.
New York, N.Y. 10017

"Beauty Fashion" was founded in 1916 by Frederick J. Pope as "Toilet Requisites." The publisher was Pope Publishing Corp. Mr. Pope was president and general manager. F. Rollin Williams was the first editor of the magazine.

Following the death of Mr. Pope in 1936, "Toilet Requisites" was sold to Frazer Sinclair and Thomas Farrell, publishers of "Drug & Cosmetic Industry." They changed the name of the magazine, in 1939, to "Beauty Fashion." The publication is oriented toward the cosmetic and toiletries sales person. Editorial content embraces the basic state of the art of the cosmetic, fragrance and toiletries industry.

"Beauty Fashion" is published ten times a year. The Summer (July/August) and Winter (December/January) editions are combined issues.

Statesman Journal Newspapers

P.O. Box 13009 • Salem, Oregon 97309 • 503/399-6611

Following the death of Frazer Sinclair in 1965, the magazine was sold to Mr. Richard Gascoigne. In July, 1974, "Beauty Fashion" was sold to a group of investors who publish the magazine under the corporate title, Beauty Fashion, Inc., 48 East 43rd Street, New York, NY 10017. John G. Ledes is editor and publisher.

Chemical Marketing Reporter

100 Church St.
New York, N.Y. 10007

"Oil, Paint and Drug Reporter," now known as "Chemical Marketing Reporter" was founded in October, 1871, by William O. Allison, a young reporter for two New York commercial daily newspapers. Allison had reported imports for these newspapers and decided that there was a place for a specialized publication concerned with this expanding sector. Thus he started publication of a weekly magazine whose editorial attention was directed to paints, oils, drugs and chemicals. Mr. Allison built his publication into a thriving, 20-page journal by 1875, then went on to absorb a number of other publications, and to start publication in 1913 of the "OPD Chemical Buyers Directory."

Harry J. Schnell joined the Allison organization in 1890 as an assistant in the mail department. By 1900 he had become general manager, and following Mr. Allison's death in 1924 continued to run the publications as executor for the estate. Mr. Schnell bought out the magazine and the buyers guide (known in the trade as the "Green Book") in 1941. When he died, just a year later, he was succeeded as editor by Hugh Craig, with William S. Auchincloss, Mr. Schnell's son-in-law becoming president of the company and publisher. Craig served as editor until his retirement in 1952. Arthur R. Kavaler, who had joined the OPD in 1947, was appointed managing editor in 1952.

Bill Auchincloss became chairman of the board in 1968, and was succeeded by his son, Stuart Auchincloss, as president and publisher. When Suart died in an accidental fall in 1971, he was succeeded by his widow, Eva S. Auchincloss, as president of the company.

Kavaler, who had been advanced to the position of editor-in-chief in 1969, was named executive vice-president of Schnell Publishing Co. Mr. Kavaler replaced Eva Auchincloss as president and publisher in 1972, and continues in these posts today, as well as that of editor-in-chief of the "Chemical Marketing Reporter," the new name for the OPD which had been adopted in January, 1972. Upon the death of William Auchincloss in July, 1973, his widow, Jean Schnell Auchincloss, succeeded him as chairman of the board.

"Chemical Marketing Reporter" continues to report the news of the chemical markets, including weekly reports on marketing developments in perfume and flavoring materials. It also publishes two in-depth special sections each year, Cosmetics and Beauty Chemicals, which offer a broad overview of new trends in the cosmetic, toiletries and fragrance industries.

Soap Perfumery and Cosmetics

42 Gerard St.
London, England W2V 72P

This British monthly magazine, published by United Trade Press, London, evolved from "The Soap Trade Review," which was started in 1928. At an early date the magazine changed its title to "Soap Trade and Perfumery Review." The first editor was John Strahan Smith. Contributors to early issues included such authorities on the cosmetic industry as Dr. Paul Jellinek, W. H. Simmons, and W. A. Poucher, a leading authority on perfumery, widely known as the author of the book, "Perfumes, Cosmetics and Soaps."

By 1934 the magazine had changed its title once more, to "Soap, Perfumery and Cosmetics Trade Review," and in that year F. V. 'Freddie' Wells took over as editor. Wells, who was also active as a perfumery consultant, was editor of the journal for 40 years. He played a significant part in helping to write the history of the modern perfumery and cosmetics industry, not only in Great Britain, but worldwide.

In 1974 Patrick Moxey took over as editor. He has been succeeded recently by Valerie Kinglake. The editorial policy of "Soap, Perfumery and Cosmetics" is to balance technical coverage of soaps, cosmetics and perfumery with strong emphasis on marketing and statistical information.

Product Marketing
124 East 40th St.
New York, N.Y. 10016

This magazine is a successor to "Drug Trade News," a magazine first published in 1925. "Drug Trade News" was started by Jerry McQuade who had started another drug trade magazine, "Drug Topics," in 1920. McQuade was later joined by Aglar Cook as a partner.

Following McQuade's death, Robert L. Swain became editor of both "Drug Trade News" and "Drug Topics." Dan Rennick, as editorial director, supervised day-to-day editorial operation of "Drug Trade News."

Both publications were sold in 1972 to Medical Economics Co., a subsidiary of Litton Industries. The new owners suspended publication of "Drug Trade News" and replaced it with a marketing magazine, "Product Management."

Within the past year "Product Management" was sold to new owners and renamed "Product Marketing." It is designed as a national news publication dealing with cosmetics, toiletries, fragrances and drugs. Paul M. Cohen is associate publisher.

Cosmetics
Box 201
Don Mills, Ontario, Canada M3C 2S2

This Canadian business magazine for the cosmetics, fragrance, toiletry and personal care industry was founded in 1972 by Edward Bodner, president of Business Industry Publications, Lt., Don Mills, Ont., Canada. The editorial purpose is to inform and educate the retailer of these products, and his staff, so they can do a better job of informing the consumer. Circulation, which is concentrated in Canada, is primarily to retail drug stores, drug chains and department stores.

The Dragoco Report

Gordon Drive
Totowa, N.J. 07512

This is actually not one but two separate publications, published by Dragoco, Inc., Totowa, N.J., whose parent company is located in Holzminden, Germany. One of the publications is directed monthly to individuals in the fragrance oriented industries, while the other goes on a bimonthly schedule to those using flavors. Both Reports have been published for over twenty years, and are printed in French, German, Spanish and Italian editions as well as in English.

The contents represent a selection of material ranging from product concepts, market evaluation and technical treatises, to industry surveys and reports from essential oil growing areas. Contributors may be Dragoco personnel or industry experts in marketing and technology. Each edition contains a section devoted to general information about Dragoco around the world.

The publications go to key people in the industries using fragrances and flavors, including people in administration, marketing, creative technology, research and development, and purchasing. Copies also go to universities and Federal agencies.

Recherches

1775 Windsor Road
Teaneck, N.J. 07666

"Recherches" is a technical journal dealing with essential oils and aroma chemicals published from time to time by Roure Bertrand Dupont, Grasse, France. Mr. P. Teisseire, technical director of RBD, and his staff at the company's research center in Grasse are the contributors to this outstanding publication of the fragrance industry.

Included in the content of "Recherches" are: reports on work on new synthetic aromatics and on the composition of natural raw materials; studies made by RBD fragrance experts on specific problems such as, for example, the application of physical chemistry to the interpretation of the behavior of perfume compounds; and analytical studies of recent reports published in specialized fields, particularly in the cosmetic field.

The first issue of "Recherches" was published in January, 1952. The most recent issue, #19, was issued in July, 1974, the next will probably appear early in 1978.

Roure Bertrand Dupont distribute "Recherches" widely to the technical people of the essential oil industry, cosmetic and fragrance manufacturers, and university libraries. Each issue is highlighted with beautiful graphics, including color photographs of the flower fields around Grasse as well as other areas of essential oil production. Methods of producing essential oils, and the equipment used in such production are discussed and illustrated to give the reader an understanding of the techniques and facilities employed in producing fragrance raw materials.

BOOKS

The Essential Oils

By Dr. Ernest Guenther
Published by D. Van Nostrand Co.

In this six-volume set, the author, one of the world's leading authorities on essential oils, had the collaboration of industry experts. Volume I (published 1948) deals with the history of essential oils, their origin in plants, and their production and analysis. Volume II (1949) covers the constituents and chemical structures of essential oils, as well as derivatives of the constituents. This volume is now being revised, and there are plans for republication, perhaps in 1978. Volumes III (1949), IV (1950), V (1952), and VI (1952), are devoted to individual essential oils from specific plant families. They cover sources, agriculture, production, components and analytical specifications of the complete list of essential oils.

This series of books is without question the premier technical reference source on essential oils. While the original edition is out of print, the plates were acquired by Robert E. Krieger, P.O. Box 542, Huntington, N.Y. 11743, from whom reprint copies may be obtained.

Technologie et Chimie des Parfums Naturels

By Yves-Rene Naves
Published (1974) by Masson & Cie.,
120 Boulevard St. Germain, Paris, France.
330 pages

One of the most recent books on natural perfuming materials, written by one of the top experts in the industry, this French volume is an invaluable addition to the literature on essential oils and aroma chemicals. The first one hundred pages are devoted to a discussion of manufacturing methods, with major attention being given to extraction processes, and a short section on distillation. The preparation of absolutes, concretes, resinoids, tinctures, infusions, etc. is covered. The balance of the book is devoted to reviewing the chemical composition and analysis of the principal perfume materials. A series of monographs on natural perfuming materials assembles the latest information available on the most important fragrance raw materials.

The Givaudan Index

Published by Givaudan Corp.,
100 Delawanna Ave., Clifton, N.J. 07014
First published in 1949. 2nd edition, 1961.
A third edition should be ready for distribution in the near future.

The Givaudan Index presents in one volume specifications and standards of basic aroma chemicals produced by Givaudan Corp. and used by the perfume and flavor industries.

The second edition contains monographs on 288 chemicals offered commercially and used commonly in perfumery. The monographs include structural formulas of all products, alternate chemical names, and physical and chemical specifications. Flash points are given, as well as color index numbers. The monographs are supplemented by descriptions of the analytical procedures and color standards for each product, indexes of odor type, chemical functional groups, and the Givaudan trade name. A foreign name index provides the French, German, Spanish and Portuguese name for each of the products listed.

Perfume and Flavor Materials of Natural Origin

By Steffen Arctander
Published privately by the author.
Now out of print.

Part I covers definitions of and methods of processing perfume and flavor materials. Part II consists of monographs on crude and processed materials of natural origin. Since stocks of the first edition were exhausted, the author has been urged to bring out a new edition. He indicated recently that he would not permit republication of the original book, which he considers out of date in many respects, but that, if the cost is underwritten by the industry, he would be willing to research and publish a new and completely revised edition. Such a possibility seems at this time to be uncertain.

Perfume and Flavor Chemicals

By Steffen Arctander
Published privately by the author (1969).
Olyphant, R.D. #1, Pennsylvania 18447.
2 Volumes — $125 per set.

A series of over 3,000 monographs reporting on 3,350 fragrance and flavor chemicals. Each monograph gives the chemical name of the material, the trade names under which it is sold, its physical and chemical characteristics, its fragrance, and its use. A few copies of this two-volume set are still available. The author has announced that no new edition will be printed, nor will photocopying or reprinting of the original edition be allowed.

Perfumery and Flavoring Synthetics

By Paul Z. Bedoukian
Published by Elsevier Publishing Co.
52 Vanderbilt Ave., New York, N.Y.
1st edition (1951). 2nd edition (1960).

This excellent technical book discusses the more important F/F aroma chemicals, their discovery, history, production, modern processes for their manufacture, specifications and uses. A technical critique of the entire aroma chemical industry. 342 pages. Copies of the 2nd edition are still available at $46.25.

Handbook of Flavor Ingredients

By Giovanni Fenaroli
2nd edition published 1975 by Chemical Rubber Co.
18901 Cranwood Pkwy., Cleveland, Ohio.
2 volumes.

Vol. I offers a fairly complete description of flavor materials and their uses. Included are definitions, history and evaluation, sensory mechanisms and chemical structure, taste perception, biogenesis, identification and evaluation of flavors. Also includes a description of sulfur and amine compounds, potentiators and sweeteners. 560 pages. $44.95 domestic, $51.75 outside U.S.A.

Vol. II lists 1,000 synthetic flavor ingredients and describes their uses in various types of goods. 944 pages. $54.95 domestic, $63.25 outside U.S.A.

Food Flavorings: Composition, Manufacture and Use

By Joseph Merory
Published 1960, 2nd edition 1968
By By AVI Publishing Co., P.O. Box 831
Westport, Conn.
$24 per copy domestic. $25 foreign.

This book contains information on the manufacture of flavoring materials. Formulas are reviewed and typical formulas presented for 350 natural and synthetic flavors for bakery products, beverages, ice cream, candy,liqueurs, etc.

The Chemistry and Physiology of Flavors

By Harold W. Schultz, Edgar A. Day and L. M. Libbey
Published (1967) by AVI Publishing Co.,
P.O. Box 831, Westport, Conn. 06880.
$20 domestic. $21 foreign.

In this symposium the authors present what was at the time of publication, the latest information on the nature of flavors and odors; on physiological aspects of olfaction and gustation; on analytical methodology; and on the nature of the flavors of such food products as bread, wine, meat, poultry, fish, dairy products, vegetables and fruits. Special attention is given to the origin and enhancement of food flavors. Based on a symposium on Flavor held at Oregon State University. 552 pages.

Cosmetics: Science and Technology

By Marvin S. Balsam and Edward Sagarin
Published by John Wiley & Sons
605 Third Ave., New York, N.Y.
2nd edition (1972-1974). Three volume set
$94.50.

Chapters on different aspects of the subject written by experts in the cosmetic field. Vol. 1 covers: creams and lotions; baby toiletries; suntan preparations; face powders; lip stick; eye makeup; rouge; dentrifices, mouthwashes.

Vol. 2 deals with: shaving preparations; depilatories; hair grooming and coloring preparations; antiperspirants and deodorants; aerosol cosmetics; bath preparations; nail preparations; and fragrances. Vol. 3 covers: structure and history of cosmetic industry; physiology of the skin; preservation of cosmetics; government regulation; color in cosmetics; and evaluation of cosmetics.

Food Chemicals Codex

Published by National Academy of Sciences,
Washington, D.C.
2nd edition 1972. 1st supplement, 1974.
2nd supplement, 1975.

The Codex is a semi-official compendium of standards for food-grade products, both natural and synthetic. It represents a compilation of the standards and specifications of products from many different categories in which a selection of the most rigid definitions of food additives describes the purest grade possible.

Perfume Album

By Jill Jessee
Published (1951) by Robert E. Krieger
Publishing Co.
645 Huntington Ave., Huntington, N.Y.
$9.50 per copy.

An informal, non-technical study of fragrance raw materials written in travelogue style. Of special interest as background material for sales people, from field representatives to retail sales personnel. Sections on: Flowers; Grasses; Spices and Herbs; Citrus products; Woods; Leaves and Roots; Gums and Balsams; Lichens; Animal Products. Includes Glossary of Terms and a Chart of Perfume Materials.

E.O.A. Book of Specifications and Standards

Published by The Essential Oil Association of
U.S.A., Inc.
60 East 42nd St., New York, N.Y. 10017
Price $75. First published in 1946. Most recent
revision in 1970.

Pooling the scientific knowledge of its member companies, the E.O.A. began publication in 1946 of monographs on analytical methods, apparatus, standards and specifications for the most important essential oils, synthetic aroma chemicals and related products. This compilation represents the latest information on products of the industry. Revised and updated periodically, this book now contains 246 monographs, 20 gas chromatographic curves and 131 infrared spectra.

To Be Published

Fragrance Chemistry

To be published by Academic Press
111 Fifth Ave., New York, N.Y.

Projected for publication, probably in 1979, is a new book dealing exclusively with the chemistry of essential oils. Dr. Ernst Theimer, former technical director of IFF, is editor. Tentative contributors include such authorities as John E. Amoree, Thomas and Marilyn Getchell, Braja Mookerjee, John Walradt, Thomas Wood, G. Ohloff, Paul Bedoukian and H. van den Dool.

Out of Print

In addition to the above, there are a number of books on essential oils and perfumery which are out of print but which are available in libraries, including the Chemists Club Library at 52 East 41st St., New York, N.Y. They include:

A Treatise on the Manufacture of Perfumes (1877)
By John Snively

The Volatile Oils (1913)
By E. Gildemeister
Published by John Wiley & Sons, New York, N.Y.

The Chemistry of Essential Oils and the Art of Perfumery (1918)
By Ernest J. Parry
Published in England

Perfumes and Cosmetics (1923)
By William A. Poucher
Published by Chapman & Hall, Ltd., London

Perfumes, Cosmetics and Soaps (2nd edition 1926)
By William A. Poucher
Published by D. Van Nostrand & Co., New York, N.Y.

Volume I is a directory of raw materials.
Volume II covers the manufacture of perfumes, cosmetics and soaps.

The Essential Oils (1926)
By Horace Finnemore
Published in England

Natural Perfume Materials
By Y.R. Naves and G. Mazuyer
Translated from the French by Ed Sagarin.
Published in the U.S.A. by Reinhold Publishing Corp.

The Chemistry and Manufacture of Cosmetics
By Maison G. de Navarre. Published (1962)
by D. Van Nostrand & Co.

Chapter XVII

People

Take away the spectrophotometers. Strip the essential oil and aroma chemical industry of its stainless steel kettles and its glass-lined storage tanks. Let the spotless research centers disappear, the plants be disassembled, and the thousands of precious formulas vanish beyond retrieval ... and what would we have left? Actually the industry's most precious element would still remain — its people! Good people are scarce in this industry, as prospective employers find every time they try to hire a top level perfumer, a flavorist, or a specially trained food technologist.

For most industries the basic problems of recruiting and handling personnel are more or less standard. The same personnel activities are common to most industries, whether the employer is manufacturing shoes, pots and pans, or rubbing alcohol. The Flavor. Fragrance and Aroma Chemical Industry, however, is different, even is unique. While companies in other industries are making and selling specific products, which can be made similarly by other companies, the Flavor and Fragrance Industry, by contrast, sells dreams, ideas, glamour and other intangibles that only certain individual companies are qualified to supply.

Assignments which firms in our industry may be given, range from creating a particular fragrance which will be compatible with the beautiful color of Elizabeth Taylor's eyes all the way to formulating a flavor which will add more zest to onion soup. The individuals who can handle such tasks must be persons of extraordinary talents, qualifications and abilities. This is what makes them different from workers in other industries. And, while the Flavor and Fragrance field is part of the overall Chemical Industry, and a chemical background is normally essential to those working in the field, there is something more required to fill a job in our industry successfully, which is usually termed an artistic temperament.

This is not to say that the sensitivities of the successful operatives in the Flavor and Fragrance Industry categorize them as difficult-to-deal-with prima donnas. For the most part nothing could be farther from the truth. It is simply that the professionals in our industry are usually permitted to employ more artistic license. They must exhibit open minds to the various ideas presented to them, and be prepared to avoid standard or routine approaches that might be acceptable in other industries where "assembly line" products are mass produced, and in which human contributions are often rigidly outlined and deliberately restricted.

Of course, not all jobs in our industry demand employees who are unusually talented or artistically inclined. There are as many routine jobs in the Flavor and Fragrance Industry as in any other. However the tone set by the top personnel in any company tends to reach down into the lower echelon, thereby influencing the thinking and attitudes of all individuals in the company.

Complicating the above picture is the supply-demand imbalance among top personnel in the Fragrance and Flavor Industry. There are very few fully trained perfumers in the country, and an even smaller number of flavorists. This situation has perhaps made some individuals, already artistically inclined by temperament, develop an exaggerated concept of their own importance. On the other hand, in terms of the true creative brilliance exhibited by some top perfumers and flavorists, perhaps in some cases they have not received the full recognition to which they are entitled.

At any rate the above factors make this industry a desirable one in which to work. Perfumers and flavorists command good salaries, and demand normally exceeds supply. You don't locate a perfumer or a flavorist by running an advertisement in the *New York Times,* the *Wall Street Journal,* or even in the trade journals. Normally you have to train your top personnel, or hope to attract an individual who has been trained by a competitor.

Even those lacking a technical degree are in demand as perfumers and flavorists. Occasionally even pseudo perfumers manage to get by on merely a smattering of training in organic chemistry, bolstered by a few years' training as a laboratory technician and a knowledge of basic formulation, provided they have that prime requisite — a "good nose." Of course an advanced degree is desirable, and a creative flair coupled with a touch of the dramatic are essential to move a perfumer or a flavorist into the top salary brackets.

But, we shouldn't talk as if there were no other jobs in this industry other than as perfumers or flavorists, or as if every perfumer spends his mornings formulating a new *Chanel 5,* and then after lunch knocks out a rival to *My Sin.* Perfumers and flavorists have the top jobs on the creative side, but there are plenty of other openings that are unique to the essential oil and aroma chemical industry, such as instrument chemist, food technologist, flavor panel evaluator, research chemist, cosmetic chemist, fragrance evaluator, etc., as well as the other obvious openings in research, production, sales, purchasing, accounting and materials handling. And, while the skilled creative artists often find themselves called upon to develop fragrances for major new perfumes, the bread-and-butter, day-after-day jobs are more likely to involve developing a low cost fragrance for a liquid soap or a floor polish, or checking the quality of incoming raw materials. No job in any industry can be all glamour.

At the end of this chapter we list a number of categories of jobs that are fairly unique to the essential oil and aroma chemical industry. We also describe the duties and responsibilities they involve, together with a word on the training and experience they require. This breakdown was supplied by a panel of leading personnel directors in the industry*. In discussing the "perfumer" and "flavorist" job descriptions, it should be pointed out that not all the major houses would agree on a common job description for these two categories. Different companies may expect a perfumer or a flavorist to perform widely different tasks.

In filling jobs of a slightly lower echelon, classified advertising is more often successfully employed. We reprint a typical advertisement from a recent issue of *New York Magazine.*

> *Fragrance Evaluation.* We are a large international company with an impressive history of leadership in the creation and manufacture of fragrance compounds, perfume specialties, and ingredients for fragrances and flavors. We are seeking individuals with experience in fragrance evaluation. A thorough working knowledge of the U.S. consumer market in perfumes, cosmetics and toiletries is absolutely essential. Good evaluation capabilities and an ability to demonstrate excellent judgment is a must. We offer a professional atmosphere that is conducive to creative instincts and an excellent opportunity for the right individual. Benefits include medical and life insurance, employee thrift and pension plans and long-term disability. Send letter and resume in strictest confidence to Box XXXXXX. An equal opportunity employer. M/F.

This advertisement is probably typical of the thousands that the industry runs each year. Incidentally, it is interesting to observe that possession of a "good nose" is

* Thomas Lotito of Givaudan Corp. was chairman of this panel. Its other members included William A. Myers of IFF, and William Rosenberger of Fritzsche D & O.

one of the prime factors in qualifying for almost any job in the flavor and fragrance industry.

Even production personnel have to do a little "sniffing" as they run their batches through various steps in the production process, to enable them to double check odor, and color quality. And the sales person presenting a fragrance or flavor formulation to a prospective buyer enjoys a definite plus if he or she is able to evaluate it accurately and objectively in comparison with alternative materials. It is of course essential that quality control personnel have dependable "noses."

Where do perfumers and flavorists come from? It is a definite "plus" if they have started by studying organic chemistry, but it comes as a surprise to some to learn how many perfumers there are who haven't been exposed to this seemingly obvious subject. Some of those who eventually get to be perfumers work up through jobs as laboratory assistants, or begin their careers in the quality control department. From such a vantage point a trainee may be called upon to help one of the perfumers who is temporarily swamped with work, then eventually get an opportunity to try the newly acquired skills by developing a formulation for some relatively minor product.

Several of the major firms operate their own schools for perfumers. Roure Bertrand Dupont operated one of the first such schools for perfumers near the town of Grasse in southern France.

Charabot & Cie. operate a full fledged school for perfumers in Grasse under the direction of Mons. George Coeur. The school, which was started about fifteen years ago as a service to the industry, occupies its own building in the Charabot plant, with separate class room, compounding and laboratory areas. Attendance is by invitation, and there is no tuition fee. There are usually at least half a dozen student perfumers in attendance, who may stay for as short a period as a month, or for as long as several years. Entrants come from all over the world, and are normally trainee perfumers already employed in the industry. The school operates like the old-fashioned one room school house, with many of the students in different grades, entering and leaving at different times, and receiving instruction on an individual basis.

Instruction emphasizes use of "natural" materials as opposed to synthetics, and has as a major purpose the training of young perfumers in the use of natural materials. The program starts with the study of basic essential oils, followed by coverage of the principal natural floral products. After a thorough grounding in "naturals," students are introduced to synthetic aromatic chemicals. Charabot manufactures a few synthetics in its own plant. For use by the students, they maintain a full stock of synthetics and specialties made by other manufacturers of aroma chemicals and most of the major aromatic houses world-wide. Thus the students have access to a wide range of perfuming materials in working up their original compositions.

After basic training has been completed, the students move on to the development of simple compositions, and next to the study of different types of perfume compositions such as single florals, floral bouquets, orientals, woody fragrances, etc. Finally they progress to the creation of original compositions. When they finish the school, they are well equipped to move on into creative perfumery, but only experience will make them full-fledged perfumers.

Givaudan has a training school for perfumers in Geneva. The course runs about a year and is under the general direction of Mr. Maurice Thiboud. It starts with work on the history and identification of natural oils, proceeds with quality appraisals of various materials and training in their selection, then moves on to basic work in creating fragrance compositions and "accords", which are the perfume concepts. During the latter part of the course, the aspiring perfumers are encouraged to work at various stages of the advanced levels of fragrance creation.

The final training is completed under the direction of a senior perfumer in one of the company's laboratories, to which the student is assigned for "on-the-job" experience after his formal training has been completed. A similar, but greatly condensed course, is offered for customers. It is generally limited to a few months.

IFF trains its perfumers in the United States, offering a program of on-the-job training for budding perfumers. The late Ernest Shiftan, IFF's senior vice-president in charge of perfuming for many years, supervised this program.

Among the few courses in the United States open to individuals other than a company's own employees are those conducted by the Continuing Education Center at the laboratories of Monsanto Flavor/Essence, Inc. in Montvale, N.J. The instructor in the courses is Henri Hoffman, vice-president and chief perfumer of Monsanto.

We prevailed upon Mr. Hoffman to put together some of his observations on "The Training of a Perfumer" which are printed at the end of this chapter.

Several courses are offered, each consisting of six or more two and one half hour sessions. The first course, titled Creative Perfumery, introduces the student to various groups of perfuming materials and classes of odors, then considers fragrance formulation, special fragrancing problems and fragrance evaluation.

A second course concentrates on the study of the principal odor groups in relation to raw materials and prototypes. A third mini-course in Creative Perfumery was given in May, 1977, for the first time. It dealt with: Perfume development and odor classification; the formulation of major commercial fragrance notes; aldehydic notes, green notes and floral notes; chypre notes, fougere notes; and the extension of perfume notes to a series of products, including shampoos, cosmetics and toiletries.

The Education Center also offers another course in Perfumery Instrumentation, which covers: the use of gas chromatography; mass spectroscopy; infrared and ultraviolet spectrophotometry; nuclear magnetic resonance spectroscopy; the sources of botanical materials used in perfumery and their quality control; psychophysical aspects of odor; classification of fragrances by odor types; and the fundamentals of perfumery formulation. The instructors in this course are Dr. Stanley K. Freeman, IFF and Christian Baude, perfumer at IFF.

Another course in "Instrument Methods in Modern Perfumery" is given at Upsala College, East Orange, N. J., by Harry Saunders, chief perfumer of Shaw Mudge & Co., Stamford, Conn.*

One of the first efforts to train perfumers, other than one's own employees, was started by A. L. van Ameringen, first president of IFF, who delivered a series of lectures on the subject at Columbia University in 1946.

At a later date, Paul Bedoukian, who was then with Compagnie Parento, started a formal course in perfumery at Columbia University. It was continued for about five years. After a short lapse, Steffen Arctander took up the idea and gave a perfumery course in the College of Engineering at Rutgers University, Newark, N.J. The curriculum dealt primarily with the source and identification of chemical compositions of fragrance materials and prototype formulations for creams, lotions, hair sprays, household products and the like. Mr. Arctander is a well known writer and fragrance consultant, who has published books on fragrance. His courses at Rutgers were discontinued several years ago.

Union College, Cranford, N.J., gave a ten-session course in perfuming in April, 1974. The course was conducted by experts from the industry, rather than by members of the college staff.

The Brooklyn College of Pharmacy at Long Island University, Brooklyn, is an institution where those interested in pharmaceuticals, cosmetics, aerosols and

*"Soap/Cosmetics/Chemical Specialties," June, 1977, pgs. 64-66.

perfumery can find special training. Dr. John Sciarra, dean of the College of Pharmacy, is fully oriented to the fragrance industry, especially aerosols, and has trained a number of students who have gone on to successful careers in this general field. Courses in pharmacy serve as a good background for those wanting to prepare themselves for the essential oil field, for the pharmacy curriculum is one of the few in which a student is exposed to botanicals and essential oils in the course of studies at the undergraduate level.

The curricula in most institutions teaching chemistry or chemical engineering generally emphasize fundamentals, anticipating that their graduates will acquire the skills they need for specific industries through on-the-job training. The basic courses they take to prepare for the fragrance and flavor field would include organic chemistry and laboratory, physical chemistry, colloid chemistry, and surface chemistry, as well as rate processes, chemical processes, and courses in materials science.

The analytical chemist in the essential oil industry is trained as other analytical chemists are, and learns to use the same analytical instruments they are taught to use. However, because in the fragrance and flavor industry they often work with minute quantities of material, there is a much greater challenge to their dexterity and patience. Organic chemists in the aroma chemical industry may make chemicals which are relatively simple compared with those made by other organic chemists, but their uniqueness and challenge lies in the ability of the chemist to predict odor properties on the basis of structural relationships, and being clever enough to find ways to manufacture these chemicals at low cost. Now, with the current emphasis on protection of the environment, it is also important that ways be found to manufacture the products the industry requires without allowing excessive quantities of reagents and intermediates to escape and thereby contaminate the environment.

Another major center that could equip students to enter the fragrance and flavor industry is the University of Pennsylvania, Philadelphia, Pa. The Monell Chemical Senses Center is located on Penn's campus. In that Center a group of graduate students work under fellowships doing research on the senses of taste and smell. Many of these grad students come from the undergraduate school where they take courses developed by the University to fit them for work in the Monell Center. Other U of P graduates go directly into the commercial field, working for one or another of the several fragrance and flavor firms that support the Center with their contributions.

For several years, back in the 60's, the Essential Oil Association tried to develop a scholarship program to encourage promising students to come into the flavor and fragrance industry. Some leaders apparently felt that the industry was becoming in-bred, and that it might benefit from an infusion of new ideas and new blood. So, a number of EOA members contributed to a Scholarship Fund designed to assist in the education of promising technical students, who would then be expected to enter the industry after graduation.

There are, as EOA discovered, problems in developing a smooth working scholarship program along these lines. Some students, after being given assistance, were not interested in going into the industry. Others the industry didn't want. Contributors to the program wanted to have a say in naming the students to be assisted. The college would not agree. They insisted the awards be made to "needy" students rather than having recipients selected on the basis of competitive tests. After trying the program for a few years, the effort was abandoned, and the funds which previously supported scholarships are now employed as a grant to the Monell Chemical Senses Center.

Apart from some basic training in chemistry, what other qualifications should be possessed by a candidate for a job in the flavor and fragrance industry? First, he or she should have an active interest in the industry. Then, he should have a well developed aesthetic sense, and above all a "good nose." Finally, he or she should possess a certain amount of sophistication. This is quite important, for in the flavor and fragrance field almost all employees have to deal with people in a "glamour industry." Telephone operators, typists, lab technicians, or members of taste panels must all present a favorable image of the company's sophistication.

This applies doubly to those in sales jobs. The industry uses a higher type sales person than the average. Men or women, they are usually better paid, and definitely not the "door bell ringer" or "drummer" type. In the fragrance industry a sales person is often more of a "customer contact person," who can meet the client on a close, personal basis, develop rapport with him or her, and hopefully be able to hold the account against the blandishments of competitors.

Sales people most often come from within the industry. A degree in chemistry is desirable, and most firms prefer candidates with prior experience in the sale of chemicals, or in the food, cosmetic or allied areas. Sales correspondents or account managers should have a chemical background, and are often brought into the inside sales force from mid-level technical positions in the laboratory, where they have gained a knowledge of products and applications.

For those who qualify, there are attractive financial rewards in this industry. And, in addition to the good pay, the flavor and fragrance field appeals to many because: the industry is an unusual blend of art, science and sociology; it offers enormous creative scope to its employees; it is in the early stages of a major growth pattern, hence there are superior growth opportunities for those engaged in it; there is an unusually stimulating working environment; and the field is international in scope, offering many opportunities for job-connected travel. But, those who enter the industry must be able to work, and make good, under pressure.

Beyond the qualifications mentioned earlier, the training of candidates for positions other than technical would include a knowledge of economics, general business practice, and commercial law, and familiarity with a foreign language, preferably French. For jobs with firms that do a considerable amount of importing and exporting, an applicant needs to understand documents and shipping, government regulations (which vary widely from one country to another), Custom's House procedure, etc. There are Import-Export Schools that train candidates for these jobs, although most such training is done on the job.

The top position in a typical essential oil dealer operation is that of a "trader," but these positions are seldom open to outsiders. The nature of the business requires that, to fill such an important role successfully, the candidate must have had years of experience with the company's customers, its suppliers, and its products. Many of those who fill these positions, where such a major element of trust is involved, tend to be members of the immediate family of the owners, or those closely related to them.

Dealers employ but few sales people, as they normally spend more time cultivating and solidifying their sources of supply than in going after new accounts. Thus the dealer segment of the industry offers only very limited openings for marketing graduates.

The essential oil and aroma chemical industry urgently needs qualified people, but up to this point has not been successful in developing a dependable source for them. Most of those now employed have worked previously for other firms in the industry, or for general line chemical firms. Some were born into the industry; others married into it. Still others just wandered into the field by chance.

There are many bright career opportunities in this field for qualified people, both men and women, and members of the industry have an active interest in encouraging qualified applicants to prepare themselves for such careers. Following are job descriptions of some typical positions in the industry.

Job Titles and Descriptions

Perfumer

In some houses this title is reserved exclusively for the creator of fine fragrances. Others in the Perfuming Department may be called "fragrance creators," research perfumers, assistant perfumers, or junior perfumers. It is worthy of note, however, that in at least one major perfume compounding house there is no grade below full perfumer, the only distinction being in the salary range. At this company they feel that no perfumer, or flavorist either, wants to be termed "junior" or "assistant." With this company, it is full perfumer or nothing. At some other houses, an instrument chemist may occasionally come close to filling the role of perfumer. There seems to be no standard practice, except that in general, at least in the USA, a trainee must have been apprenticed to a perfumer for a period of years, then elected a member of the American Society of Perfumers before he or she may claim the title of perfumer.

Whatever the title, the man or woman filling this important creative post must be able to evaluate competitive or customer products, and to create new and exciting fragrances. He or she must have an artistic and creative sense, have a good sense of smell (a good "nose") and possess as good a knowledge as possible of the vast array of perfuming materials on which it may be necessary to draw in formulating a target fragrance. A background in chemistry is certainly helpful, although not all perfumers have a formal technical education. More important, perhaps, is that special aesthetic feeling for aromas. The individual who wants to be a perfumer may be born, but must still acquire an aesthetic olfactory sense. Lacking that, the applicant will never become a perfumer.

Other essential assets are an active imagination and a good sense memory. After training, a master perfumer must have learned how to make the best use of the myriad fragrance materials which may be used to create a formulation which will please the buyer In preliminary training, the perfumer must learn to distinguish, and remember, four to five hundred of the more common aroma chemicals and natural products. Later, this knowledge will have to be extended to as many as three thousand fragrance materials.

The basic function of the perfumer is to create new fragrances and revise existing formulae as required by customers. Depending on the area to which he or she is assigned, a perfumer may be asked to put fractions of an aroma chemical or essential oil together to produce a target fragrance, and also may have the responsibility for checking the odors of incoming materials against standards. Typical functional responsibilities of the perfumer would include:

Creating new fragrances.

Assisting in solving technical problems relating to perfumery.

Evaluating stability of created fragrances as to compatibility with specific bases.

Evaluating new research chemicals, olfactively in perfumes or cosmetic formulations.

Evaluating new and existing raw materials and other perfume additives.

Conceiving applications for new fragrances.

Serving on panels to determine suitability of purchased raw materials.

Calling on select customers to discuss technical matters related to their fragrance requirements.

The top perfumer is the *creative perfumer*. He has to do all the tasks listed above from time to time, but his essential role is to create new perfumes — combining natural and synthetic products to create fragrances for those hundreds of new products that are introduced every year in the perfume, cosmetic, soap, chemical specialty and household product fields.

Cosmetic Chemist

Should have degree in chemistry or related field. Must have knowledge of emulsions, anti-perspirants, shampoos, lotions, creams, aerosols, etc. Should be familiar with the application of fragrances to all cosmetic products. Responsibilities include:

Stability testing of fragrances in shelf-life cosmetic products

Preparation of finished cosmetic packages for sampling to customers

Preparing cosmetic formulations for general use and stability and compatibility studies

Performing required cosmetic applications for customer technical service requests: colognes, cosmetics, extracts, etc.

Assisting in formulation and development work on all types of cosmetics

Maintaining laboratory

Fragrance Evaluator

A knowledge of the consumer market in perfumes and cosmetics is necessary to fill this post. A keen sense of smell and a retentive memory for hundreds of fragrances is also highly important. Also invaluable is accurate knowledge of products and trends, and an insight into the history and future of the industry.

Fragrance Quality Control Chemist

Is responsible for odor quality and approval, rejection or clean-up recommendations on all fragrance raw materials, intermediates, keys and finished products. In addition to unacceptable odor, materials may also be rejected for differences in clarity, color, physical state, viscosity or fluidity. A good nose is required for this post, as well as a well rounded knowledge of all fragrance materials.

Recommendations Assistant (Fragrance)

The qualifications for this job are a certain amount of fragrance experience, an ability to discern odor characteristics of fragrances, and a knowledge of commercial fragrances. Principal responsibilities are:

Recommending fragrances in response to requests from marketing for potential new accounts

Making recommendations for major customers, either on own or working with a perfumer

Conducting meetings with the product manager and/or salesmen on requests not requiring a major review meeting

Keeping records on all fragrance information involving performance in specific products, stability, etc.

Aerosol Chemist

This member of the staff must be able to formulate, develop and test all forms of aerosol products. In addition, it is the responsibility of the aerosol chemist to:

Provide aerosol samples and service to the sales department, perfumers and other laboratories

Maintain sample records, inventory, etc.

Supervise stability testing and evaluation of aerosol products

Keep informed on developments in aerosol industry

Investigate new raw materials and aerosol package components

Perfume and Cosmetic Lab Technician

High school or equivalent required. Degree advantageous. Under direct supervision, weighs, measures, mixes liquids, essential oils and other ingredients. Fine opportunity for those who are pursuing a chemistry degree in night school.

FLAVOR DEPARTMENT

Flavorist (Flavor Chemist)

Like the perfumer, the flavorist is an artistic, scientific worker who creates finished formulations (flavors) from basic aroma chemicals and essential oils. Like the perfumer, the flavorist evaluates customer and competitive products. He or she must be capable of duplicating competitive material and developing new and unique flavors as required for customer's new products. Training in chemistry is desirable, preferably a degree, and the successful flavorist must be imaginative and creative. It is an important talent to be able to recognize a large number of aroma chemicals and essential oils, and a good memory is a valuable asset.

There are a number of special categories of flavorists such as new product flavorists, beverage flavorists, meat flavorists, etc. The basic function of the new product flavorist is to provide liaison between development and basic flavor research. The principal functional responsibilities of this individual would be to:

Assist flavorists in application and development of GRAS (generally recognized as safe) chemicals

Screen and evaluate potential GRAS chemicals

Establish use levels and use data

Coordinate submission to panels of experts

Participate in flavor project teams

Take flavor chemist problems to basic research

Collate chemical evaluation findings for retrieval

Create flavors

Senior flavorist

By contrast, the Senior Flavorist would concentrate more directly on creating and developing flavors for a wide range of products. This individual would also:

Evaluate new aroma chemicals for possible use

Review incoming requisitions

Prepare technical reports

Provide functional supervision and/or direcction

Assist sales with specific technical problems

Advise food technologists on application of flavor formulae in consumer products

Assign responsibilities to other flavorists and provide supervision

Food Technologist (equivalent to cosmetic chemist in the fragrance plant)

People who work in the application of flavors to foods are known as flavor application chemists or food technologists. They must be familiar with the formulation of a wide range of food products such as snacks, sauces, meats, soups, dairy products, beverages, candy, baked goods, ice cream, etc. A college degree in food science is desirable. Food technologists may be thought of as applied scientists, as opposed to perfumers or flavorists, who are in the main artistically oriented.

Food technologists are usually broken down into various categories such as candy technologist, dairy technologist, meat technologist, chewing gum technologist, vegetable oil technologist, etc. The flavor industry cannot normally afford to have specialists in every category, but a good food technologist can usually function quite successfully in applying flavors in the above categories.

The food technologist working in the flavor industry must have very good sensory perception, especially as regards odor and flavor (taste). It is important that this individual have the capacity and ability to explain the functioning of flavor to the customer. It is highly important, too, to have good taste in selecting the appropriate flavor to use, and an ability to estimate the optimum use level. The food technologist must be able to devise new concepts or prototypes in conjunction with new and varied flavors.

Taste Panel Chief

The taste panel chief prepares materials for panel testing under the supervision of a food technologist who is responsible for evaluating the results of the panel's testing. The taste panel chief develops prototype food bases for flavor application testing, and helps to develop flavor usage levels for various food products on all new flavors that are developed. Prepares products for organoleptic panel testing, conducts panel tests, and tabulates results.

Flavor Quality Control Chemist

This position requires an experienced taste expert who is responsible for the flavor (taste and aroma) of raw materials, in-process, intermediate and finished products. Responsibility includes evaluating, approving or rejecting the material. Such characteristics as color, clarity, fluidity, sedimentation, etc. are important considerations for this operative.

CHEMICAL DEPARTMENT

Chemist

There are many types of chemists who work in essential oil and aroma chemical companies. Cosmetic chemists, flavor chemists, research chemists, aerosol chemists, all have special responsibilities. Under the basic title, chemist, the principal function of the individual in a fragrance oriented house would be to assist the efforts of the perfumery group by operating a fragrance instrumentation laboratory, and to provide liaison with the fragrance research group or the research department. Functional responsibilities for this individual would include:

Conducting fragrance analyses

Maintaining reference files on natural materials, specialties and aroma chemicals, including products of competitors

Assisting perfumers in olfactory analyses

Assisting in solution of quality control problems

Conducting other technical services for the fragrance division, including extractions of commercial products, and small scale distillation

Obviously a different set of responsibilities would be involved for a chemist working with a flavor group.

Research Chemist

The basic function of the research chemist in the fragrance field is to analyze, identify, improve and/or purify natural products and aroma chemicals, and to synthesize aroma chemicals. Specifically:

Prepare new compounds, isolate and identify them and summarize their properties.

Search chemical journals for new materials, stay abreast of technical literature and patents.

Run analytical instruments to check purity of samples, separate and identify natural components.

Keep notebooks on experiments and analyze results.

Write up lab processes.

Run stability tests, determine reactivity and compatibility of materials under test, and establish specifications.

Chemist - Process Development

Major function is to scale up and develop products from the R&D laboratories to production level in a manner both practical and economical. This individual performs analyses, tests products that have been developed by R&D, and operates small-scale plant equipment, including some instrumentation. Makes literature searches and writes technical reports.

Chemist - Instrumental

A candidate for this position should have at least a B.S. in college chemistry, with some physics and math. Should be familiar with lab procedure, apparatus and instrumentation. A knowledge of mechanics and electronics is an important asset. The instrumentalist detects and determines components in natural and synthetic products, checks quality by strictly objective methods, and is a chemical investigator for all products.

Chemist - Quality Control

This staff member evaluates incoming and outgoing materials for compliance with specifications; improves processes, and uses instruments as well as the senses to check quality of raw material and finished product.

The Training of a Perfumer

By Henri Hoffman
Monsanto Flavor/Essence, Inc.

In my years in the perfuming business I don't believe I have ever found anyone who lacked the basic sense of smell. Practically everyone has at least a passable "nose." Those who say they can't smell probably just don't know how to translate what they smell into words.

When a candidate decides he or she wants to become a perfumer, a fragrance evaluator, or a participant in a recommendation group, the first step in training is to become aware of odor everywhere. You must find out, by experiment, how sensitive your nose is to begin with. Then, as training progresses, you can trace how your sense of odor perception develops.

Next, you must study the raw materials of your craft. Become familiar at first hand with the fragrance of the natural materials such as rose, jasmine, orangeflower, tuberose, cassia, etc., and don't neglect the less expensive naturals. There are many useful lower cost naturals which some modern perfumers forget about. Touches of these materials in a well-balanced fragrance are often very helpful.

Whenever possible, the student perfumer must go to the shelf and smell the materials. This repeated study of the aromas of the different materials assists the trainee, just as exercising and playing scales aids the student pianist. And the importance of this perfumery exercise is not so much to train the individual in recognizing a family of fragrances as it is to keep the nose tuned to smelling, and to reinforce the memory. Beginners should use a blotter in sniffing, rather than smelling the material direct from the bottle. It is further recommended that the trainee avoid staying with one type odor too long, or for that matter continuing to sniff for hours even with frequent changes in the fragrance. After perhaps fifteen minutes, the sense of smell becomes dulled, and the results of the sniffing cannot be depended upon.

The trainee must be aware that a single fragrance material can have more than one odor. This point is made by Steffen Arctander in his book, "Perfume and Flavor Chemicals." He observes that "perfuming chemicals are rarely found in such a state of purity that they are absolutely free from traces of other components." The impurities present may have an odor, sometimes consistent with the odor of the basic material, but at other times distinctly different. When the odor of the impurity is different, it may blend in to produce a more complex odor which will be distinctly different from the odor of the pure material. Or, it may have a fragrance note which will be apparent either before or after the odor of the major ingredient. Because of this factor, it is important for the trainee to sniff the odor of many samples of the same material in order to get used to the range of fragrance which an odoriferous substance may have. He must also learn to recognize the top note, the middle note and the dry out note (residual character) of a fragrance.

The student must also be familiar with the mostwidely used synthetic aromatics — the aldehydes, esters, alcohols, ketones, etc. He or she must learn which materials are used in large amounts, which normally in small amounts, or perhaps only in traces. Which materials are used in solution. Solutions in what? Which materials are used in dilution, and how dilute? All the materials you are going to use must be studied in various dilutions and solutions, so you will know what result to expect when they are employed in different forms and concentrations.

As training progresses, the student perfumer will find it very useful to build a card file, broken down in some logical fashion — by odor group or type of product to be perfumed. Within groups, the cards should be arranged alphabetically. Separate cards for the same material can be included in more than one major group, wherever they belong. Then, every time you work with an aromatic material, such as rose crystals, linalool or heliotropine, you make up a card on that material for your file. Give yourself a quota of at least five cards a day to be filled out, and don't let the work pile up, expecting to find time to do it later. If you are swamped with other work, you may postpone making up your cards for a day or two, but you should at least make notes immediately in a notebook, then transfer the data to a card when you get the chance. Your card file will be an important tool for years, and you cannot afford to neglect it.

Enter on the cards all the information you can find about the materials with which you work — the trade name of the product, chemical name, type fragrance, cost if available, how it performs in different solvents, stability, tendency to discolor, purity, etc. When a material is sold under more than one name, it is a good idea to make up cross-reference cards, filing the extras under the name of the product rather than under the group of which it is a member.

When working with a specific aldehyde or ester, for example, its character should be checked when used in various solvents, and notes made on how it performs in each. This will illustrate why choice of solvent is such an important factor in the use of many fragrance materials.

When checking out a new aroma chemical, a card should be made up on it and added to the file. Whenever you work with an internal base, your own company's product, a card should be prepared on it. Spend some time each week looking at samples of new materials received. Prepare cards incorporating all the information you get from your supplier and from your chemist. Every detail, no matter how trivial it may seem, should be put on the cards, because at least in your early years as a perfumer you will find it difficult to remember details.

All the information that may help in your work goes into your card file, for you will not be able to remember it otherwise. It takes years of experience before you can develop such a dependable fragrance memory that you will not have to refer to the cards. Of course, after a few years, you may be able to start pruning the file. The obvious material can be eliminated, keeping cards on only the less frequently used materials.

You should be personal in the details and opinions you include. The file must be arranged by odor or product groups, because it must be built on a system, but within the broad groups your comments should be personal. Be sure to include your own ideas about the material. If you have doubts, make up two cards. When you don't know what to relate an odor to, relate it to the name of the product. Don't just routinely go to the Givaudan Index to describe it. In this way you limit yourself too much, and the file will reflect some one else's ideas rather than your own. If you feel that the particular material you are working with gives the same effect as another, or smells like it under certain specific conditions, say so

Preparation of a card file is also helpful to those involved in fragrance recommendation. The fragrance should be described in your own words, your own ideas.

After your card file is well on its way, you may be ready for some basic fragrance formulation. The first thing to bear in mind is not to get involved with too many materials.

Don't try to balance a hundred ingredients, at least in your early stages of training. When a fragrance is over-complicated, we might say the perfumer has lost his original idea, that is if he had one to begin with.

Take four or five materials and from them select one major note, the fragrance around which you will build your formulation. Then, balance around it three or four related notes. When you think you have a good balance, select a couple which you will play around with as minor tones and then balance one or two aromatics or naturals on the formulation, trying not to lose the original character selected, or destroying the balance of the blend. Balance up to eight or ten materials, and leave it there. Don't make your early formulations too complex. And, don't overuse the same materials. Start your next formulation with completely different products.

At this point, we can add another training suggestion from Steffen Arctander, who has such an excellent reputation both as a teacher of the art of perfumery and as a writer on the subject. In his book, "Perfume and Flavor Materials of Natural Origin," he mentions that a good exercise for the advanced apprentice perfumer is the development of a lily-of-the-valley perfume. He indicates that this problem in formulation is a family severe test, and that "next to rose, jasmine and lilac, the creation of a lily-of-the-valley perfume type is one of the first problems presented to the advanced apprentice perfumer."

As your training progresses, you will learn that balance is perhaps the most important characteristic in a successful formulation. The difference between a successful formulation created by an experienced perfumer and a bungled one by an amateur is often merely the fact that the expert's formulation is well balanced, while the fragrance made by the beginner is too complicated. And, once the basic error has been committed, there is not too much that can be done to save a formula. Loading a bad fragrance with "naturals" is expensive, but it is not necessarily going to make it good. If you had a bad balance to begin with, you can put in all the jasmine, rose, orange flower or tuberose you want, and it still won't make it a good fragrance. But, if you have started with a well balanced formulation, the addition of naturals can improve it immensely.

When we mention naturals, incidentally, too many perfumers think automatically of the expensive materials — rose, jasmine, cassia, tuberose, etc. They tend to forget that there are many relatively inexpensive naturals that are very useful. We mean such products as mimosa absolute for floral fragrances, lavender, labdanum, etc. Touches of these materials added to a well balanced fragrance improve the aroma greatly.

Whenever you work on a formula you should learn something from it. If you smell it after making each addition, you may be surprised. Often you can add a material which by itself seems to have little odor value, but which will produce a definite aromatic value in the finished compound. There are quite a few materials of this type — such as the isophytols, nerolidol, and triethylcitrate — which are weak in odor by themselves, but can prove very important in your finished formulation. Don't neglect them.

The Training of a Flavourist

Excellent advice on "The Training of a Flavourist" is contained in an article by E. Cowley of Bush Boake Allen Ltd., London, in the January, 1973, issue of "The Flavour Industry." In selecting trainees, Mr. Cowley counsels, a most important factor is that the candidate indicate an interest in flavor technology. High ratings in the natural sciences and the visual arts are also important. Elementary tests in flavor assessment are given to all candidates to insure that they are not "flavor blind." For an association of ideas the candidates are shown allyl caproate, amyl butyrate and 6-methylcoumarin and asked to pair them up with the flavors, pineapple, banana and coconut. They are also given tests on sweetness, bitterness and acidity.

Following a general introduction to laboratory procedure, the trainees then go through a 24-week training module which introduces them to flavor quality control, flavor development and applications of flavors in finished products. A condensed course of lectures takes up the subjects of: quality control, confectionery flavors, beverage flavors, food flavors, bakery, colors, and analytical work. Following this introductory period, more specialized training follows. It consists of such subjects as:

Laboratory organization, procedures, legalities, profitability, size of business, etc.

Flavoring agents, subdivided into
(a) Synthetics, further subdivided into chemical groups
(b) Naturals, further subdivided into tinctures, oils, oleoresins, concretes, juices, etc.

The design of flavors to meet specific end uses such as ice cream, yogurt, confectionery, etc.

Quality control of raw material and finished products

Instrumental aids to flavor development, technical services and quality control

Flavor research

This advanced instructional period continues over a period of about two years, and includes a series of lectures at 2-week intervals on such subjects as esters at one session, flavoring agents at another, and perhaps cloudy beverages at the following session. During this period the trainee gains considerable experience with the standard raw materials, both natural and synthetic, and while working in quality control gains experience in the reformulation which occasionally becomes necessary when raw material variations make it necessary to rework "wrong" batches.

"After a period of one to three years in this area," says Mr. Cowley, "we generally find the trainee is ready for moving into the development area as an assistant to a flavourist. He will leave the quality control area with the foundation of a library of tastes and smells and a smattering of knowledge as to the effect of blending and formulation work. He will also have picked up the basic terminology which will allow him to converse on a more intelligent level with the creative flavourist."

Trainees are expected to continue their education independently, and encouraged to obtain a degree in chemistry or food science. In his next step forward the trainee becomes an assistant to a flavorist, moving him to an area where the emphasis is on creativity. At this point, the strategy of Bush Boake Allen management is to encourage self expression. Too much instruction during this period, Mr. Cowley suggests, "can stultify the inagination. We are trying to produce a creative individual who will produce original concepts, and if the flavorist is directed too forcibly into another person's channel of thinking, then we may defeat the objective we are trying to achieve. Hence you have to strike a balance between complete freedom and good supervision."

During the remainder of his development into a full-fledged flavorist, the trainee works closely with the company's top flavorists. While they are obviously in a position to help him greatly, the Bush Boake Allen view is that precautions must be taken to see that they do not influence him or her unduly. Says Mrs. Cowley, "top management must keep a fairly close watch on this stage of development. In other words, certain set rules of flavouring are communicated, *e.g.*, vanilla or rose for rounding flavour off, esters for brightening them, increasing the volatiles to compensate for losses during processing; but we do not inhibit the trainee in his experimental work, providing common sense prevails. The worst thing for any flavour house, in my opinion, is for its flavours to be typed by the ideas of one man. Obviously all flavourists have their recognizable type of profile to a certain degree; but ideally if a company has ten flavourists, these ten should be individual creators, each with his own individual style. In other words I cannot stress too strongly that formalized training in *actual creativity* is, in my opinion, a very bad thing. The training should really consist of bringing the individual along to the point where creativity can take place by that individual. Obviously all intelligent beings will discuss and talk about their flavour problems and there is a tendency for a terminology to be perpetuated in a given flavour house. Generally speaking we find that at the beginning terms like green, fatty, roasted, bright and full-bodied are used. After a while chemical terms like acidic and aldehydic begin to creep in and in finality the terminology does come down to specific chemical names. We involve all trainees in the development area and creative flavourists in the assessment of new raw materials, of which, as well as standard raw materials, a master is maintained. These are also card indexed and each flavourist's comments on the description of the profile and its possible uses are maintained on the index.

"The trainee's knowledge is enlarged by participation, but interpretation and application are not rigidly channelled. We encourage flavourists and their assistants to co-ordinate most strongly with the basic research grop if there are common objectives, which there invariably are. This is giving the trainee and flavourist an insight into the technical feasibility of future potential creations providing certain key factors of natural products can be separated and identified. We insist that all experiments are kept in laboratory books and when their final creations are accepted for the market place then that flavourist assumes the responsibility for piloting his creation through to the point of the market launch. In other words before it is offered to the sales department he has to make sure about the application, description and the claims which can be made in the application, the suitability in various countries throughout the world, the compounding instructions, the raw materials availability, etc."

The above training procedure, Mr. Cowley believes, produces finished flavorists who will be competent in:

(1) Testing and evaluation.
(2) Possession of a vast library of tastes and smells
(3) Ability to communicate with other functions and customers.
(4) Possession of a complete understanding of the problems involved when flavors are used in processed foods.
(5) Ability to create without supervision.
(6) Possession of the knowledge to assess the technical feasibility of a project with the marketing department.

Chapter XVIII

Statistics

The basic figure about any industry is, of course, what is its dollar volume? In the case of the essential oil and aroma chemical industry it is rather difficult to arrive at a dependable total, because information about the figures of individual companies is closely held, and there are no U.S. Census figures on which to base even an educated guess. The firms manufacturing and importing fragrance and flavor raw materials generally report to the Census Bureau as an insignificant part of the chemical industry and there are no break-downs to show their production or sales separately.

Several surveys made recently have some claim to reasonable accuracy, provided they are read by those sufficiently familiar with the fragrance and flavor industry to be able to read between the lines and sift out the basic facts from the admixture of misinformation that always creeps into such surveys.

A rather over-generous estimate of the size of the industry is contained in a recent "Study of Flavors, Fragrances and Essential Oils" published in 1976 by Delphi Marketing Services, 400 East 89th St., New York, N.Y. They report an overall figure of $2.4 billion for the value of the flavor and fragrance industry in 1974, including raw materials as well as perfume oils and flavour compounds. The industry, they say, has experienced an annual growth rate of 7% since 1970.

We have no quarrel with their assessment of the growth rate, but their grand total for the industry has been inflated by the inclusion of $1.7 billion of flavoring extracts, and an additional $146.8 million of spices. Perhaps the spice figure may belong in the total, although only a few firms such as Norda, Fritzsche D & O, Glidden-Durkee and the Meer Corp., which are active in the spice field, are also factors in the perfuming materials market. Although a few companies like these do handle both type products, we don't believe that sales totals for flavoring extracts should be bulked in with fragrance and flavor sales figures. Flavoring extracts are finished products. It seems as inaccurate to add the sales value of flavoring extracts to the essential oil and aroma chemical total as it would be to add in the retail value of all the other fragrance-using products such as cosmetics and

Table I

VALUE OF THE FLAVOR & FRAGRANCE INDUSTRY

Item	1974 millions of $	1975 estimated
Essential Oils (import value)	75.8	35.6
Essential Oils (shipment value)	100.4	109.0
Synthetic Aroma Chemicals (sales value)	136.6	146.0
MSG (sales value)	30.1	35.0
Synthetic Sweeteners	not available	not available
Spices (value imports)	146.8	155.0
Flavor Extracts (value of shipments)	1706.0	1844.0
TOTAL	2195.7	2325.0

household items — perfumes, cosmetics, soaps, hair sprays, etc.

Delphi's figures for essential oils and aroma chemicals (Table I) on the other hand, seem incomplete. They list $75.8 million for 1974 for the value of essential oil imports, and $100.4 for the "shipment value" of what are presumably domestically produced essential oils, then add another $136.6 million for the sales value of synthetic aroma chemicals. The sum of these three items gives a total of only $312.8 million, which is certainly much too small a figure. The top ten companies in the U.S. industry alone account for at least that volume of sales.

The Delphi study states that the leading four companies in the U.S. essential oil and aroma chemical industry do 40% of the industry's total business. In this appraisal we feel that they are quite close to the actual facts as reported to us informally and confirmed by other estimates.

Discussing the end markets for manufacturers of fragrances, Delphi recognizes that the cosmetic market is the most important outlet for fragrance materials. The value of cosmetic sales, they say, was $5 billion in 1975 at the consumer level, which represented a growth of 8% over 1974. Delphi assigns the largest portion of the market for fragrances among cosmetic items to bath products — bubble baths, gels, and some soaps. Bath products, they estimate, account for 24% of the total volume of fragrance raw materials.

The second largest market for fragrances, says Delphi, is for "handkerchief perfumes," which, they estimate, amount to 20% of the total market. Men's fragrances take 10%, and toilet soaps 6% of the fragrance raw materials sold, according to their break-down. The balance, they believe, is accounted for by sales of fragrances to manufacturers of household products, industrial products, air deodorizers, paints, clothing, industrial soaps and cleaners, rubber, plastic, etc. They cite no authority for this apportioning of the market.

Delphi also offer estimates on the sales volume of leading companies in the flavor and fragrance industry.

Table II

MAJOR PRODUCERS AND THEIR SALES in 1975

Company	Estimated Sales (millions of $)
Chem-Fleur, Inc.	$4.5
Felton International, Inc.	25.0
Florasynth, Inc.	25.0
Fritzsche, Dodge & Olcott, Inc.	not avail.
Givaudan Corp.	50.0
International Flavors & Fragrances, Inc.	270.0
Naarden, Inc.	4.5
Norda, Inc.	45.0
Polaks' Frutal Works, Inc.	not avail.
A.M. Todd Co., Inc.	33.0
Ungerer & Co., Inc.	not avail.
Universal Oil Products	936.0

Sales figures are total corporate sales, not sales of flavor and fragrance materials alone.

Figures are probably quite accurate except for the inflated figure of $936 million they assign to UOP, which as they indicate is the total for the company's overall corporate activities (which include the manufacture of chemicals, plastics, hose, tubings, the building of refineries, etc.). UOP's sale of flavor and fragrance materials alone would obviously amount to only a small percentage of the corporate sales total.

Another study of Sales of Flavors and Fragrances which offers estimates of the individual figures of eleven of the leading companies dealing in flavor and fragrance raw materials has come to our attention recently. The study was made by a market analyst whom we are not free to identify, but his findings seem fairly accurate, at least in the opinion of some top industry executives who have seen them. We offer them in the accompanying Table III for what they may be worth. It should be noted that no figures are included for such major suppliers of fragrance and flavor materials as PFW, Monsanto, Ungerer & Co., Haarmann &

Reimer, Glidden-Durkee, Union Camp, and other such firms.

We believe the sales estimates for the eleven companies reported by this investigator are reasonably accurate. However, on the basis of our own investigations, we think the estimate of the overall size of the market by this analyst is on the low side. Once we get past the ten or eleven market leaders, we believe the remaining 100 to 150 firms which constitute the balance of the industry do a total business closer to $200 million than the $138 million estimate this investigator offers for "other." This would put the overall total dollar value of the essential oil and aroma chemical market at $600 million rather than the $510 million figure reported.

Table III

ESTIMATED U.S. SALES of the MAJOR SUPPLIERS OF FLAVORS AND FRAGRANCES — 1976

Millions of Dollars

Company	Fragrance	Flavors (a)	Aroma Chemicals (b)	Other	Total
IFF	47	15	28	C	90
Givaudan	14	5	31	—	50
F D & O	15	18	3.5	11.5	48
Norda	12	10	13	10	45
Firmenich	16.5	7	2	—	25.5
Roure Dupont	13	—	0.5	1	14.5
Felton	12.5	14	1.5	2	30
Florasynth	11	5	1	—	17
Dragoco	6	8	1	1	16
Naarden-UOP	17	3	—	—	20
Rhodia	5	2	8	2	17
Other	21	23	30.5	62.5	138
Total	190	110	120	90	510

(a) Excludes essential oils, spices, vanilla and MSG
(b) Includes large commodity chemicals
(c) Included in aroma chemicals

Our own analysis of the size of the market was made by relating figures we collected on number of employees working for a selected group of companies to the dollar volume of sales these companies are believed to enjoy, then applying these "dollar-per-employee" factors to the known number of employees of a large number of the remaining firms in the industry where we know how many employees they have but have little clue as to their dollar sales. This statistically valid procedure supports a $600 million estimate of the overall size of the fragrance and flavor raw material market.

Still another estimate of the size of this market was made in 1974 by C. H. Kline, Inc., Fairfield, N.J., in the 2nd edition of "The Kline Guide to the Chemical Industry" (pgs. 150-157). This study puts the value of domestic U.S. production of aroma and flavor materials at $420 million for the year 1973. To this total they add $25 million for flavor enhancers and $5 million for synthetic sweeteners, reaching a grand total of $450 million.

Of the $420 million reported for flavors and fragrances, $235 million was reported to have been in blended compounds, $80 million in aroma chemicals, and $105 million in essential oils and other natural products. Of the $235 million of blended compounds, $135 million was in fragrances, and about $100 million in flavors. Value of the overall sales of flavors, fragrances, and related products has been growing at an average annual rate of 6.6% over the past ten years, this source reports. Essential oils have exhibited the highest growth rate, says the Kline Guide, largely as a result of price increases in the producing areas.

The Kline study predicts that in the future there will be a more rapid growth in the value of sales of blended compounds than for raw materials. They suggest that domestic output of blended compounds will increase at a rate of 6.9% a year, reaching $375 million in 1980, and $525 million by 1985. Synthetic aroma chemicals, they say, will grow at a much slower rate — 3.8% a year — reaching $105 million in 1980 and $125 million in 1985. They predict that essential oils will grow at an average rate of 5.1% a year, reaching a total of $150 million in 1980, and $190 million by 1985.

The Kline Guide reports that, in 1972, the top eleven U.S. producers of flavors and fragrances enjoyed approximately 70% of the overall market. The top four firms had 37% of the total, and the next four firms an additional 22%. It is pointed out that the fragrance and flavor field is one industry in which small, privately owned companies still thrive. Of the ten leading companies only two are publicly owned. "The business is based to a high degree on personal contacts not only between suppliers and customers, but also among suppliers who often buy and sell among themselves."

Another factor in the success of the smaller, privately-owned firms is that the success or failure of an individual fragrance or flavor depends almost entirely upon the success of the creative staff. "With few exceptions, the larger, diversified chemical companies," says Kline, "have not done well in the compounding business."

In our study of significant statistics of the fragrance and flavor industry we also had an opportunity to see data on the number of employees of essentially every company in the industry. There is a wide range in size of the companies in this field. There are quite a few relatively small operations, where the number of employees may be no more than 8 or 10. On the other end of the spectrum a sizeable number of firms employ upwards of one hundred, and annual sales run well up into the millions of dollars. The top 80 to 100 companies in the industry can be broken down into five groups as follows:

A	those having ten or fewer employees	20% of total
B	those having 11 to 25 employees	26%
C	those having 26 to 50 employees	17%
D	those having 51 to 100 employees	17%
E	those having over 100 employees	19%

From the figures reported by these firms on number of employees, it appears that there are approximately seven to eight thousand people overall employed in the essential oil and aroma chemical business. The number of employees ranges from as few as three or four in the smaller firms up to approximately eleven hundred (domestically) for the largest company in the industry.

The companies were studied as to their comparative interest in the sale of fragrances vs. flavors. We found that about 64.4% of the total number of employees in the industry are concerned with production and sale of fragrance materials, compared with 35.6% for flavors.

Between 80 and 85% of the firms in the industry are interested in both flavors and fragrances. The remainder sell only one type product, either fragrances or flavors. The break-down among the single-interest group is about 40% who are concerned solely with fragrance materials, and 60% interested only in flavors.

Of those concerns that handle both fragrance and flavor materials, fragrance is the dominant product for 74%, and flavors for 20%. The remaining 6% report their interest is about evenly divided between flavors and fragrances.

Interest in flavors seems to be more heavily concentrated in the larger firms. In terms of number of employees concerned with flavor operations, the top ten flavor firms employ about 70% of the total number of flavor workers. This offers a contrast with the fragrance materials side, where the top ten fragrance houses (based on number of workers concerned with fragrance products) employ only about 54% of the total number of fragrance workers in the industry.

U.S. Trade Commission Data

Data on U.S. production and sale of Flavor and Perfume Materials is published annually by the United States International Trade Commission, Washington, D.C. The latest figures, for the year 1975, show that total domestic production of flavor and perfume materials in 1975 amounted to 101.5 million pounds. Sales totaled 83.5 million pounds, with a value of $143.6 million. This represented a substantial decline from 1974 when production totaled 107.4 million pounds, valued at $166.7 million.

The totals do not include benzyl alcohol which is no longer reported in the flavor and perfume materials list.

The overall total is broken down into cyclic and acyclic materials. In 1975, production of cyclic materials amounted to 44.7 million pounds; sales were 33.0 million pounds; and the value of those sales was $91.7 million. Output of acyclic flavor and perfume materials in 1975 was 56.8 million pounds; sale of these materials amounted to 50.5 million pounds; and the value of the materials sold was $51.9 million.

Imports of Essential Oils — 1976

The U.S. Department of Commerce reports the following figures for United States imports of essential oils for the year 1976.

Almond Oil, Bitter	-(Natural & Imitation)	161,870	264,326
Anise Oil	-(Natural & Imitation)	123,598	547,331
Bergamot Oil		156,742	1,041,739
Bois de Rose Oil		134,785	517,052
Camphor Oil		219,877	128,311
Caraway Oil		37,557	302,297
Cassia Oil	-(Natural & Imitation)	171,563	2,416,028
Cedar Leaf Oil		22,305	250,449
Cinnamon Oil		311,321	361,292
Citronella Oil		2,282,146	2,650,580
Clove Oil	-(Leaf Bud Stem)	1,803,768	3,410,836
Eucalyptus Oil		475,891	845,760
Geranium Oil		229,444	4,409,181
Grapefruit Oil		26,080	10,704
Lavender Oils		241,277	2,588,275
Lemon Oil		634,004	3,350,407
Lemongrass Oil		244,314	469,662
Lime Oil	-(Distilled & Expressed)	966,327	6,559,647
Neroli or Orange Flower		1,834	230,083
Nutmeg Oil		251,366	1,117,126
Orange Oil		747,196	569,690
Orris Oil		1,743	612,519
Palmarosa Oil		21,316	168,418
Patchouli Oil		480,861	2,547,041
Peppermint Oil (Mentha Arv.)		582,388	1,924,634
Peppermint Oil (Mentha Pip.)		32,935	225,507
Petitgrain Oil		439,813	1,603,588
Pine Needle Oil		69,420	302,760

Rose Oil	(Natural & Imitation)	33,873	1,452,713
Rosemary Oil		300,069	766,907
Sandalwood Oil		91,391	2,765,829
Thyme Oil		24,336	186,606
Vetiver Oil		246,417	5,798,892
Ylang Ylang		143,624	2,651,487
Other Misc. Distilled or Essential Oils		3,949,199	18,230,539
Total dollar value of essential oils imported during 1976			$71,278,216

Index of Proper Names

Subject Index

The Fragrance and Flavor Industry

published by
Wayne E. Dorland Publishing Company
Mendham, N.J.

Illustration and design by
Leonard Frank

Body copy set in Baskerville by
Benway, Maxwell & Smith, Chatham, N.J.

Title page in Bulmer italic

Cover printed by offset lithography on
Process Material Corporation's
Multicolor Antique Willow Green

End papers on
Multicolor Antique Thistle

Text pages on
Westvaco's Pinnacle Offset

by
Litho-Four, Butler, N.J.

Bound by
Tapley-Rutter Company
Moonachie, N.J.